COORDINATION CHEMISTRY IN PROTEIN CAGES

COORDINATION CHEMISTRY IN PROTEIN CAGES

Principles, Design, and Applications

Edited by

TAKAFUMI UENO
YOSHIHITO WATANABE

Published by John Wiley & Sons, Inc., Hoboken, New Jersey
Published simultaneously in Canada

For general information on our other products and services or for technical support, please contact our
Customer Care Department within the United States at (800) 762-2974, outside the United States
at (317) 572-3993 or fax (317) 572-4002.

Wiley also publishes its books in a variety of electronic formats. Some content that appears in print may
not be available in electronic formats. For more information about Wiley products, visit our web site at
www.wiley.com.

Library of Congress Cataloging-in-Publication Data:

Coordination chemistry in protein cages principles, design, and applications / edited by Takafumi Ueno,
Yoshihito Watanabe.
     pages cm
   Includes index.
   ISBN 978-1-118-07857-0 (cloth)
  1. Protein drugs.   2. Protein drugs–Physiological transport.   3. Carrier proteins.   I. Ueno, Takafumi,
1971– editor of compilation.   II. Watanabe, Yoshihito, editor of compilation.
   RS431.P75C66 2013
   615.1'9–dc23

                                                                                            2012045122

Printed in the United States of America

ISBN: 9781118078570

10 9 8 7 6 5 4 3 2 1

CONTENTS

PART V   APPLICATIONS IN NANOTECHNOLOGY

11 Protein Cage Nanoparticles for Hybrid Inorganic–Organic Materials   275

Shefah Qazi, Janice Lucon, Masaki Uchida, and Trevor Douglas

FOREWORD

The field of biological inorganic chemistry has taken many new directions in the first part of the 21st century [1]. One area of great current interest deals with the role of outer sphere interactions in tuning the properties of active sites in metalloproteins. Investigators are building on early ideas of "the entatic state" and "the rack effect" in interpreting the modulation of active site properties in the cavities of folded polypeptide structures [2, 3]. Others are designing and constructing synthetic molecular cavities for studies of host-guest interactions. The cavity in ferritin, the iron storage protein and the superstar of nanocages, is the subject of the opening chapter in this timely collection of reviews edited by my good friends Yoshi Watanabe and Takafumi Ueno. In their review, Theil and Behera set the tone for the book in their thorough discussion of ferritin structures and mechanisms as well as uses of this natural nanocage in imaging, drug delivery, catalysis, and electronic devices. I have a soft spot in my heart for ferritin, as John Webb and I worked on the nature of iron coordination in its cavity many years ago [4]. And in the late 1980s, Bill Schaefer and I designed the ferritin fountain that decorates the courtyard of the Beckman Institute at Caltech. The cavity in this fountain is a real megacage![1]

The collection of reviews put together by Watanabe and Ueno is most impressive. Synthetic cavities are discussed by Nastri et al., and designed protein assemblies that allow systematic investigation of molecular interactions at interfaces are treated by Tezcan. Incorporation of oxometal species in natural protein cages is reviewed by Müller and Rehder; manipulation of myoglobin and other heme protein cavities for ligand binding and chemistry is treated by Hayashi; engineering both copper and

---

[1] See: http://ww2.cityofpasadena.net/arts/images/molecule.JPG

heme proteins for altered redox functions is discussed by Marshall et al.; catalysis of organic transformations in both natural and artificial protein cavities are the subjects of reviews by Ueno and Abe and by Praneeth and Ward. Two sections of the book deal with applications in biology and nanotechnology, with chapters on optical imaging (Kurishita and Hamachi), magnetic materials (Arakaki et al.), hybrid inorganic–organic materials (Qazi et al.), nanoelectronic devices (Yamashita et al.), and nanostructured inorganic materials (Dennis et al.). The role of coordination chemistry in the assembly of metal-organic cages is highlighted by Fujita and Sato, who also discuss reactions both inside and outside such structures in the final chapter of the book.

The book is comprehensive and up-to-date. What is more, it captures the excitement of the protein nanocage field. In my view, it is a must-read for all investigators working in biological inorganic chemistry, biological organic chemistry, and nanoscience.

HARRY B. GRAY
California Institute of Technology
Pasadena, California, USA

REFERENCES

[1] H. B. Gray, *Proc Natl. Acad. Sci. USA* **2003**, *100*, 3563.

[2] H. B. Gray, B. G. Malmström, and R. J. P. Williams, *J. Biol. Inorg. Chem.* **2000**, *5*, 551.

[3] J. R. Winkler, P. Wittung-Stafshede, J. Leckner, B. G. Malmström, and H. B. Gray, *Proc. Natl. Acad. Sci. USA* **1997**, *94*, 4246.

[4] J. Webb and H. B. Gray, *Biochim. Biophys. Acta* **1974**, *351*, 224.

PREFACE

Today is a most exciting time to be working in coordination chemistry, in particular at the interface of biology and materials science. Until recently, most coordination chemistry related to biology was used for the reconstruction of metal-binding sites in proteins and the elucidation of the mechanisms of protein functions involving metal ions. Bioinorganic systems have evolved for the storage of metal ions, biominerals such as teeth, bones, and various complex framework structures for carrying out different functions. These processes occur at the nano-, meso-, and micro-scales. With the rise of nanotechnology, the role played by bioinorganic chemistry has changed from fundamental understanding to providing manipulation tools consisting of large proteins with numerous subunits. In some cases, situations may arise in which we cannot design the protein functions for the coordination chemistry or cannot design the coordination chemistry for the protein functions. We, therefore, think that there is a need for a guide book covering the interesting, rapidly developing areas of bioinorganic chemistry of protein cages, which are fundamentally important and inspire applications in biology, nanotechnology, synthetic chemistry, and other disciplines.

In this book, we focus on protein cages, which have attracted much attention as nanoreactors for coordination chemistry because of their unique internal molecular environments and few of these have been constructed, even with modern organic and polymer synthetic techniques. The book is divided into six major sections: (1) Coordination Chemistry in Native Protein Cages, (2) Design of Metalloprotein Cages, (3) Coordination Chemistry of Protein Assembly Cages, (4) Applications in Biology, (5) Applications in Nanotechnology, and (6) Coordination Chemistry Inspired by Protein Cages. Each part contains articles by experts in the relevant area. The contributors to all the sections of the book are extremely well-known in their areas of research and have made significant contributions to coordination chemistry in biological systems. In the first part, the principles of coordination reactions in natural protein cages are explained. The second part focuses on one of the emerging areas of bioinorganic chemistry, and deals with the fundamental design of coordination sites of small artificial metalloproteins as the basis of protein cage design. The third part explains the supramolecular design of protein cages and assembly for or by metal coordination. Parts IV and V, respectively, consist of dedicated sections on applications in biology and nanotechnology; these are extremely important and the most recent work by experts in these fields is covered. Part VI describes the principles of coordination chemistry governing self-assembly of numerous synthetic cage-like molecules; similar principles are often applicable in biology. We believe that this book, which has a different scope from previously published books on bioinorganic chemistry, supramolecular chemistry, and biomineralization, will be of interest not only to specialists but also to readers who are not familiar with coordination chemistry.

Finally, a large number of people have helped to put this book together, so that it ultimately provides an invaluable resource for all those working on the principles, design, and applications of coordination chemistry in protein cages. I would especially like to thank all the contributors, who have all produced excellent manuscripts, and made improvements to this book. We also thank Anita Lekhwani and Cecilia Tsai at Wiley who have helped to put this together, and Dr. Tomomi Koshiyama for her dedicated efforts in drawing an excellent cover image representing this book's concept.

TAKAFUMI UENO
YOSHIHITO WATANABE
January 2013

CONTRIBUTORS

Satoshi Abe Graduate School of Bioscience and Biotechnology, Tokyo Institute of Technology, Nagatsuta-cho, Yokohama, Japan

Atsushi Arakaki Division of Biotechnology and Life Science, Institute of Engineering, Tokyo University of Agriculture and Technology, Koganei, Tokyo, Japan

Rabindra K. Behera Children's Hospital Oakland Research Institute, Oakland, CA

Rosa Bruni Department of Chemistry, Complesso Universitario Monte S. Angelo, University of Naples Federico II, Via Cintia, Naples, Italy

Patrick B. Dennis Nanostructured and Biological Materials Branch, Materials and Manufacturing Directorate, Air Force Research Lab, WPAFB, OH

Trevor Douglas Department of Chemistry and Biochemistry and Center for Bio-Inspired Nanomaterials, Montana State University, Bozeman, MT

Makoto Fujita Department of Applied Chemistry, School of Engineering, University of Tokyo, Bunkyo-ku, Tokyo, Japan

Harry B. Gray Division of Chemistry and Chemical Engineering, Beckman Institute, California Institute of Technology, Pasadena, CA

Itaru Hamachi Department of Synthetic Chemistry and Biological Chemistry, Graduate School of Engineering, Kyoto University, Katsura, Kyoto, Japan

Takashi Hayashi Department of Applied Chemistry, Osaka University, Japan

Kenji Iwahori JST PRESTO, Laboratory of Mesoscopic Materials and Research, Nara Institute of Science and Technology, Takayama, Ikoma, Nara, Japan

Shinya Kumagai Department of Advanced Science and Technology, Toyota Technological Institute, Nagoya, Japan

Yasutaka Kurishita Department of Synthetic Chemistry and Biological Chemistry, Graduate School of Engineering, Kyoto University, Katsura, Kyoto, Japan

Angela Lombardi Department of Chemistry, Complesso Universitario Monte S. Angelo, University of Naples Federico II, Via Cintia, Naples, Italy

Yi Lu Department of Chemistry, University of Illinois at Urbana-Champaign, Urbana, IL

Janice Lucon Department of Chemistry and Biochemistry and Center for Bio-Inspired Nanomaterials, Montana State University, Bozeman, MT

Ornella Maglio Department of Chemistry, Complesso Universitario Monte S. Angelo, University of Naples Federico II, Via Cintia, Naples, Italy

Nicholas M. Marshall Department of Chemistry, University of Illinois at Urbana-Champaign, Urbana, IL

Tadashi Matsunaga Division of Biotechnology and Life Science, Institute of Engineering, Tokyo University of Agriculture and Technology, Koganei, Tokyo, Japan

Kyle D. Miner Department of Chemistry, University of Illinois at Urbana-Champaign, Urbana, IL

Achim Müller Fakultät für Chemie, Universität Bielefeld, Bielefeld, Germany

Rajesh R. Naik Nanostructured and Biological Materials Branch, Materials and Manufacturing Directorate, Air Force Research Lab, WPAFB, OH

Flavia Nastri Department of Chemistry, Complesso Universitario Monte S. Angelo, University of Naples Federico II, Via Cintia, Naples, Italy

Michiko Nemoto Division of Biotechnology and Life Science, Institute of Engineering, Tokyo University of Agriculture and Technology, Koganei, Tokyo, Japan

V. K. K. Praneeth Department of Chemistry, University of Basel, Basel, Switzerland

Shefah Qazi Department of Chemistry and Biochemistry and Center for Bio-Inspired Nanomaterials, Montana State University, Bozeman, MT

Dieter Rehder Fachbereich Chemie, Universität Hamburg, Hamburg, Germany

Sota Sato Department of Applied Chemistry, School of Engineering, University of Tokyo, Bunkyo-ku, Tokyo, Japan

Joseph M. Slocik Nanostructured and Biological Materials Branch, Materials and Manufacturing Directorate, Air Force Research Lab, WPAFB, OH

F. Akif Tezcan Department of Chemistry and Biochemistry, University of California, San Diego, La Jolla, CA

Elizabeth C. Theil Children's Hospital Oakland Research Institute, Oakland, CA

Masaki Uchida Department of Chemistry and Biochemistry and Center for Bio-Inspired Nanomaterials, Montana State University, Bozeman, MT

Takafumi Ueno Graduate School of Bioscience and Biotechnology, Tokyo Institute of Technology, Nagatsuta-cho, Yokohama, Japan

Thomas R. Ward Department of Chemistry, University of Basel, Basel, Switzerland,

Tiffany D. Wilson Department of Chemistry, University of Illinois at Urbana-Champaign, Urbana, IL

Ichiro Yamashita Panasonic ATRL, Laboratory of Mesoscopic Materials and Research, Nara Institute of Science and Technology, Takayama, Ikoma, Nara, Japan

Bin Zheng Laboratory of Mesoscopic Materials and Research, Nara Institute of Science and Technology, Takayama, Ikoma, Nara, Japan

PART I

COORDINATION CHEMISTRY IN NATIVE PROTEIN CAGES

1

THE CHEMISTRY OF NATURE'S IRON BIOMINERALS IN FERRITIN PROTEIN NANOCAGES

ELIZABETH C. THEIL AND RABINDRA K. BEHERA

1.1 INTRODUCTION

Ferritin protein nanocages, with internal, roughly spherical cavities ∼5–8 nm diameter, and 8–12 nm external cage diameters, synthesize natural iron oxide minerals; minerals contain up to 4500 iron atoms, but usually, in normal physiology, only 1000–2000 iron atoms are present; in solution the cavity is filled with mineral plus buffer. The complexity of protein-based iron-oxo manipulations in ferritins, which are present in contemporary archaea, bacteria, plants, humans, and other animals, is still being discovered [1]. Such new knowledge indicates that current applications exploit only a small fraction of the ferritin protein cage potential. To date, applications of ferritin protein cages to nanomaterial synthesis have been mainly as a template [2–4], and as a catalyst surface [5, 6]; some applications of ferritins as nutritional iron sources and targets or chelators in iron overload are also developing [3, 4].

There are two biological roles of the ferritins: concentrating iron within cells as a reservoir with large concentrations of iron for rapid cellular use, as in red blood cells (hemoglobin synthesis) or cell division (doubling of heme and FeS protein content), and for recovery from oxygen stress (scavenging reactive dioxygen and ferrous iron released from damaged iron proteins). Equation 1.1 shows the simplified forward reaction of ferritin protein nanocages.

$$2Fe(\text{II}) + O_2 + 2H_2O \rightarrow Fe_2O_3 \cdot H_2O + 4H^+ \tag{1.1}$$

Coordination Chemistry in Protein Cages: Principles, Design, and Applications, First Edition.
Edited by Takafumi Ueno and Yoshihito Watanabe.
© 2013 John Wiley & Sons, Inc. Published 2013 by John Wiley & Sons, Inc.

In ferritin protein nanocages, the Fe/O reactions occur over distances within the cage of ~50 Å (ion entry channels/oxidoreductase sites/nucleation channels/ mineralization cavity) and over time spans that vary from msec (ferrous ion entry, active site binding, and reaction with dioxygen) to hours (mineral nucleation and mineral growth). Moreover, the reaction can be reversed by the addition of an external source of electrons, to reduce ferric to ferrous, and rehydration to release ferrous iron from the mineral and the protein cage. Exit of ferrous iron from dissolved ferritin minerals is generally slow (minutes to hours) but is accelerated by localized

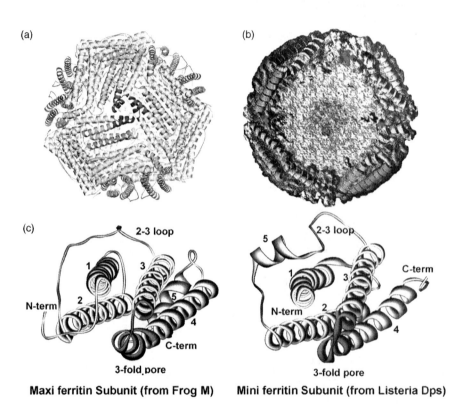

FIGURE 1.1 Eukaryotic ferritin protein nanocages and structural comparison of maxi- and mini-ferritin subunit. (a) An assembled 24-subunit ferritin protein with symmetrical Fe(II) entry/exit site at threefold pores (dark gray) that connects the external medium to inner protein cavity. Reprinted with permission from Reference 24. Copyright 2011 Journal Biological Chemistry. (b) Cross-section of ferritin protein cage (PDB:1MFR); sketch (gray cavity) showing filled ferric oxide mineral; arrow pointing ion channels. Reprinted with permission from Reference 12. Copyright 2010 American Chemical Society. (c) The 4-α-helix bundle (subunit) of maxi- (left) and mini-ferritin (right) drawn by us using PYMOL and the PDB files indicated; the fifth short helix is at the end of 4-α-helix and in 2-3 loop in maxi- (PDB:3KA3) and in mini-ferritin (PDB:2IY4), respectively, that defines fourfold and twofold symmetry upon self-assembling. The threefold pore regions that are conserved in both maxi- and mini-ferritin are shown in dark gray.

protein unfolding around the cage pores (Fig. 1.1), mediated by amino acid substitution of key residues in the ion channels or adding millimolar amounts of chaotrope [1, 7].

Ferritin proteins are so important that the genetic regulation is unusually complex. For example, in addition to ferritin genes (DNA), mRNA is also regulated by metabolic iron, creating a feedback loop where the catalytic substrates ferrous and/or oxidant are also the signals for ferritin DNA expression, ferritin mRNA regulation, and ferritin protein synthesis [8]. The iron concentrates in ferritins are used when rates of iron protein synthesis are high or, in animals, after iron (blood) loss. Rates of synthesis of iron proteins are high, for example, in preparation for eukaryotic cell division (mitochondrial cytochromes), plant photosynthesis (ferredoxins), plant nitrogen fixation (nitrogenase and leghemoglobin synthesis), red blood cell maturation (synthesizing 90% of cell protein as hemoglobin), immediately after birth, hatching, or metamorphosis (animals liver ferritin), and germinating legume seeds. After oxidant stress, ferritins play an important role in recovery, and in pathogenic bacteria in resistance to host oxidants by removing from the cell cytoplasm potent chemical reactants, ferrous ion and dioxygen or hydrogen peroxide. Protein-based catalytic reactions use $Fe(II)$ and O to initiate mineralization of hydrated ferric oxide minerals.

The family of iron-mineralizing protein cages is apparently very ancient since they are found in archaea as well as bacteria, plants, and animals as advanced as humans. All ferritins share the following:

1. unique protein cavities for hydrated iron oxide mineral growth;
2. unusual quaternary structure of self-assembling, hollow nanocages with extraordinary protein symmetry (Fig. 1.1);
3. catalytic sites where di-$Fe(II)$ ions react with dioxygen/hydrogen peroxide to initiate mineralization;
4. subunit protein folds in the common 4-α-helix bundle motif;
5. ion entry pores/channels;

Ferritin protein cages have variable properties as well:

1. cage size (subunit number: $n = 12$ or 24);
2. amino acid sequence: conserved among eukaryotes but divergent among eukaryotes, bacteria, and archaea (>80%) [9];
3. hydrated iron oxide mineral crystallinity and phosphate content;
4. catalytic mechanism for mineral initiation;
5. location/mechanism of mineral nucleation.

In this chapter we discuss protein cage ion channels, ferritin inorganic catalysis, and protein mineral nucleation and growth to provide fundamental understanding of ferritin structures and functions. The protein nanocage itself and the protein subdomains can be developed for templating nanomaterials, synthesizing nanopolymers with fixed organometallic catalysts, producing nanodevices, and delivering nanosensors and medicines. The current status of such applications of ferritin cages are discussed extensively in Sections 1.2, 1.3, and 1.4.

1.2   FERRITIN ION CHANNELS AND ION ENTRY

1.2.1   Maxi- and Mini-Ferritin

Ferritins are hollow nanocage proteins, self-assembled from 24 identical or similar subunits in maxi-ferritins and 12 identical subunits in mini-ferritins that are also called Dps protein (*DNA-binding proteins from starved cells*) [10]. Cavities in the center of the protein cages account for as much as 60% of the cage volume. Cage assemblies from the 4-α-helix bundle protein subunits have 432 symmetry (24 subunits) or 23 symmetry (12 subunits) [11]. The unusual symmetry has functional consequences for ion entry and exit, and in the larger ferritins for protein-based control over mineral growth. Ferritin subunits are 4-α-helix bundles, ~50 Å long and 25 Å wide (Fig. 1.1), that use hydrophobic interactions for helix and cage stability; charged residues on surfaces enhance cage solubility in aqueous solvents and ion traffic into, out of, and through the protein cage.

Ion channels with external pores are created around the threefold cage axes by sets of helix-loop-helix segments at the N-terminal ends of three subunits in the assembled cages (Fig. 1.1). The channels, analogs to ion channels in membranes, are lined with conserved carboxylate groups for cation entry and transit from the outside to the inside; ion distribution to multiple active sites occurs at the interior ends of the channels [12]. Each channel, eight in maxi-ferritin cages and four in mini-ferritin cages, is funnel shaped [1, 12]. In maxi-ferritins, constrictions in the channel center may be selectivity filters. (Fig. 1.2). At the C-terminal end of ferritin subunits in maxi-ferritins, a second cage symmetry axis occurs, created by the junctions of four subunits that can form pores [13]. The region is primarily hydrophobic and the fourfold arrangement of nucleation channel exits contributes to mineral nucleation. In mini-ferritins, by contrast, the C-termini form pores with threefold symmetry [11, 14] that in some mini-ferritins can participate in DNA binding [15].

Maxi-ferritins are found in all types of higher organisms (animals and plants including fungi), and bacteria (ferritins and heme-containing bacterioferritins (BFRs)). Mini-ferritins, by contrast, are restricted to bacteria and archaea and may represent the transition of anaerobic to aerobic life, because of the ability to use dioxygen and/or hydrogen peroxide as the oxidant in the Fe/O oxidoreductase reaction. The secondary and quaternary structures of ferritins are very similar [10] in spite of very large differences in primary amino acid sequences.

The fifth helix in ferritin subunits, attached to the four-helix bundle, distinguishes maxi-ferritins (animal and plant H/M chains and animal L chains, bacterial ferritins, and the heme-containing bacterioferritins) from the smaller mini-ferritin (Dps proteins). In maxi-ferritins, helix 5 is at the C-terminus, positioned along the fourfold symmetry axis, at ~60° to the principal helix bundle. A kink in helix 4 occurs at

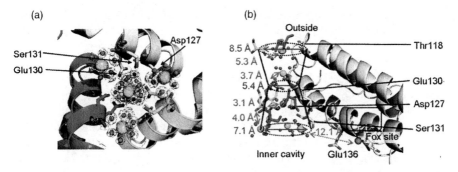

FIGURE 1.2   Iron entry route in ferritin. Fe(ii) ion channels showing a line of metal ion (Big spheres) connecting from the outside to inner protein cavity (PDB:3KA3). Adapted with permission from Reference 12. Copyright 2010 American Chemical Society. (a) A view from inside the cavity through the channel toward the cage exterior, showing three metal ions at the exit into the cavity; they are symmetrically oriented toward each of the catalytic centers in the subunits that form the channel by binding to one of the Asp127 residues in each subunit. The fourth metal ion in the center is the end of the line of metal ion stretching from the external pore through the channel to the cavity. (b) The conserved negatively charged residues in one subunit that defines the ion channel are shown as sticks. Big spheres are Mg(ii) and small spheres are water molecules; a ~4.5 Å diameter constriction is at the center of the channel at Glu 130.

the position of an aromatic residue (Tyr/His/Phe 133), which sterically alters the helix at the point where localized pore unfolding stops [16, 17]. Mini-ferritins (Dps proteins), by contrast, lack the fifth helix at the C-terminus; some have helices in the loop connecting helices 1 and 2 as in the mini-ferritin cage of *Escherichia coli*, and Dps proteins (Fig. 1.1) where helix 5 participates in subunit–subunit interactions along the twofold axes of the cage. In most of the maxi-ferritins, the di-iron oxidore-ductase (ferroxidase or F_{ox}) center is located in the central region of the four-helix subunit bundle and is composed of residues from all four helices of the bundle. In spite of the structural differences between maxi- and mini-ferritins and the large sequence differences (>80%), all ferritins remove Fe(ii) from the cytoplasm, catalyze Fe(ii)/O oxidoreduction, provide protein-caged, concentrated iron reservoirs, and act as antioxidants [10, 11].

1.2.2   Iron Entry

Central to the understanding of ferritin iron uptake and release is the route of iron ion entry and exit through the protein cage. Both the theoretical (electrostatic calcu-lations) and experimental (x-ray crystal structure, site-directed mutagenesis) studies of different ferritins converge on the ion channels around the threefold cage axes as the iron-uptake route (Fig. 1.2). Early x-ray crystal structures [18, 19] and elec-trostatic calculations indicated iron entry along the 12 Å long hydrophilic threefold channels. Recently, a line of metal ions in the channel was observed in ferritin chan-nels (Fig. 1.2b) much like those in other ion channels [20] that connect the external threefold pores to the internal cavity [12]. Electrostatic potential energy calculations describe a gradient along the threefold channels of maxi-ferritins that can drive metal ions toward the protein interior cavity [10, 21, 22]. Site-directed mutagenesis studies confirmed the role of the negatively charged residues both on oxidation [23] and on diferric peroxo (DFP) formation, where selective effects of different carboxylate groups were also observed [24]. In the mini-ferritin, *Listeria innocua* Dps (*Li*Dps), channel carboxylates also control iron oxidation rates that, in contrast to maxi-ferritins, also control iron exit rates [25]. A functional role of the threefold channel in transit of Fe(II) ions through the ferritin cage to and from the inner cavity of ferritin was indicated by crystallographic studies on many ferritins (human, horse, mouse, frog) where divalent metal ions (Mg^{2+}, Co^{2+}, Zn^{2+}, and Cd^{2+}) are observed in the channels [11, 12, 26, 27]. However, how the hydrated Fe(II) with diameter of 6.9 Å passes through the ion channel constrictions, diameter about 5.4 Å, (Fig. 1.2) is not known. A possible explanation might be fast exchange of labile aqua ligands of Fe(II) with the threefold pore residues [28] or dynamic changes in the ion channel structure or both.

1.3 FERRITIN CATALYSIS

Ferritin nanocage proteins sequester and concentrate iron inside the cell by iron/O_2 chemistry. In bacteria and plants, ferritins are usually found as homopolymers composed of H-type subunits whereas in vertebrates, they are heteropolymers of 24 subunits, comprising two types of polypeptide chains called the H and L subunits [10]. The ratios of H to L subunits are tissue specific, with more H subunits present in the heart whereas L subunit contents are more in the liver [10]. Ferritin H subunits possess catalytic, di-Fe(II)/O_2 oxidoreductase sites (ferroxidase/F_{ox}) at the center of the 4-α-helix bundle of each eukaryotic maxi-ferritin subunit that rapidly oxidize Fe(II) to Fe(III). By contrast, the L subunit possesses amino acid residues known as the nucleation sites that provide ligands for binding Fe(III); in this case, initiation of crystal growth and mineralization occurs on the inside surface the ferritin protein nanocage. Another type of ferritin subunit is found in the amphibians known as "M" type that closely (85% sequence identity) resembles the vertebrate H type [12, 29]. In this section structure–function relationships in ferritin catalysis in eukaryotic maxi-ferritins and characterizations of reactive intermediates are described. Detection of reaction intermediates in heme-containing bacterioferritins and mini-ferritins (Dps proteins) with hydrogen peroxide as the oxidant has been elusive.

1.3.1 Spectroscopic Characterization of μ-1,2 Peroxodiferric Intermediate (DFP)

Fe(II) ions from the solution move to the inner cavity of the ferritin nanocage through hydrophilic threefold channels. At the channel exits, on the inner surface of the protein cage, the ions are directed to each of the 24 oxidoreductase sites where oxidization to Fe(III) by dioxygen is completed. Transit and oxidation requires only milliseconds, whereas solid hydrated ferric oxide mineral formation takes many hours [13]. The

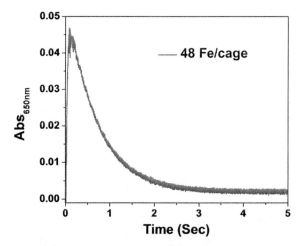

FIGURE 1.3 A typical time course of DFP formation and decay. The reaction conditions for 48 Fe(II)/cage are mixing of equal volumes of protein solution (4.16 μM wild-type frog M ferritin in 200 mM Mops, pH 7.0, 200 mM NaCl and 0.2 mM FeSO$_4$ solution in 1.0 mM HCl) in air (mixing time <10 msec) at 20°C (Π-Star, Applied Photophysics); absorbance values were measured at 650 nm, the λ_{max} for the DFP intermediate in the Fe(II)/O$_2$ oxidoreductase.

process of iron oxidation catalysis and biomineralization of ferritin core proceeds via complex mechanism of formation and decay of transient intermediate DFP (Eq. 1.2); a blue complex is observed by rapid-mixing, UV-vis spectrometry [30–32].

$$\text{Ferritin} + 2\text{Fe(II)} + \text{O}_2 \longrightarrow [\text{Fe(III)}-\text{O}-\text{O}-\text{Fe(III)}]^* \xrightarrow[-\text{H}_2\text{O}_2]{\text{H}_2\text{O}} [\text{Fe(III)}-\text{O(H)}-\text{Fe(III)}]$$

$$\longrightarrow (\text{Fe(III)}-O-\text{Fe(III)})_{2-8} \longrightarrow (\text{Fe}_2\text{O}_3) \cdot (\text{H}_2\text{O})_x + 4\text{H}^+ \qquad (1.2)$$

DFP ($\lambda_{max} = 650$ nm) is the first detectable intermediate during iron oxidation at the F$_{ox}$ site of the ferritin. The complex has been characterized by a variety of spectroscopic analyses in frog M ferritin where DFP formation and decay kinetics are particularly favorable for DFP accumulation and spectroscopy (see Fig. 1.3) [33, 34]. Mössbauer parameters ($\delta = 0.62$ mm/s, $\Delta E_Q = 1.08$ mm/s), typical for ferric species but with diamagnetic ground state, indicated the presence of anti-ferromagnetic interaction [35]. Peroxodiferric complexes with similar Mössbauer parameters are also formed as early intermediates in the reaction of O$_2$ with human H ferritin and the catalytic di-iron non-heme proteins (Table 1.1). The formation and decay profiles of this transient intermediate obtained both by rapid freeze-quench (RFQ) Mössbauer and stopped-flow absorption spectroscopy were identical, and confirmed the formation of only single transient species in the solution mixture during that time regime. The molar extinction coefficient for this species was estimated to be ~1000 mM^{-1}cm^{-1} at 650 nm, using the results obtained by the two different methods mentioned above [33].

Information on the O-O bridge in DFP was obtained from RFQ resonance Raman and RFQ X-ray absorption studies. The detection of O-isotope sensitive bands

TABLE 1.1   Spectroscopic Parameters for the DFP Intermediates in Ferritin and Di-iron Carboxylate Proteins

	Optical Parameter		Mössbauer Parameter		Peroxide Binding	
Protein	λ_{max} (nm)	ε (cm^{-1}M^{-1})	δ (mm/s)	ΔE_Q (mm/s)	Mode	Reference
Frog M ferritin	650	1000	0.62	1.08	μ-1,2	[30]
Human H ferritin	650	850	0.58	1.07	μ-1,2	[29]
sMMOH	725	1500	0.66	1.51	–	[36]
RNR-R2 D84E	700	1500	0.63	1.58	μ-1,2[a]	[37]
Δ^9-desaturase	700	1200	0.68; 0.64	1.90; 1.06	μ-1,2	[38]

[a]To trap DFP, amino acid substitutions D84E/W48F variant were used [37].
MMOH, methane monooxygenase; RNR-R2, ribonucleotide reductase R2; Δ^9-desaturase, Δ^9-stearoyl-acyl carrier protein desaturase.

assigned to ν(Fe-O) at 485 and 499 cm^{-1} and ν(O-O) at 851 cm^{-1} is consistent with a μ-1,2-bridging mode of the peroxide ligand as reported for different di-iron proteins [39]. An unusually short Fe-Fe distance (2.53 Å) in the μ-1,2 peroxodiferric intermediate was detected in the early steps of ferritin biomineralization by RFQ EXAFS spectroscopy [40], which is significantly shorter than those in other di-iron proteins such as R2 and sMMOH (3.1 to 4.0 Å). This short Fe-Fe distance (2.53 Å) was proposed to have stronger O-O bond and requires a small Fe-O-O angle (106° to 107°). The unique geometry in the ferritin catalysis was suggested to favor the decay of the DFP intermediate by the release of H_2O_2 and μ-oxo or μ-hydroxo diferric biomineral precursors instead of forming high-valent Fe(IV)-O species as in R2 and sMMOH that oxidizes organic substrate (Fig. 1.4).

Both ferritin and di-iron oxygenases are members of the di-iron-carboxlyte protein family [41]. While both share the same DFP, the decay paths are different. In ferritins, diferrous is a substrate, whereas in oxygenases, diferrous is a cofactor and is retained at di-iron sites throughout catalysis. The lability of diferrous ions in ferritin in air prevents direct identification of the F_{ox} site ligands in typical ferritin protein crystals. In Ca or Mg (ferrous analogs) ferritin cocrystals, longer metal–metal distances [11, 12, 29] were observed compared to iron–iron distances or ligands from solution EXAFS [40]; the ligand sets were also slightly different than for di-ferrous binding by VTVH MCD/CD [42].

Using protein chimeras, the iron sites (1 and 2) at the active centers required for DFP formation were positively identified when iron ligands proposed to be important for catalysis were inserted into catalytically inactive L ferritin [43]. Four of the six active site residues are the same in ferritins and di-iron oxygenases (except in heme-containing BFR with all six di-iron cofactor site ligands); ferritin-specific Gln137 and variable Asp/Ser/Ala140 (at Fe2) substitute for Glu and His, respectively, at di-iron cofactor active sites (Fig. 1.4a). The contrasting properties of ferritin or di-iron oxygenases, toward Fe(II) as substrate or cofactor, reflect differences in the amino acid residues at one of the di-iron sites in the catalytic centers. Amino acid residues (ligands) at Fe1 site are shared by ferritins and di-iron cofactor proteins. However, different and weaker Fe(II) ligands are present at Fe2 site in ferritin di-iron catalytic

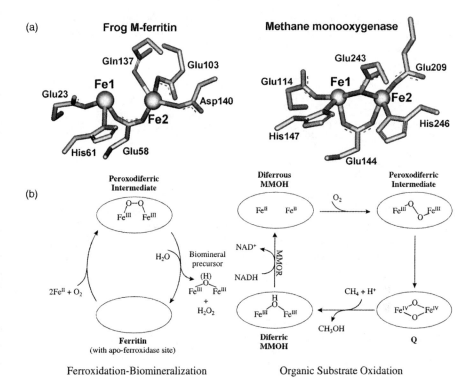

FIGURE 1.4 Comparison of the active site residues in the fate of peroxodiferric-protein (DFP) complexes during biomineralization (ferritin) and oxygen activation (MMOH). Reprinted with permission from Reference 40. Copyright 2000 AAAS. (a) Structures of the di-iron active sites in frog M-ferritin (1MFR) crystallized with Mg(ii) as a Fe(ii) homolog (left) and in di-iron cofactor site in MMOH (1FYZ) (right). Note the activity-specific ligands in the Fe2 site; ferritin specific: Gln (137) and Asp (140) are replaced by Glu (243) and His (246), respectively, in MMOH that decides the function of DFP intermediate. (b) Comparison of the fate of peroxodiferric-protein complexes during biomineralization (ferritin) and oxygen activation (MMOH). Modified from Reference 44. In ferritin, the eukaryotic ferritin peroxodiferric complex decays to diferric oxo or hydroxo mineral precursors that move from catalytic sites into nucleation channels with the release of hydrogen peroxide (left). In MMOH, DFP decay produces a high valent oxidant (Q) that oxygenates organic substrates and forms the diferric cofactor which is reduced to the initial diferrous cofactor by the redox partner, MMOR (right).

centers which permit the release of diferric oxo products and entry into the nucleation channels. Residue Ala26, identified by covariation analysis, is between large numbers of catalytically active and inactive ferritin subunits proximal to the active site. Ala26 is critical in normal turnover and release of catalytic products (diferric oxo) into the nucleation channels of eukaryotic ferritins [24] and is replaced by Ser in some L subunits, which are catalytically inactive. The loss of flexibility around the active sites also coincides with the loss of activity [12, 44].

Within the ferritin superfamily, two alternatives to the eukaryotic Fe1 and Fe2 sites at catalytic centers are known. In mini-ferritins, catalytic centers have fewer iron ligands. The substrate iron is usually bound between by ligands contributed by two different subunits at the subunit dimer interface [15, 25]. In heme bacterioferritins, the Fe2 site is similar to the di-iron oxygenase cofactor sites; iron serves both as substrate and as cofactor in BFR.

1.3.2   Kinetics of DFP Formation and Decay

The iron oxidation product in all the ferritins absorbs in the range 300–420 nm; due to the Fe(III) oxy absorption, an unresolvable mixture of DFP, Fe(III) oxo/hydroxo dimer, and Fe(III) biomineral results. However, some ferritins, in addition to this absorption band, also exhibit a broad band in the range 600–720 nm, which arises due to the formation of DFP with absorption maxima around 650 nm [10, 31, 45, 46].

In the chimeric ferritin, the Fe1 site (e, ExxH) was insufficient for ferroxidase activity [43]. Both Glu-Glu-xx-His and Glu-Gln-xx-Asp were required. In wild-type ferritins, when both the ligand set of di-iron oxygenase cofactors were created by amino acid substitution (Fe2: Gln137→Glu or Asp140→His), catalytic activity (DFP formation) was destroyed and the Fe(III)O formation rate decreased 40-fold [44]. The kinetics of ferroxidase activity in ferritin with di-iron cofactors residues illustrates the importance of ferritin-specific residues in the Fe2 site in ferritin catalytic centers for the DFP formation and its decay to different species in ferritin and di-iron cofactor proteins as described earlier.

Conserved amino acid residues located at the threefold axis (Asp127 and Glu130) and residues distal to the active sites influence the kinetics of DFP formation and iron oxidation, in addition to the Fe site ligands at the active sites [24]. Positive cooperativity, sigmoidal behavior with increasing Fe(II) concentration, occurs in the formation of DFP and Fe(III)/O species. However, the Hill coefficient for the formation of DFP and Fe(III)/O species ($n \sim 1.7 \pm 0.2$) [43, 44] is smaller than that reported ($n \sim 3$) [42] for the binding of Fe(II) at F_{ox} site in the absence of oxygen [42–44]. The Hill coefficient of 3 indicates binding of Fe(II) at three oxidoreductase sites directed by three clustered carboxylate residues (Asp127) at the opening of the Fe(II) entry channels into the cavity of the ferritin protein cage around threefold axes of the cage. Since the cooperativity of Fe(II) binding was independent of di-Fe(II) binding at each of the active sites and occurred in the absence of oxygen, protein–protein interactions involving three subunits were indicated. In the presence of dioxygen, cooperativity might differ. However, in Asp127A ferritins, cooperativity is eliminated ($n \leq 1$) indicating the role of Asp127 in the previously observed cooperativity (R. Behera and E.C. Theil, to be published). The mechanism of iron oxidation is well studied, but the DFP decay pathway/kinetics are unknown; changes in decay kinetics by amino acid substitutions occur by changing residues at the active sites—residue 140 [44], iron transport residues 26, 42, and 149 [24], and Fe(II) substrate concentrations (R. Behera and E.C. Theil, to be published).

H_2O_2 generated during initial F_{ox} reactions, when the total ferritin iron content is low, is released into solution; diferric oxymineral precursors leave the active sites traveling through the protein to the protein cavity. The generation of free radicals by the so-called Fenton reaction between the DFP decay products, such as H_2O_2, and substrates, such as Fe^{2+}, is very low in ferritin as they are spatially separated by the protein during turnover. The mechanism of DFP decays to diferric mineral precursors and H_2O_2 is not known, but hydrolysis of DFP has been proposed to be involved. At low ratios of iron : protein, H_2O_2 generated during DFP decay was correlated quantitatively with the amount of DFP accumulated at 70 ms, determined by RFQ Mössbauer spectroscopy [47]; DFP decayed to H_2O_2 with 1 : 1 stoichiometry at or below 36 Fe(II)/cage. Multiple diferric oxo species were detected by RFQ Mössbauer spectroscopy and found to decay very slowly to mineral, suggesting the existence of multiple stages in the pathway between DFP and the diferric hydroxo and H_2O_2 products.

Protons ($2H^+/Fe$) generated by hydrolysis of water, coordinated to ferric ions during catalysis/mineral nucleation/mineral growth, diffuse from the protein to the solution. Still unknown is how the diferric oxo reaction products are propelled into and along the nucleation channels (Fig. 1.5a). Possibly changes in acidity of water coordinated to iron during oxidation contribute to di-Fe(III)O release. The long catalytic turnover times [30, 48] and the hours required for the stabilization of ^{13}C–^{13}C NMR NOESY spectra after each addition of saturating amounts of Fe(II) substrate [13] both suggest slow, reversible, likely local, conformational (helix unfolding/folding?) changes are occurring during diferric oxo product release and mineral nucleation.

In some bacterial ferritin, such as the Dps proteins that protect DNA against the oxidative damage, the putative F_{ox} site is on the inner surface of the protein cavity at the dimer interfaces, rather than in the center of each subunit, as seen in eukaryotic ferritin. Although both types of ferritin make iron minerals, a property of Dps proteins is the preferential use of H_2O_2 as the iron oxidant (Eq. 1.3) during ferritin core formation [10].

$$2Fe^{2+} + H_2O_2 + (H_2O)_{x+1} \rightarrow [?] \rightarrow Fe_2O_3(H_2O)_x + 4H^+ \qquad (1.3)$$

1.4 PROTEIN-BASED FERRITIN MINERAL NUCLEATION AND MINERAL GROWTH

Ferritin minerals naturally vary in average size, phosphate content, and crystallinity depending on the cytoplasmic environment. The average mineral size in ferritins isolated from natural tissues represents a distribution of mineral sizes, which coincides with a distribution of densities within a s ample of the protein mineral complex. The number of iron atoms/cage mineral can vary from zero ("apoferritin") to many thousands. Sub-fractions of ferritin, varying in both mineral size and cross-links between protein cage sub units, can be purified from natural ferritin preparations by sedimentation in sucrose gradients. When iron mineral from natural tissue ferritin isolates is dissolved and the iron mineral is reconstituted in apoferritin (empty protein cages) *in vitro*, the distribution of ferritin mineral sizes is more narrow than the mineral sizes present in natural tissue ferritins [49, 50]. Why mineral sizes vary among ferritins

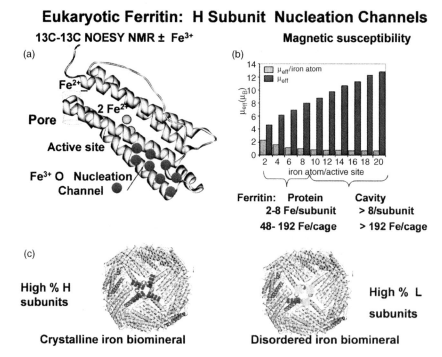

FIGURE 1.5   Protein-based control of iron biomineral order in eukaryotic ferritins. (a) Location of Fe(III) during multiple oxidoreductase turnover cycles at ferritin active sites, determined by residue broadening in $^{13}C–^{13}C$ solution NOESY; note that four Fe(II) catalytic cycles/active sites, are required for iron to reach the cavity entrances Adapted with permission from Reference 1. Copyright 2011 Curr Opin Chem Biol. (b) Changes in Fe(III)O magnetic susceptibility of growing ferritin mineral precursors inside the nucleation channels during multiple turnovers at the active sites. Reprinted with permission from Reference 13. Copyright 2010 Proc. Nat'l Acad. Sci. U.S.A. (c) An illustration of proximity effects for multiple H (catalytically active) or multiple L (catalytically inactive) animal ferritin subunits on organized mineral nucleation and biomineral growth, drawn using Pymol and PDB file 1MFR.

from *in vivo* sources remains unknown, but is likely the results of complex variations in iron delivery and iron utilization as well as subcellular distributions of ferritins in different tissue/cell types.

Ferritin minerals with a high phosphate content relative to iron, as in bacteria and plants (Table 1.2), are amorphous. The phosphate content of the bacterial cytoplasm and plant plastids, which have endosymbiotic ancestors that were single-celled organisms [51], is relatively high. If ferritin protein cages isolated with more ordered minerals are demineralized and reconstituted in solution with minerals in the presence of large amounts of phosphate, the minerals are disordered [52, 53], indicating the role of the mineralization environment on mineral structure. In the case of plants, ferritin is synthesized in the cytoplasm but is targeted to the high-phosphate plastids. Such observations, combined with the amorphous, high-phosphate structure of plant

TABLE 1.2   Natural Variations in Ferritin Average Mineral Sizes and Phosphate Content

Ferritin Source	Mineralized Ferritin Iron, (Average Number Atoms/Cage)	Ferritin Mineral-Iron : Phosphate (Moles/Mole)	Reference
Human thalassemic liver	2500	2 : 1	[37]
Horse Spleen	2000	8–10 : 1	[55]
Pea[a]	1800–2000	2.8 : 1	[51, 52]
Azotobacter Vinelandii	800–1200	1.5–1.9 : 1	[47]
Pseudomonas Aeruginosa	700	1.7 : 1	[56]
E. coli heme-ferritin (BFR)	25–75	1.4–2.2 : 1	[57]

[a]Plant ferritins are sometimes called phytoferritins because of the biological source [58]. However, structurally, plant ferritins vary in sequence only 40% from animals and are clearly in the eukaryotic ferritin family, contrasting with the 80% sequence divergence between bacterial and eukaryotic ferritins [9].

ferritin minerals has led to the conclusion that in plants, ferritins are mineralized inside the plastid [52, 53].

Minerals in ferritin protein cages from animal tissues can be either disordered or relatively ordered even though the mineral phosphate content is relatively constant [52, 54]. Recent evidence suggests that the control of mineral order in animal ferritins is an inherent property of the protein cages. In a study of the amino acids that were near Fe(III) during ferritin mineral nucleation, a previously unknown channel was discovered [13]. In the channel, the diferric oxy products of ferritin catalysis from each active site interact during multiple catalytic turnovers, forming tetrameric ferric oxy species and larger. The mineral nuclei are dispersed within the channels, apparently linearly along the 20 Å channel, and emerge at exits into the cavity around the fourfold symmetry axis of the cage. As many as eight Fe(III) ions are required to reach the end of the cavity, suggesting that iron mineral nuclei of significant size emerge into the cavity of the protein cage (Figs. 1.1, 1.5). Each nucleation channel exit is near the exits of three other subunits around the fourfold symmetry axes of the protein cage (Figs. 1.1, 1.4), which facilitate the ordered interactions of mineral nuclei from four subunits and the buildup of highly ordered ferritin minerals.

If mineral nuclei from each active site are directed through intra-cage nucleation channels to exits from four subunits that are symmetrically clustered to form ordered minerals, how then do animals form less ordered ferritin mineral observed [54] in some tissue? The answer appears to reflect a unique gene product in animal ferritins, the L subunit which is encoded in an apparently duplicated/modified H ferritin gene; L subunit deficiencies are related to changes in cell proliferation [59]. L subunits are catalytically inactive and lack some of the residues required for catalysis, such as Glu23 and Gln137 (Fig. 1.4a), and also the residues required for protein-channel nucleation, such as Ala26 [24]. H and L subunits spontaneously self-assemble into cages with the ratio of H : L subunits genetically controlled and varied for each animal tissue. In the liver, for example, ferritin protein cages have a large number of catalytically inactive L subunits and a relatively small number of catalytically active H subunits, whereas in the heart, ferritin proteins predominately have H subunits. Since there are few catalytic sites and few nucleation channels in a ferritin rich in L subunits, as in liver, the ferritin mineral should be poorly ordered (Fig. 1.5), as is, in fact, the case [54]. In the high H subunit ferritin from heart, where there are many catalytic sites and nucleation channels in each cage, multiple mineral nuclei will emerge in the cavity near mineral nuclei from many other subunits (Fig. 1.5c,d), leading to a highly crystalline mineral, which indeed it is [54]. The physiological advantage of tissue-specific ferritin minerals in animals with different degrees of order is not understood, but may relate to the rates of mineral dissolution and/or local dioxygen concentrations. Liver ferritin is a reservoir of iron for the entire body, recruiting iron quickly for distribution to other tissue, for example, after blood loss when increased synthesis of new red blood cells (hemoglobin and iron-rich) is required. A disordered mineral in the L-subunit-rich liver ferritin may dissolve more quickly in response to the sudden biological iron need. Conversely, heart ferritin, which accumulates highly crystalline minerals in the H-subunit-rich heart ferritin may reflect the antioxidant properties of ferritin minerals in cells that have high, local dioxygen concentrations and need to sequester iron more tightly until regulated opening/unfolding of the ion channel exit pores [1].

1.5 IRON EXIT

Regulating iron entry/exit channels can be important in the application of ferritin cages to nanochemistry and nanotechnology, and is also important for the growth and the survival of living organisms. *In vivo* iron release from ferritin and turnover of the empty protein cages is complex; it occurs in at least two sites within the cells: proteasomes, the supramolecular protein complexes that mediate ferritin cage and other protein degradation in the cytoplasm after iron is released; and lysosomes, acidic, subcellular compartments with hydrolytic enzymes and membrane pumps to transport to the cytoplasm after degradation of the protein cages [10, 60]. Required to dissolve ferritin minerals under physiological conditions, based on solution studies, are reductants, thought to be $FMNH_2$, protons, water for ion rehydration, and external Fe(II) transporters that are yet to be identified.

Localized changes in ferritin protein cage structure contribute significantly to the rates of Fe(II) exit from the caged ferritin minerals and regulate the rates of ferritin mineral reduction and rehydration. The Fe(II) ion entry channels, formed by helix-loop-helix motifs at the threefold symmetry axes of the cage, are also the Fe(II) exit channels. However, in contrast to the use of negatively charged residues and constrictions to propel cations into the cage [10, 22], residues controlling the rates of mineral dissolution in eukaryotic ferritins are more involved with helix folding/unfolding. Localized unfolding in ferritin protein cages is controlled by hydrophobic amino acids such as leucine, an arginine–aspartate salt bridge [16, 17, 61, 62], and by the addition of external chaotropes such as urea or guanidine [7]. Heptapeptides, identified as binding, from a combinatorial library can either increase or decrease localized folding and Fe(II) release from ferritin minerals [63]. Conserved Arg72 and Asp122, which form salt bridges at the openings of the ion channels, and conserved Leu110 interact with conserved Leu134 deep in the channels; all have a significant impact on folding of the threefold channel; Glu130 and Asp127 by contrast have no effect on iron exit but control ion entry.

The N-terminal extension of the 4-α-helix bundle lies near the external opening of the ion entry pores. Substitutions Arg72Asp and Asp122Arg increase mineral dissolution and Fe(II) exit [62]. In protein crystal structures of ferritin M Arg72Asp, the N-terminal extension is completely disordered apparently because of disruption the H-bond network between Gln5/Arg72, which is linked to loop Asp122 by a salt bridge; protein crowding restores normal function [64]. Such studies show that the N-terminal extensions of the ferritin four-helix bundles act as exit gates for the ion pores in eukaryotes and contrast sharply with mini-ferritins where negatively charged residues appear to control both cation entry and exit [25]. Such variety in ion channel function and control between maxi- and mini-ferritins should find useful applications of ferritin cages in nanoscience.

1.6 SYNTHETIC USES OF FERRITIN PROTEIN NANOCAGES

Ferritin protein nanocages are currently used in a number of applications that are briefly summarized here and described in detail elsewhere in the book. Examples include catalysis [5, 65, 66] (Chapter 7), templating of nanoparticles/materials [2, 4, 6, 67], delivery of imaging agents [3, 4, 67] (Chapter 11), and nanoelectronics [68, 69] (Chapters 11–12). Among the metal ions used in ferritin-based catalysts or materials are Au, Pd, Rh ions Pt, Ni, Cr, CdS, Ti, Eu, Ti, Co, and Fe [2, 67].

The unusual protein stability of ferritin cages means ferritin protein cages can be used under relatively extreme conditions. Ferritin is naturally stable in aqueous solutions (heat up to 80 °C), chaotropes such as 6 M urea or guanidine (at pH 7), and detergent such as 1% sodium dodecyl sulfate (SDS) [70]. However, coupling of long chain hydrocarbons (C9, C12, C14) allows reversible dissolution of intact ferritin proteins cages in organic solvents such dichloromethane, ethyl acetate, and toluene [71], and low pH/high pH disassembly/assembly has allowed entrapment in aqueous systems [72].

Availability of many natural metal-binding sites in ferritin, provided by the disproportionately large fraction of amino acid side chains with negative charges (over 700/24 subunit maxi-ferritin cage), makes ferritin cages particularly attractive for metallocatalysts and nanomaterials. During the natural function of ferritin, such sites are important in cage transport of iron ions (Fe^{2+}), binding the catalytic substrate (di-Fe^{3+})O, and binding the catalytic product and mineral precursors. Many nanochemistry/nanomaterial studies use a commercially available ferritin rich in L subunit [2] catalytic centers and mineral nucleation channels and is specific to animals.

Ferritin L subunits, specific to animals, lack residues for the natural Fe^{2+}/O_2 catalysis and for $Fe^{3+}O$ transport and appear to be important in regulating iron mineral order [1, 24]. In addition, new functional iron-ferritin protein sites are still being discovered in ferritin [13, 24]. Thus, the full potential of ferritin metal-binding sites for catalysis and protein-controlled nanoparticle growth remains to be reached.

1.6.1   Nanomaterials Synthesized in Ferritins

The oldest use of ferritin cages in nanochemistry has been as a nanomaterial template [3, 4, 67]. When used to template nanomaterials, cations enter the cavity through the ion entry channels (Figs. 1.1, 1.4) and self-assemble into materials inside the protein cage. Ferritin protein nanocages contain arrays of ion channels, which, because of the protein properties, are uniquely soluble in aqueous solvents, contrasting with the related membrane ion channel proteins. Disassembled/reassembled ferritin cages have been used to encapsulate a DNA probe and fluorescence reporters for picomolar bioassays and immunoassays [3, 4].

Among the materials made using ferritin protein nanocage templates are semiconductors (CdS nanodots) gold nanospheres, and a variety of iron particles with magnetic properties [3, 4, 67]; such iron minerals when used for imaging are likely dissolved by the biological pathways for natural ferritin iron biominerals.

When ferritin protein cages are used as a template, the uniform nanoparticles are limited by the interior cavity of the protein cage. Thus, the sizes are <8 nm diameter when 24 subunit, maxi-ferritin cages from bacteria, plants, and animals are used; and <5 nm in diameter when 12 subunit mini-ferritin cages are used (See Section 1.2). Ferritin templates for larger structures have been developed by modification of cage surface residue that enhance ferritin aggregation [3, 4, 67] by stabilization of 3D ferritin crystals with silica in the interstitial spaces of the protein crystals [73], or as 2D lattices on various substrates or aggregates, for example, [74, 75]. When empty ferritin protein cages are used as starting material, they are sometimes called "apoferritin" because they have no mineral. Variant protein cages for templating are produced as recombinant proteins from wild-type mammalian sequences. Negatively charged amino acids, naturally on the interior of the cage/surface of the cavity, create local cation concentrations high enough to initiate nanoparticle formation. In some cases design changes that facilitate metal binding are introduced by protein engineering, exemplified by the production of silver and gold nanoparticles inside ferritin [76].

The outer surfaces of ferritin cages have also been recruited for nanosyntheses, for example, [77]. Here, negatively charged caged residues normally on the outside of a mini-ferritin protein cage from the bacterium *Listeria innocua* were made available as a templating surface by turning the cage "inside out" using immobilization, two-step modifications, and elution techniques. Other approaches use modifications of the outer surface. The wide range of amino acid sequences available among ferritin protein cages, which can vary as much as 80% [9], as well as the two different cages sizes, indicate the vast untapped potential of ferritins cages in nanoparticle synthesis.

1.6.2   Ferritin Protein Cages in Metalloorganic Catalysis and Nanoelectronics

Ferritin protein nanocages have been used as catalysts to produce polymers of defined molecular weight, with a narrow size distribution. Palladium and rhuthenium metalloorganic complexes bind at amino acid side chains in the cage, for example, [5, 65, 66]. Changes in organic metallic binding densities, stabilities, and location have recently been achieved with engineered amino acid substitutions, such as selective insertion of cysteine or histidine, especially on the interior surfaces of the cage, facing the cavity. Reactants diffuse through the ferritin protein cage to the nanocavity (Fig. 1.1) where polymerization occurs. Manipulation of ferritin protein cage pore/channels to enhance reactant entry through pore unfolding by amino acid substitutions or with added chaotropes [7] or to control polymer properties using modified eukaryotic sites/extrusion channels for controlled polymerization order and crystallinity are subjects for future exploration and development.

Larger, inorganic catalysts synthesized inside ferritin nanocages have been used to produce with a variety of spatial distributions. In this approach, empty protein cages are distributed in the desired array using surface properties of the protein [6]. Then catalytic nanoparticles are produced inside the cages. Next, the protein cages are selectively removed leaving the nanoparticles exposed. Finally, the nanoparticles deposited in arrays controlled by ferritin protein cage properties catalyze carbon nanotube growth in arrays determined by the controlled pattern of ferritin deposition. Single Pt nanoparticles, synthesized inside ferritin and embedded inside NiO film, have demonstrated low power and stable operation in a memory cell [69]. Future manipulations of ferritin cage surface structure should expand an array of deposition possibilities, and manipulation of ferritin protein natural metal-binding sites should produce more selective control of synthesized nanoparticle properties.

1.6.3   Imaging and Drug Delivery Agents Produced in Ferritins

Entrapment inside ferritin protein cages has been used to deliver hydrophilic drugs, imaging agents such as fluorophores radionuclides, nuclear medicine such as ^{177}Lu, ^{90}Y, and antitumor drugs such as doxorubicin [3, 4, 67], cisplatin, carboplatin, and ozaloplatin [78]. Dissociation/association of ferritin protein cages at low/high pH has also been used to trap DNA probes or probes for bioassay and immunoassay at picomolar concentrations [3].

Recently, modified ferritin targeted to alpha-v beta-3 integrins have found use in imaging vascular inflammation and angiogenesis, and influencing metabolism in C32 human melanoma cells [79]. Ferritin-containing magnetic nanoparticles or conjugated to Cy5.5 has been successfully used to the image cells (vascular macrophages) by magnetic resonance or fluorescence that are key to understanding atherosclerotic plaques and heart disease [80].

The natural ability of living cells to selectively recognize and incorporate exogenous ferritin [38, 81, 82] reflect the intrinsic and complex surfaces of ferritin protein cages that themselves are only beginning to be understood [1]. Coupled with modifications of ferritin protein cages that modulate interactions with specific disease cells, the vast potential of using ferritin protein cages lies mostly untapped in nanomedicine. Further, the emphasis of many ferritin-based nanomaterial studies on L subunit maxi-ferritin without active sites, or mini-ferritins (Dps proteins) with active sites on the inner cage surface, leaves underutilized ferritin protein-based control of mineral nucleation and growth associated with buried catalytic sites and mineral nucleation channels (Section 1.4, Fig. 1.5). Exploitation of the inherent properties of H ferritin protein cages that influence the order of ferritin nanomineral in Nature remains for future development of ferritin-based synthetic nanomaterials.

1.7 SUMMARY AND PERSPECTIVES

Ferritins are water-soluble, highly symmetrical, protein nanocages with ion channels for Fe(II) entry to Fe/O oxidoreductase sites that initiate the synthesis of hydrated ferric oxide minerals inside large (60% v/v) cage cavities. The protein-caged, iron biominerals provide required intracellular iron concentrates for cell growth and division. Ancient in origin, ferritins self-assemble from four α-helix bundle protein subunits to achieve a very stable structure that is soluble in aqueous solvents. Each funnel-shaped ion channel in ferritin cages has external pores and directs ferrous substrate through the cage to multiple protein oxidoreductase sites that initiate biomineral nucleation; ferrous ion exit from dissolved ferritin biomineral also occurs through the external pores ion channels, dependent on a different set of residues than ion entry. Recent identification in ferritin protein cages of sites that control mineral order/crystallinity (nucleation channels) provides a new design dimension for ferritin-based nanotechnology.

Current uses of ferritin protein nanocages in nanotechnology are as follows: templates for nanomaterials, arrays used in nanodevice production, catalysts for nanopolymer synthesis, reagents for cellular imaging and picomole bioanalytical chemistry; these use ferritin protein cages in relatively native states.

Design modifications have mainly been of nonconserved residues for catalysts binding or cell recognition. An exciting frontier is the modification of ferritin functional subdomains, such as the external pores, ion channels, and protein-based mineral nucleation/crystallization controls to increase the sophistication of ferritin-based nanomaterials, nanocatalysts, and nanodevices. Nature has manipulated ferritin protein cages to produce iron particles with defined properties. Scientists have just begun to explore the potential of engineered ferritin protein cages in nanotechnology.

ACKNOWLEDGMENTS

The authors are grateful for the financial support of the NIH and the CHORI Foundation during the preparation of this chapter and the intellectual contributions of colleagues and members of the Theil Group.

REFERENCES

[1]   Theil, E. C. *Curr. Opin. Chem. Biol.* **2011**, *15*, 304–311.

[2]   Mann, S. *Nat. Mater.* **2009**, *8*, 781–792.

[3]   Maham, A.; Tang, Z.; Wu, H.; Wang, J.; Lin, Y. *Small* **2009**, *5*, 1706–1721.

[4]   Uchida, M.; Kang, S.; Reichhardt, C.; Harlen, K.; Douglas, T. *Biochim. Biophys. Acta* **2010**, *1800*, 834–845.

[5]   Takezawa, Y.; Bockmann, P.; Sugi, N.; Wang, Z.; Abe, S.; Murakami, T.; Hikage, T.; Erker, G.; Watanabe, Y.; Kitagawa, S.; Ueno, T. *Dalton Trans.* **2011**, *40*, 2190–2195.

[6]   Yamashita, I.; Iwahori, K.; Kumagai, S. *Biochim. Biophys. Acta* **2010**, *1800*, 846–857.

[7]   Liu, X.; Jin, W.; Theil, E. C. *Proc. Natl. Acad. Sci. USA* **2003**, *100*, 3653–3658.

[8]   Theil, E. C.; Goss, D. J. *Chem. Rev.* **2009**, *109*, 4568–4579.

[9]   Bevers, L. E.; Theil, E. C. Maxi- and mini-ferritins: minerals and protein nanocages. In *Progress in Molecular and Subcellular Biology*; Muller, W. E. G., Ed.; Springer-Verlag: Berlin-Heidelberg, **2011**; pp 18.

[10] Liu, X.; Theil, E. C. *Acc. Chem. Res.* **2005**, *38*, 167–175.

[11] Crichton, R. R.; Declercq, J. P. *Biochim. Biophys. Acta* **2010**, 1800, 706–718.

[12] Tosha, T.; Ng, H. L.; Bhattasali, O.; Alber, T.; Theil, E. C. *J. Am. Chem. Soc.* **2010**, *132*, 14562–14569.

[13] Turano, P.; Lalli, D.; Felli, I. C.; Theil, E. C.; Bertini, I. *Proc. Natl. Acad. Sci. USA* **2010**, *107*, 545–550.

[14] Ilari, A.; Stefanini, S.; Chiancone, E.; Tsernoglou, D. *Nat. Struct. Biol.* **2000**, *7*, 38–43.

[15] Chiancone, E.; Ceci, P. *Biochim. Biophys. Acta* **2010**, *1800*, 798–805.

[16] Takagi, H.; Shi, D.; Ha, Y.; Allewell, N. M.; Theil, E. C. *J. Biol. Chem.* **1998**, *273*, 18685–18688.

[17] Theil, E. C.; Liu, X. S.; Tosha, T. *Inorganica Chim. Acta* **2008**, *361*, 868–874.

[18] Banyard, S. H.; Stammers, D. K.; Harrison, P. M. *Nature* **1978**, *271*, 282–284.

[19] Stefanini, S.; Desideri, A.; Vecchini, P.; Drakenberg, T.; Chiancone, E. *Biochemistry* **1989**, *28*, 378–382.

[20] Jiang, Y.; Lee, A.; Chen, J.; Ruta, V.; Cadene, M.; Chait, B. T.; MacKinnon, R. *Nature* **2003**, *423*, 33–41.

[21] Douglas, T.; Ripoll, D. R. *Protein Sci.* **1998**, *7*, 1083–1091.

[22] Takahashi, T.; Kuyucak, S. *Biophys. J.* **2003**, *84*, 2256–2263.

[23] Levi, S.; Santambrogio, P.; Corsi, B.; Cozzi, A.; Arosio, P. *Biochem. J.* **1996**, *317 (Pt 2)*, 467–473.

[24] Haldar, S.; Bevers, L. E.; Tosha, T.; Theil, E. C. *J. Biol. Chem.* **2011**, *286*, 25620–25627.

[25] Bellapadrona, G.; Stefanini, S.; Zamparelli, C.; Theil, E. C.; Chiancone, E. *J. Biol. Chem.* **2009**, *284*, 19101–19109.

[26] Lawson, D. M.; Artymiuk, P. J.; Yewdall, S. J.; Smith, J. M.; Livingstone, J. C.; Treffry, A.; Luzzago, A.; Levi, S.; Arosio, P.; Cesareni, G.; Thomas, C. D.; Shaw, W. D.; Harrison, P. M. *Nature* **1991**, *349*, 541–544.

[27] Toussaint, L.; Bertrand, L.; Hue, L.; Crichton, R. R.; Declercq, J. P. *J. Mol. Biol.* **2007**, *365*, 440–452.

[28] Barnes, C. M.; Theil, E. C.; Raymond, K. N. *Proc. Natl. Acad. Sci. USA* **2002**, *99*, 5195–5200.

[29] Ha, Y.; Shi, D.; Small, G. W.; Theil, E. C.; Allewell, N. M. *J. Biol. Inorg. Chem.* **1999**, *4*, 243–256.

[30] Treffry, A.; Zhao, Z.; Quail, M. A.; Guest, J. R.; Harrison, P. M. *Biochemistry* **1995**, *34*, 15204–15213.

[31] Fetter, J.; Cohen, J.; Danger, D.; Sanders-Loehr, J.; Theil, E. C. *J. Biol. Inorg. Chem.* **1997**, *2*, 652–661.

[32] Bou-Abdallah, F.; Papaefthymiou, G. C.; Scheswohl, D. M.; Stanga, S. D.; Arosio, P.; Chasteen, N. D. *Biochem. J.* **2002**, *364*, 57–63.

[33] Pereira, A. S.; Small, W.; Krebs, C.; Tavares, P.; Edmondson, D. E.; Theil, E. C.; Huynh, B. H. *Biochemistry* **1998**, *37*, 9871–9876.

[34] Moënne-Loccoz, P.; Krebs, C.; Herlihy, K.; Edmondson, D. E.; Theil, E. C.; Huynh, B. H.; Loehr, T. M. *Biochemistry* **1999**, *38*, 5290–5295.

[35] Krebs, C.; Bollinger, J. M., Jr.; Theil, E. C.; Huynh, B. H. *J. Biol. Inorg. Chem.* **2002**, *7*, 863–869.

[36] Bollinger, J. M.; Krebs, C.; Vicol, A.; Chen, S.; Ley, B. A.; Edmondson, D. E.; Huynh, B. H. *J. Am. Chem. Soc.* **1998**, *120*, 1094–1095.

[37] Tinberg, C. E.; Lippard, S. J. *Acc. Chem. Res.* **2011**, *44*, 280–288.

[38] Li, L.; Fang, C. J.; Ryan, J. C.; Niemi, E. C.; Lebron, J. A.; Bjorkman, P. J.; Arase, H.; Torti, F. M.; Torti, S. V.; Nakamura, M. C.; Seaman, W. E. *Proc. Natl. Acad. Sci. USA* **2011**, *107*, 3505–3510.

[39] Moenne-Loccoz, P.; Krebs, C.; Herlihy, K.; Edmondson, D. E.; Theil, E. C.; Huynh, B. H.; Loehr, T. M. *Biochemistry* **1999**, *38*, 5290–5295.

[40] Hwang, J.; Krebs, C.; Huynh, B. H.; Edmondson, D. E.; Theil, E. C.; Penner-Hahn, J. E. *Science* **2000**, *287*, 122–125.

[41] Nordlund, P.; Eklund, H. *Curr. Opin. Chem. Biol.* **1995**, *5*, 758–766.

[42] Schwartz, J. K.; Liu, X. S.; Tosha, T.; Theil, E. C.; Solomon, E. I. *J. Am. Chem. Soc.* **2008**, *130*, 9441–9450.

[43] Liu, X.; Theil, E. C. *Proc. Natl. Acad. Sci. USA* **2004**, *101*, 8557–8562.

[44] Tosha, T.; Hasan, M. R.; Theil, E. C. *Proc. Natl. Acad. Sci. USA* **2008**, *105*, 18182–18187.

[45] Bou-Abdallah, F.; Zhao, G.; Mayne, H. R.; Arosio, P.; Chasteen, N. D. *J. Am. Chem. Soc.* **2005**, *127*, 3885–3893.

[46] Bou-Abdallah, F.; McNally, J.; Liu, X. X.; Melman, A. *Chem. Commun.* **2011**, *47*, 731–733.

[47] Jameson, G. N.; Jin, W.; Krebs, C.; Perreira, A. S.; Tavares, P.; Liu, X.; Theil, E. C.; Huynh, B. H. *Biochemistry* **2002**, *41*, 13435–13443.

[48] Waldo, G. S.; Theil, E. C. Ferritin and Iron Biomineralization. In *Comprehensive Supramolecular Chemistry, Bioinorganic Systems, Vol. 5*; Suslick, K. S., Ed.; Pergamon Press: Oxford, UK, **1996**; pp 65–89.

[49] Mertz, J. R.; Theil, E. C. *J. Biol. Chem.* **1983**, *258*, 11719–11726.

[50] May, C. A.; Grady, J. K.; Laue, T. M.; Poli, M.; Arosio, P.; Chasteen, N. D. *Biochim. Biophys. Acta* **2010**, *1800*, 858–870.

[51] Gould, S. B.; Waller, R. F.; McFadden, G. I. *Annu. Rev. Plant Biol.* **2008**, *59*, 491–517.

[52] Wade, V. J.; Treffry, A.; Laulhere, J.-P.; Bauminger, E. R.; Cleton, M. I.; Mann, S.; Briat, J.-F.; Harrison, P. M. *Biochim. Biophys. Acta* **1993**, *1161*, 91–96.

[53] Waldo, G. S.; Wright, E.; Whang, Z. H.; Briat, J. F.; Theil, E. C.; Sayers, D. E. *Plant Physiol.* **1995**, *109*, 797–802.

[54] Pierre, T. St.; Tran, K. C.; Webb, J.; Macey, D. J.; Heywood, B. R.; Sparks, N. H.; Wade, V. J.; Mann, S.; Pootrakul, P. *Biol. Met.* **1991**, *4*, 162–165.

[55] Andrews, S. C.; Brady, M. C.; Treffry, A.; Williams, J. M.; Mann, S.; Cleton, M. I.; de Bruijn, W.; Harrison, P. M. *Biol. Met.* **1988**, *1*, 33–42.

[56] Moore, G. R.; Mann, S.; Bannister, J. V. *J. Inorg. Biochem.* **1986**, *28*, 329–336.

[57] Zhao, G. *Biochim Biophys Acta* **2010**, *1800*, 815–823.

[58] Aitken-Rogers, H.; Singleton, C.; Lewin, A.; Taylor-Gee, A.; Moore, G. R.; Le Brun, N. E. *J. Biol. Inorg. Chem.* **2004**, *9*, 161–170.

[59] Cozzi, A.; Corsi, B.; Levi, S.; Santambrogio, P.; Biasiotto, G.; Arosio, P. *Blood* **2004**, *103*, 2377–2383.

[60] De Domenico, I.; Ward, D. M.; Kaplan, J.; *Blood* **2009**, *114*, 4546–4551.

[61] Hasan, M. R.; Tosha, T.; Theil, E. C. *J. Biol. Chem.* **2008**, *283*, 31394–31400.

[62] Jin, W.; Takagi, H.; Pancorbo, B.; Theil, E. C. *Biochemistry* **2001**, *40*, 7525–7532.

[63] Liu, X. S.; Patterson, L. D.; Miller, M. J.; Theil, E. C. *J. Biol. Chem.* **2007**, *282*, 31821–31825.

[64] Tosha, T.; Behera, R. K.; Ng, H.-L.; Bhattasali, O.; Alber, T.; Theil, E. C. *J. Biol. Chem.* **2012**, *287*, 13016–13025.

[65] Wang, Z.; Takezawa, Y.; Aoyagi, H.; Abe, S.; Hikage, T.; Watanabe, Y.; Kitagawa, S.; Ueno, T. *Chem. Commun.* **2011**, *47*, 170–172.

[66] Abe, S.; Hirata, K.; Ueno, T.; Morino, K.; Shimizu, N.; Yamamoto, M.; Takata, M.; Yashima, E.; Watanabe, Y. *J. Am. Chem. Soc.* **2009**, *131*, 6958–6960.

[67] Flenniken, M. L.; Uchida, M.; Liepold, L. O.; Kang, S.; Young, M. J.; Douglas, T. *Curr. Top. Microbiol. Immunol.* **2009**, *327*, 71–93.

[68] Iwahori, K.; Yamashita, I. *Nanotechnology* **2008**, *19*, 495601.

[69] Kobayashi, M.; Kumagai, S.; Zheng, B.; Uraoka, Y.; Douglas, T.; Yamashita, I. *Chem. Commun.* **2011**, *47*, 3475–3477.

[70] Santambrogio, P.; Levi, S.; Arosio, P.; Palagi, L.; Vecchio, G.; Lawson, D. M.; Yewdall, S. J.; Artymiuk, P. J.; Harrison, P. M.; Jappelli, R. *J. Biol. Chem.* **1992**, *267*, 14077–14083.

[71] Wong, K. K.; Colfen, H.; Whilton, N. T.; Douglas, T.; Mann, S. *J. Inorg. Biochem.* **1999**, *76*, 187–195.

[72] Liu, G.; Wang, J.; Wu, H.; Lin, Y. *Anal. Chem.* **2006**, *78*, 7417–7423.

[73] Lambert, E. M.; Viravaidya, C.; Li, M.; Mann, S. *Angew. Chem. Int. Ed.* **2010**, *49*, 4100–4103.

[74] Srivastava, S.; Samanta, B.; Jordan, B. J.; Hong, R.; Xiao, Q.; Tuominen, M. T.; Rotello, V. M. *J. Am. Chem. Soc.* **2007**, *129*, 11776–11780.

[75] Saha, K.; Bajaj, A.; Duncan, B.; Rotello, V. M. *Small* **2011**, *7*, 1903–1918.

[76] Butts, C. A.; Swift, J.; Kang, S. G.; Di Costanzo, L.; Christianson, D. W.; Saven, J. G.; Dmochowski, I. J. *Biochemistry* **2008**, *47*, 12729–12739.

[77] Suci, P. A.; Kang, S.; Young, M.; Douglas, T. *J. Am. Chem. Soc.* **2009**, *131*, 9164–9165.

[78] Yang, Z.; Wang, X.; Diao, H.; Zhang, J.; Li, H.; Sun, H.; Guo, Z. *Chem. Commun.* **2007**, 3453–3455.

[79] Kitagawa, T.; Kosuge, H.; Uchida, M.; Dua, M. M.; Iida, Y.; Dalman, R. L.; Douglas, T.; McConnell, M. V. *Mol. Imaging Biol.* **2011**.

[80] Terashima, M.; Uchida, M.; Kosuge, H.; Tsao, P. S.; Young, M. J.; Conolly, S. M.; Douglas, T.; McConnell, M. V. *Biomaterials* **2011**, *32*, 1430–1437.

[81] San Martin, C. D.; Garri, C.; Pizarro, F.; Walter, T.; Theil, E. C.; Nunez, M. T. *J. Nutr.* **2008**, *138*, 659–666.

[82] Li, J. Y.; Paragas, N.; Ned, R. M.; Qiu, A.; Viltard, M.; Leete, T.; Drexler, I. R.; Chen, X.; Sanna-Cherchi, S.; Mohammed, F.; Williams, D.; Lin, C. S.; Schmidt-Ott, K. M.; Andrews, N. C.; Barasch, J. *Dev. Cell* **2009**, *16*, 35–46.

2

MOLECULAR METAL OXIDES IN PROTEIN CAGES/CAVITIES

ACHIM MÜLLER AND DIETER REHDER

2.1 INTRODUCTION

Two questions may be posed at the beginning: Why does Nature construct hybrid systems exhibiting metal oxides integrated in protein cavities? And why is the investigation of these structures of interest for scientists, especially chemists? The first question is simple to answer: As specific metal cations are important for biological processes, they are stored in a form that they can be easily taken up and released which, in many cases, is simply achieved by changing the metal oxidation state. This assertion mainly applies to oxides of 3d elements such as iron. On the other hand, scientists can learn from Nature how the mechanism of nucleation processes leading to molecular oxides proceeds in view of the fact that, in natural systems, metal oxides could definitely be shaped by internal functionalities at the cavity surfaces, acting as templates (see, e.g., Reference 1). For example, this is the case for the molybdenum and tungsten storage proteins containing Mo/W polyoxometalates (POMs) (Section 2.3) [2], where the release and uptake can be directed mainly by pH changes, as well as for oxidovanadium storage proteins (Section 2.2), the manganese cluster in photosystem II (PSII) (Section 2.4), and iron oxide clusters in iron storage and related proteins (Section 2.5).

Generally speaking, the chemist can learn from Nature how metal–metal oxide polyhedra can be linked under the confined conditions of the cavity—template or non-template controlled. Since the time of Linus Pauling's famous book "The Nature of the Chemical Bond" (cf. the chapter "The Sharing of Polyhedron Corners, Edges

Coordination Chemistry in Protein Cages: Principles, Design, and Applications, First Edition.
Edited by Takafumi Ueno and Yoshihito Watanabe.

and Faces"), assembly/nucleation processes based on simple AB_n-type polyhedra have been important research subjects, not only in inorganic chemistry, but also in fields such as discrete mathematics, crystal chemistry (discussed by Pauling in detail for silicates), materials science, and geoscience. POM chemistry in particular [3], which is dominated by the elements Mo and W, should be mentioned, as it is based on this type of assembly/nucleation processes. This research area, which represents an extreme structural versatility and includes the largest structurally well-defined inorganic discrete species known to date, is directly related to the biological scenario of the structure of Mo/W storage proteins, exemplified for the protein loaded with molybdenum and tungsten [1]. Here, the uniqueness roots in the fact that, for the first time, different POMs are independently formed, and exist independently, in one and the same cavity. No other case is known for this type of unique scenario. For the relevant POM research, two instructive discoveries should be mentioned here: (i) a protein-sized cluster, containing 368 Mo ions, with a cavity of $25 \times 25 \times 40$ Å^3 and (ii) porous capsules with internal functionalities in which the pores can be stepwise opened and closed and which allow to model biological ion transport, as well as cell response [4–7]. As a matter of fact, the sizes and structures of the metal oxides considered differ tremendously. This is due to their extremely different functions in the proteins.

In the following discussion, the term "metal oxides" encompasses neutral oxides (ferrous and ferric oxides) as well as anionic oligo- and polyoxometalates (vanadates, molybdates, tungstates) and species derived from cationic fragments ($\{V^{IV}O\}_n$, $\{Mn^{III-V}O\}_n$). Further, we will address—in the context of "protein cages/cavities"— (vacant) sites in apoproteins which are able to accommodate a metal-oxido cluster (the $\{Mn_4CaO_5\}$ cluster in the oxygen-evolving center (OEC) of PSII) and protein sites which bind oligovanadates (e.g., myosin and tubulins) or can coordinate a sizable number of partially interacting VO^{2+} ions (the vanabins).

We only refer to purely biological systems and exclude those where the protein/ virus shells/capsoids act as containers or are used for reactions in small spaces.

2.2 VANADIUM: FUNCTIONAL OLIGOVANADATES AND STORAGE OF VO^{2+} IN VANABINS

At ambient physiological conditions, vanadium mainly occurs as vanadate(v) $H_2VO_4^-$ and vanadyl (oxidovanadium(IV)) VO^{2+}, the latter affording stabilization against precipitation (due to the formation of sparingly soluble $VO(OH)_2$) by coordination to biogenic ligands. Vanadate tends to aggregate, at concentrations $>$ ca. 10 μM, to form di-, tetra-, and pentavanadates (pH > 7) and decavanadate (pH $<$ 6.3). Decavanadate $H_nV_{10}O_{28}^{(6-n)-}$ ($n = 0$–3), V_{10}, which can form by condensation of vanadate under slightly acidic physiological conditions in environments where the vanadate concentration becomes locally sufficiently high, is also kinetically comparatively stable at micromolar concentrations and in the pH regime where it should degrade, thermodynamically driven, to vanadates of lower nuclearity. V_{10} is also stabilized upon binding to (target) proteins. Decavanadate has been shown to

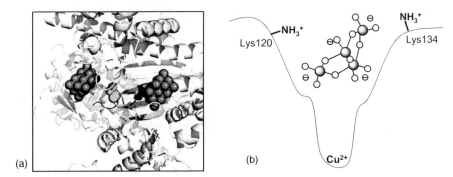

FIGURE 2.1   (a) Docking of decavanadate (red spheres) to the "back-door" region of the S1 domain of *Dictyostelium discoideum* myosin. Segments forming the phosphate-binding loop are represented in green (serine), cyan, and orange. Reprinted with permission from Reference 9. © Elsevier. (b) The interaction of tetravanadate ($V_4O_{12}{}^{4-}$; gray circles, V; empty, O) with two lysines in the pocket of the solvent channel of bovine Cu,Zn superoxide dismutase. One of the possible conformations of tetravanadate is shown (modified from Reference 10). *See insert for color representation of the figure.*

interact with Ca^{2+} homeostasis by docking to the nucleotide binding site of the calcium pump and to bind to myosin (Fig. 2.1a)—a protein which plays a pivotal role in muscle contraction—and thus inhibit actomyosin ATPase activity [8, 9]. Tetravanadate is incorporated by various phosphate-metabolizing enzymes and, particularly efficiently, by Cu,Zn superoxide dismutase, CuZn-SOD [10] (Fig. 2.1b), an enzyme which is responsible for the elimination of the superoxide radical. Both, deca- and tetravanadate also interact strongly with microtubule-associated proteins [11]. Microtubules serve as structural components in cells and are involved in various cellular processes.

The vanadyl ion VO^{2+} is intermittently formed in vanadium-accumulating ascidians (sea squirts). Sea squirts can take up vanadate from sea water and sequester vanadium, by a factor of up to 10^7 mainly in the form of aqua complexes of V^{3+}, in the highly acidic vacuoles of special blood cells, so-called vanadocytes. The intermediate VO^{2+} is loaded to and transported by small proteins termed vanabins (for *vana*dium-*bin*ding proteins) [12, 13], AsGST (*A*scidia *s*ydneiensis **g**lutathione **S** transferase) [14] and AsNramp (*A*scidia *s*ydneiensis **n**atural **r**esistance-**a**ssociated **m**acrophage **p**rotein) [15]. GSTs catalyze the coupling of glutathione to, for example, xenobiotic electrophiles and thus promote detoxification. Nramp is an important stimulating component of the immune system.

Vanabins, which are bow-shaped proteins consisting of four α-helices connected by disulfide bonds, can accommodate up to 20 VO^{2+} ions, exclusively located on the same face of the protein. According to EPR results [16a], about 84% of the VO^{2+} is present in a mononuclear form, while the remaining 16% are EPR-silent and hence magnetically interacting. Vanabins have been located in the cytoplasm of the vanadocytes (vanabins 1–4) and in the blood serum (vanabinP, VBP-129). The well-characterized vanabin2 has a molecular weight of 10,467 Da, corresponding to 91 amino acids, 14 of which are lysine and 5 of which are arginine. Vanabin2 is bow shaped; four α helices are linked by nine disulfide connectivities. These cystines provide a reducing potential, that is, vanadbin2 also acts as a vanadate reductase [16b]. The Lys and Arg are primarily on one side of the protein; binding of the vanadyl ions occurs essentially via the NH_2 functions of Lys and Arg [17]. Aspartate and other O-functional amino acids are also involved, providing an equatorial N_2O_2 donor set for VO^{2+}. The medium binding constant, at about neutral pH, is ca. 3 × 10^4 M^{-1} per VO^{2+}. The highest affinity binding site involves the amino acids Lys10 and Arg60.

The action of myosin in muscle contraction is coupled to the interaction with another muscle protein, actin, and the hydrolysis of adenosine triphosphate (ATP) to ADP and phosphate HPO_4^{2-}. Decavanadate, at concentrations of $c(v) \approx 1$ μM, tightly binds at the "back door" of the myosin–ATP complex, preventing the association of myosin and actin and thus the release of phosphate necessary for initiating muscle contraction [9]. Tetravanadate $V_4O_{12}^{4-}$, V_4, is a cyclic vanadate not subjected to protonation/deprotonation. It interacts with bovine CuZn-SOD to form (V_4)SOD and $(V_4)_2$SOD, with binding constants of $\approx 10^7$ M^{-1} [10]. There is no direct, that is, covalent, interaction between V_4 and SOD; rather, V_4 is electrostatically linked to the ammonium groups of two lysines (Lys120 and Lys134) in the solvent channel otherwise allowing access of superoxide to the copper center (Fig. 2.1b). Oligomeric vanadates, tetra- and decavanadate in particular, also bind to tubulin proteins and thus affect microtubule formation in the cytoskeleton [11]. ^{51}V NMR studies suggest that, at least in the case of tetravanadate, direct (i.e., covalent) interaction with the protein matrix occurs.

We have discussed here "vanadium oxides" in a broader context, referring to partially interacting VO^{2+} ions bound to a protein surface (in the vanabins) and vanadates docking to proteins (the decavanadate–myosin system) or interacting electrostatically with amino acids in a protein pocket (tetravanadate and Cu,Zn superoxide dismutase). The oncoming sections on Mo, W, Mn, and Fe will provide a more direct connection to metal oxides in protein matrices. Polyoxovanadates, and POMs, containing both Mo/W and V, which have so far not been investigated in this context, may be expected to behave in a manner comparable to what is being displayed in the oncoming section.

2.3 MOLYBDENUM AND TUNGSTEN: NUCLEATION PROCESS IN A PROTEIN CAVITY

During recent years, tremendous progress has been made in the understanding of the very versatile role of Mo in fundamental biological processes in which it becomes functional in protein-associated form [18]. Interestingly, biologically relevant Mo can, in a few cases, be replaced by W. Examples are enzymes containing the tungstopterin cofactor in some hyperthermophilic organisms [19]. The reason is that the chemical behavior of the two transition elements, at least in the highest oxidation state, that is, +VI, is similar. A rather unique situation was found for the Mo-storage (MoSto) protein from the nitrogen-fixing bacterium *Azotobacter vinelandii*. With respect to its function, this protein is formally comparable to the iron storage protein ferritin. The MoSto is not only capable of binding a large number of, approximately 100, biologically relevant Mo ions (see References 2 and 20) but alternatively can also bind an equivalent amount of tungsten atoms [21] (therefore termed here Mo/WSto) in the form of aggregates of the polyoxotungstate type, which has recently been structurally characterized [1]. The release of Mo and W depends on parameters such as the pH value and temperature, a behavior which can be traced back to the properties of "normal" POMs [22]. The biological process thus exhibits, as expected, similarities with the corresponding degradation processes of POMs under bulk conditions [3a, 22] (for biological control of the important biological metal-storage/release processes of the protein ferritin, which contains a nano-sized iron oxide fragment, see References 23 and 24, and Section 2.5). In the case of the Mo/WSto, we are dealing with appropriately functionalized protein pockets within the cavity of the cage-type storage protein. These pockets harbor different—and this is unique—individual polynuclear (molybdenum and) tungsten oxide aggregates. The fascinating aspect anchors in the fact that different pocket functionalities can be specifically correlated with differently directed assembly processes [1]. In the present context, we can refer to a "biological supramolecular chemical system" [25] as the pockets in the protein-cavity shell act as a polytopic host for the noncovalently or weakly bonded polyoxotungstate guests. The process is definitely analogous to the molybdate case, but has not been seriously investigated until now.

We refer here to the real biological situation in a bacterium where Mo is partially released from its storage protein to keep the nitrogen-fixation process running under conditions where there is insufficient supply of Mo from the soil: The Mo/WSto protein containing tungstates is organized as an $(\alpha\beta)_3$ hetero-hexamer consisting of a trimer of $\alpha\beta$ dimers (Fig. 2.2). The two structurally similar subunits, α and β, represent an open α,β structure whose architecture is similar to that of the amino acid kinase family [26], the most closely related member being the hexameric uridine monophosphate (UMP) kinase [27]. The subunits are composed of an α,β core and two lobes attached above the edges of the central β sheet of the α,β core, thereby forming a groove which serves as an ATP-binding site in the α subunit [28].

An important aspect here roots in the fact that the hetero-hexamer contains a large central cavity of about 7250 Å^3 (i.e., comparable in size to the functionalized cavities of the giant polyoxomolybdate clusters [4, 6, 7], especially of $\{Mo_{368}\}$ [5]). The cavity (Fig. 2.3) is endowed with functionalized/appropriate pockets that serve as binding sites for the 14 polynuclear tungsten oxide clusters, while the rest is filled with solvent/water molecules. The clusters are of the following types: one W_3, three W_6, three W_7, three $W_2(\text{I})$, three $W_2(\text{II})$, and one $W_{7+x} \equiv (W)W_{6+x}$ (Fig. 2.3). Owing to the molecular symmetry of the Mo/WSto protein, each cluster occurs three times with the exception of the two located on the C_3 axis which therefore are found only once. Remarkably, 60–70 W-atom sites could be identified, based on the anomalous difference Fourier electron-density map, which was calculated from data collected at the Se and W absorption edges. The W–W distances of the aggregates are in the range of 3.3±0.4 Å, comparable to the situation in a large number of polyoxotungstates

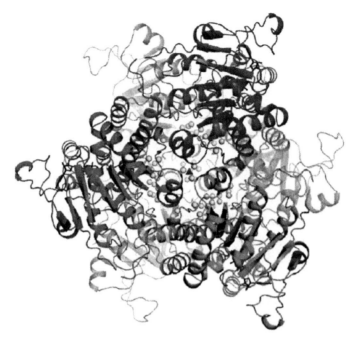

FIGURE 2.2 Structure of the Mo/WSto protein of *A. vinelandii*. The hetero-hexameric $(\alpha\beta)_3$ protein complex has a size of about $100 \times 100 \times 70$ Å^3 and pseudo *32* (D_3) symmetry. The monomers α_1 (red), α_2 (blue), and α_3 (green), and β_1 (light red), β_2 (light blue), and β_3 (light green) assemble as trimers on either side (called interfaces α and β) of the hetero-hexamer. The three monomers of each trimer are related by the crystallographic threefold axis (▲). The ATP molecules are shown as black stick models and the W atoms of the polynuclear tungsten oxide clusters as yellow spheres (see Reference 1). *See insert for color representation of the figure.*

(see Reference 3a, and references therein). Unfortunately, the complete structure containing all of the O atoms could only be resolved for the well-defined W$_3$ cluster (Fig. 2.4). All of the other polyoxotungstates show only fractional occupation due to the fact that there is partial tungstate release during the crystallization process.

The well-defined {W$_3$O$_{10}$N$_3$}-type cluster is composed of three edge-sharing octahedra that are related by the crystallographic threefold axis (Figs. 2.3 and 2.4). Each W atom is coordinated to five O atoms and to the imidazole N$_{\varepsilon2}$ of Hisα139. The W$_3$O$_{10}$ unit corresponds (formally) to a constituent of the well-known interesting [W$_{12}$O$_{40}$H$_2$]$^{6-}$ Keggin-type ion with encapsulated H$^+$ ions [3a]. As the O atoms could be clearly identified, an accurate description of hydrogen bonds and van der Waals interactions with the protein matrix is feasible (Fig. 2.4b).

The W positions of the three W$_7$ clusters correspond to the related positions of the known heptanuclear POMs [M$_7$O$_{24}$]$^{6-}$ (M = Mo, W) [3a] (Fig. 2.3) that are formed with high formation tendency in aqueous solution at intermediate pH values. The {M$_7$O$_{24}$} building block is also found in the largest polyoxomolybdate,

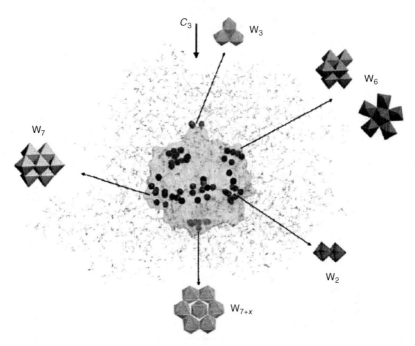

FIGURE 2.3   Surface representation of the Mo/WSto protein. The (roughly) ellipsoidal protein shell (gray) and the nanocavity are shown as a yellow surface. The cavity protein surface forms several well-distributed pockets, which harbor the polynuclear tungsten oxide clusters. The W_3 (yellow) and $W_{7+x} \equiv (W)W_{6+x}$ (green) clusters are positioned on the threefold axis, in front of the interfaces α and β, respectively, and are embedded in a small hydrophilic and a larger hydrophobic pocket, respectively. The binding sites of the three W_6 clusters (red; two idealized structural options are shown) are essentially built up by segments of the α subunit forming a flat and hydrophilic pit. The three W_7 clusters (blue) lie between subunits α and β in a deep pocket, while the W_2 clusters I and II (brown) are located in the vicinity of the W_7 cluster. The clusters are represented as idealized polyhedra constructed by assuming the O positions on the basis of the knowledge of POM chemistry [3a] and the present results; the individual W centers are shown as spheres of the corresponding color in the cavity of the Mo/WSto protein; see Reference 1. *See insert for color representation of the figure.*

$[Mo_{36}O_{112}]^{8-}$, obtained under nonreducing conditions [3a]. Interestingly, the W positions are homogeneously occupied to about 30%, in agreement with homogeneous cluster degradation during the release process.

The W_6 cluster, the formation of which is directed by the asymmetrical arrangement of the adjacent polypeptides, can be formally (!) correlated with a distorted pentagonal pyramidal $(W)W_5$ unit (Fig. 2.3). The arrangement of the W atoms is approximately comparable to the central metal part in the (flatter) building unit $(Mo)Mo_5$, which appears 12 times in the spherical porous capsules/artificial cells of the type $\{(Mo)Mo_5\}_{12}\{linker\}_{30}$ (linker, e.g., $\{Mo^V_2O_4(SO_4)\}$) [4, 6, 7], in the hedgehog-shaped Mo_{368} cluster [5], as well as in different wheel-shaped Mo_{154}- and

Mo_{176}-type clusters [29]. The pentagonal units are mentioned here because they are of importance for the chemistry of giant POMs. But we have to realize that the peak heights of the W sites of the W_6 cluster are different. This might suggest that the superimposed clusters result from nonhomogenous degradation processes. The W_6 cluster might in principle be a degradation product of the above-mentioned W_7 cluster, as it resembles the W_6 part of the latter (Fig. 2.3).

Regarding the uptake of small tungstate ions, structural-comparison studies suggested that the related hexameric Mo/WSto and UMP kinase [27] should exist in both a closed and an open conformation. In the open state, pores across the protein shell are formed, which are, in the present case, sufficiently wide to facilitate the entrance of small tungstate ions, such as $[WO_4H]^-$ and $[WO_4]^{2-}$. A consequence of the small protein pores is that the cluster syntheses can only proceed inside the cavity. An important issue in this context is the fact that the W_{7+x} and W_6 clusters, which are not stable in bulk solution, are necessarily synthesized in a protein-assisted process whereby the protein not only serves as a template, but also protects these clusters against hydrolysis.

The nucleation and growth processes for the different clusters differ considerably. The template-directed formation (e.g., see References 30 and 31) of the W_3 and $W_{7+x} \equiv (W)W_{6+x}$ species primarily follows geometry constraints in agreement with the threefold molecular symmetry (in contrast, e.g., to the W_6 species). The comparatively large $W_{7+x} \equiv (W)W_{6+x}$ cluster shows typical multiple contacts to a complementary shaped protein pocket and the W_3 cluster correspondingly weak bonds to the appropriately positioned Hisα139 (for further details see Fig. 2.4a). The special growth mechanism of the W_{7+x} cluster might be correlated with a cascade-directed process postulated for the Mo_{37}-type cluster [31]; the process proceeds in such a way that the "older" parts direct the formation of the next fragments and subsequently react with them. The subsequent part of the (essentially protein-independent)

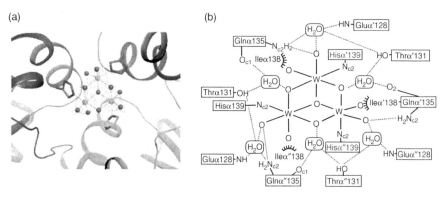

FIGURE 2.4 (a) The structure of the W_3 cluster and its protein surroundings. (b) Amino acid residues and water molecules in the direct environment of the W_3 cluster. The three triangularly arranged W atoms (yellow) are 3.4 Å from each other. Each W atom is linked to five O atoms and the $N_{\epsilon 2}$ atom of Hisα139 (light green residue in (b); see Reference 1). *See insert for color representation of the figure.*

growth is accompanied by a (dramatic) decrease in the occupancy of the W sites because of a lack of protection at the conditions of crystallization.

The three W_6 clusters are specifically hydrogen-bonded to three positively charged histidine residues, the electrophilic surface of which attracts tungstate ions and supports the assembly processes by decreasing the repulsion between the negatively charged small tungstate ions. The most important aspect here is that nucleation always starts at the protein pocket with the largest degree of protection while the cluster de-aggregation proceeds in the reverse direction. The three rather stable W_7 clusters should, according to our knowledge of bulk-type POM chemistry [3a], form spontaneously in the cavity in a nondirected self-assembly process and subsequently become bound to the protein surface.

The protein exhibiting its polynuclear tungsten oxide system combines, in a unique manner, macromolecular biochemistry with an unprecedented type of inorganic chemistry and supramolecular chemistry (for chemical examples, see References 32 and 33) under confined conditions. The term "supramolecular" is correct in this context because the protein cavity acts as a polytopic host for several different inorganic clusters, that is, for weakly or noncovalently bonded guests. All of the aggregates found exhibit structural details well known from POM chemistry [3a, 22]. Interestingly, some of the clusters have not been detected in bulk media, because of their instability under those conditions. On the other hand they could, at least in principle, serve as precursors for bulk-type species. Though nature has developed the Mo/WSto protein system for the storage of Mo in the form of compact polynuclear molybdates, the process for the formation and release of the polyoxotungstate species discussed here must be considered as analogous, because the POM chemistry of Mo^{VI} and W^{VI} is very similar [3a, 22, 34, 35]. Most importantly, from the results discussed here we can learn much about cluster syntheses based on tailor-made protein-based templates. Unprecedented aspects we refer to are nucleation/condensation processes on a single molecular level in the sense of single molecule chemistry, as we can study with physical methods one (!) capsule in which different nucleation processes occur. The option to use the Mo/WSto protein as a suitable "nano test container" can probably be explored. Variation of the aqueous solutions should lead to different encapsulated polyoxotungstates, while modification of the relevant protein pockets—that is, modification of the template conditions—by site-directed exchange of amino acids can create new functionalities or patterns with directing functions.

2.4 MANGANESE IN PHOTOSYSTEM II

Substantial photosynthesis (by cyanobacteria), known as the "Great Oxygen Event," started on our planet about 2.4 billion years (Ga) ago. It led to the oxidation of Fe^{2+} in the primordial water and the formation of the iron-banded ores with less soluble Fe^{3+}. In a second phase of oxygen supply, 0.9–0.5 Ga ago, the present equilibrium with about 21% (by volume) O_2 in our atmosphere became established. Photosynthetic organisms—green plants, algae, protozoa containing symbiotic chloroplasts, many photosynthetic bacteria, and cyanobacteria—all employ an oxygen evolving center

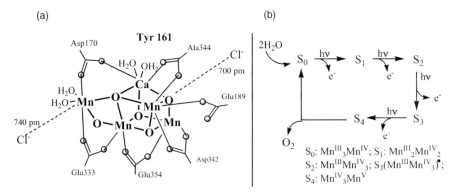

FIGURE 2.5   (a) The $\{Mn_4CaO_5\}$ cluster of the oxygen-evolving center (OEC) in photosystem II, based on a recent X-ray structure determination [36]. Only direct bonds to amino acids at a distance ≤ 2.4 Å are drawn. The redox-active Tyr161 is not in direct contact with the cluster. The chlorides are embedded in-between Asn, Asp, and Glu (not shown) and additionally coordinated to H_2O [37]. (b) The Kok cycle of S-state transitions in the OEC [38], including proposed oxidation states of the Mn ions. $S_4(Mn^{III}Mn^{IV}_3)^\bullet$ refers to a Mn^{IV}-oxyl radical. An alternative for the S_0 state is $Mn^{III}_3Mn^{IV}$, and for the S_4 state $Mn^{III}Mn^{IV}_2Mn^V$.

(OEC) as an integral constituent of their photosystem II (PSII). PSII is a multi-subunit complex which forms an integral part of the thykaloid membrane of the chloroplasts. The OEC is a chemically and structurally interesting oxidomanganese cluster of composition $\{Mn_4CaO_5\}$ embedded in a protein matrix, with at least six amino acids directly linked to the metal centers (Fig. 2.5). According to the present stand, based on EXAFS and X-ray diffraction data (down to 1.9 Å resolution [36, 39–43]), three of the manganese centers and the Ca^{2+} ion are in a cuboidal arrangement of composition $\{Mn_3CaO_4\}$, which is linked, via two oxido bridges, to a fourth manganese, the "dangler Mn." In the vicinity of this $\{Mn_4CaO_5\}$ core, there is a chloride ion, which may take over a role in proton removal from, or substrate access to, the OEC [37a]. A functional water-oxidizing complex forms, in a two-step process, as the apoprotein of PSII is treated with Mn^{2+}, Ca^{2+}, and Cl^-, accompanied by photo-oxidation of Mn^{2+} to Mn^{3+} [37b].

With the use of light energy, the OEC powers the splitting of water according to Equation 2.1, providing oxygen O_2, protons, and reduction equivalents which are employed in CO_2 fixation. The light is harvested, and its energy equivalents eventually delivered to the $\{Mn_4CaO_5\}$ cluster, by chlorophyll-a and β-carotene bound to the protein domain of the OEC [42b]. The redox-active Tyr161 in Figure 2.5a is one of the mediators in this transduction of energy. During the water-splitting process, the OEC cluster runs through five successive states, S_0 to S_4 (Fig. 2.5b), four of which consume light energy, while the fifth, $S_4 \rightarrow S_0$, is associated with the liberation of molecular oxygen. Manganese changes between different oxidation states, likely involving the oxidation states +III, +IV, and +V during turnover, while the oxygen in water may be oxidized stepwise via $^\bullet OH/H_2O_2$ and $^\bullet O_2^-/HO_2$ (Eq. 2.2)

[44]. A possible inorganic progenitor for the OEC is the MnO_2-derived, layer-type mineral rancieite $(CaMn^{II})Mn^{IV}O_9 \cdot 3H_2O$ [38], found in cavities in limonite $FeO(OH) \cdot nH_2O$. An evolutionary origin for the Mn_4 cluster in PSII from manganese superoxide dismutase (containing one Mn center) and manganese catalase (with two Mn centers bridged by carboxylate and water, plus a structural Ca^{2+}) has also been proposed [45].

$$2H_2O + 4h\nu \rightarrow O_2 + 4H^+ + 4e^-, \tag{2.1}$$

$$H_2O \rightarrow \cdot OH + H^+ + e^-; 2 \cdot OH \rightarrow H_2O_2; H_2O_2 \rightarrow HO_2 + H^+ + e^-;$$

$$HO_2 \rightarrow O_2 + H^+ + e^-. \tag{2.2}$$

2.5 IRON: FERRITINS, DPS PROTEINS, FRATAXINS, AND MAGNETITE

The immediate embedding of nutritional iron into a large variety of different functional units such as heme, mitochondrial frataxins, and iron–sulfur clusters, or the storage in ferritins, is imperative, since free $Fe^{2+/3+}$ is toxic due to its potential to produce reactive oxygen radicals. The iron storage proteins, present in all cell types of all organisms, consist of a hollow protein sphere, the apo-ferritin. Human apo-ferritin is a 450 kDa protein, made up of 24 subunits of ca. 170 amino acids each [46] (Fig. 2.6). The outer diameter is 130 Å and the inner diameter 75 Å. The inner surface of this protein capsule is lined with carboxylate functions (Glu and Asp) for the coordination of the first incoming Fe^{3+} ions. Up to 4500 Fe^{3+} can be incorporated, interconnected by bridging oxido and hydroxido groups, akin to the mineral ferrihydrite $FeO(OH) \cdot 0.4H_2O$ or goethite $FeO(OH)$. The overall composition of the iron nucleus comes close to $8FeO(OH) \cdot FeO(H_2PO_4)$, that is, some phosphate is also

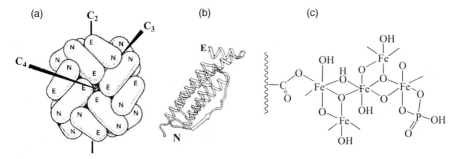

FIGURE 2.6   The iron storage protein ferritin. (a) The 24 subunits and the fourfold, threefold, and twofold symmetry axes are shown. The threefold axes correspond to the direction of the ion transport channels. (b) Subunit of ferritin; N and E refer to the N-terminal lobe and the α helical moiety. (c) Schematic view of a section from the iron core. The phosphate can further link to adenosinemonophosphate.

built in. Pores and channels of threefold symmetry in the protein envelope, furnished with Asp and Glu residues, allow for an exchange of iron ions between the interior and exterior. For the primary uptake, iron has to be in the oxidation state +II (soluble iron). Uptake through the channels and assembly into the core structure of ferritin is accompanied by oxidation, by oxygen, to Fe^{3+}. Interestingly, the effective magnetic moment per Fe^{3+} center in ferritin is only 3.85 Bohr magnetons, and hence clearly less than the ideal value of 5.92 for high-spin ferric iron. This points to a super-exchange (partial antiferromagnetic coupling) in the highly ordered, pseudo-crystalline ferritin core.

In addition to ferritins, other systems also constitute protein cages with iron oxide cores, namely, the Dps proteins, the frataxins, and nanoclusters of magnetite embedded in magnetosomes (vide infra). While ferritins function only as iron storage proteins, the related Dps proteins (_D_NA-binding _p_roteins from _s_tarved cells) also attain ferroxidase activity. They protect DNA from oxidative damage by OH radicals, which are generated from Fe^{2+} and hydrogen peroxide, known as Fenton reaction. Dps proteins counteract the Fenton reaction in that they catalyze the oxidation, by H_2O_2, of Fe^{2+} to Fe^{3+}, which is then taken up into the protein cage. Frataxins are mitochondrial proteins which play a role in iron–sulfur cluster assembly in the mitochondrial machinery. Magnetite Fe_3O_4 (and greigite Fe_3S_4), in the magnetosomes of magnetotactic bacteria and in the magnetic field sensors of a variety of other organisms, including fruit flies, bees, fish, homing pigeons, and salamanders, are employed for orientation in the ca. 50 μT magnetic field of our planet.

Dps proteins [47, 48], sometimes also termed Dpr proteins (for _D_ps-like _p_eroxide _r_esistance [49]), are restricted to archaea and, mainly, bacteria, and their main (but not exclusive) function is protection against oxidative stress. Dps proteins reveal structural homology to ferritins, constituting, however, 12 subunits only and a storage capacity for Fe^{3+} of ca. 500. The Dps monomer, a four-helix bundle (ferritin: five helix), has a molecular weight of ca. 20 kDa. The 12 subunits are arranged in a hollow sphere of _23_ symmetry, with an outer diameter of ca. 95 Å and an inner diameter of ca. 45 Å. There are four C_3 axes, connecting 4×2 pores/channels of threefold symmetry, four of which—the N-terminal ones—are negatively charged and hydrophilic and thus serve as entry (and exit) channels. The antioxidant protective function involves the binding of Fe^{2+} to ferroxidase sites, which is followed by oxidation of Fe^{2+} to Fe^{3+} and, finally, mineralization [50]. The ferroxidase sites are located at the interface of two symmetry-related protein monomers, thus contrasting ferritins where the two Fe^{2+} bind to _intra_-subunit sites of the ferroxidase center. The carboxylate groups of Asp and Glu and two His (directly, and indirectly via a water molecule) are involved in iron binding [48]. The ferric oxide core formed in the Dps proteins after the nucleation process in the hollow cavity comes close in overall composition to ferrihydrite or goethite and contains, again in accordance with the mineral core of ferritins, varying amounts of phosphate. Additional metal ions M^{2+} (M, e.g., Zn, Mn, Ni, Co, Cu, Mg [51]) may be taken up.

Frataxins, which are present throughout the kingdoms of bacteria (where they are termed CyaY) and eukarya, but not in archaea, are proteins essential for the delivery of iron to those scaffolds which are responsible for the assemblage of transiently formed iron–sulfur clusters (Isu), such as the Isu assembly proteins [52]. In eukarya, frataxins are mainly responsible for the FeS cluster biogenesis not only in mitochondria but, to some extent, also in extra-mitochondrial machineries. In humans, a (genetically inherited) lack of frataxin leads to progressive symptoms of diminished coordination of muscle movement and sensory loss (Friedreich ataxia) through degeneration of nerve tissue, accompanied by a deficiency of FeS clusters and accumulation of iron in the form of iron phosphate nanoparticles in the mitochondria [53].

The buildup of frataxin complies to a 14 kDa α,β sandwich (2 parallel α-helices, 5 antiparallel β-strands) with binding sites for up to 7 ferrous ions in the N-terminal α-helix, typically dominated by the carboxylate residues of aspartate and glutamate, and supplemented by histidine [53]. The coordination sphere of (high-spin) Fe^{2+} is supplemented by water ligands; the overall binding affinity, 10–55 μM in human frataxin [54], is comparatively low.

Frataxin subunits can form multimeric assemblies in the course of oxidation of ferrous to ferric iron, and hence resemble Dps proteins and ferritins in this respect. In contrast to the latter, they disassemble again into monomers on reduction of Fe^{III} to Fe^{II}, followed by the delivery of Fe^{2+} to proteins engaged in the synthesis of FeS clusters and possibly also certain heme-type proteins. In the case of a yeast frataxin, 24-mers with the *234* symmetry established for ferritins have been characterized (Fig. 2.7), although with a lower iron loading and less ordered $\{FeO(OH)\}_n$ core (with n up to 2400) than in the case of ferritins [55].

A prominent example for biologically controlled biomineralization is the formation of the mixed-valent magnetic materials magnetite Fe_3O_4 ($Fe^{II}Fe^{III}_2O_4$) and greigite Fe_3S_4 ($Fe^{II}Fe^{III}_2S_4$) by magnetotactic bacteria, allowing for their orientation

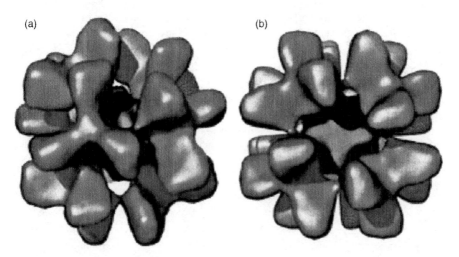

(a)   (b)

FIGURE 2.7   Reconstruction of an iron-loaded oligomeric (24 subunits) yeast frataxin viewed close to the threefold axis (a) and along the fourfold axis (b). The iron oxide core is shown in red, the protein shell in blue. Reprinted with permission from Reference 55; © American Chemical Society. *See insert for color representation of the figure.*

in the Earth's magnetic field. Particularly pure magnetite crystallites are usually considered as "biomarkers," that is, they point toward a biological origin.

Magnetotactic bacteria produce, in many cases, mono-domain magnetite or greigite, enveloped by a lipid bilayer membrane associated with interacting proteins which are involved in, *inter alia*, biomineralization. These specific cell organelles, termed magnetosomes, exhibit species-dependent morphologies and sizes [56], typically with a diameter of around 50 nm and a membrane thickness of 7 nm [57]. Crystals of this size are permanently magnetic. The overall process of magnetosome formation includes four successive steps: (1) formation of the vesicle; (2) iron transport into the vesicle; (3) anchoring of iron ions to membrane-associated proteins; and (4) crystallization of magnetite within the vesicle. The amino acids Asp, Glu, Ser, and Tyr are involved in iron anchoring [58].

2.6   SOME GENERAL REMARKS: OXIDES AND SULFIDES

Metal oxides (molecular and solid state type) [59] are of importance in several disciplines. Metal oxides present in metal protein cavities are of special interest, for instance, regarding their unique structures and biosyntheses (as in the ferritin case) and also for aspects of catalysis. "Biological" metal sulfides, especially the larger ones present in nitrogenases, should also be mentioned in this context [60]. Interestingly, besides the usual molybdenum nitrogenases, alternative nitrogenases exist which do not contain the molybdenum–iron–sulfur clusters (FeMocos) but vanadium–iron–sulfur clusters (FeVcos) or pure polynuclear iron sulfides (FeFecos) [61]. A comparison of the three different nitrogenases is of interest with respect to evolutionary aspects, in particular regarding their very different remarkable catalytic properties [12, 62]. The interdisciplinary interest for metal sulfides—including molecular ones [63]—corresponds to that for oxides as they have properties in common. A fascinating point: the FeFe cofactor can be replaced by the FeMo cofactor in the iron-only nitrogenase [64]; this means there is the option to replace metal clusters containing different metals in protein cavities [65].

Changes in the redox state of iron initiate important inputs in its physiological function. Multicopper-containing proteins can be involved in the oxidation of ferrous to ferric iron [66]. Ceruloplasmin, a ferroxidase and copper storage protein in the blood plasma, attains an intriguing role in this respect. Ceruloplasmin is a 134 kDa glycoprotein, which contains six copper centers (three type 1, one type 2, and two type 3 centers) embedded in the protein matrix in a binding sphere dominated by histidine and cysteine residues. A paper published after the submission of this review refers to the pure molybdenum storage protein [67].

REFERENCES

[1] Schemberg, J.; Schneider, K.; Demmer, U.; Warkentin, E.; Müller, A.; Ermler, U. *Angew. Chem. Int. Ed.* **2007**, *46*, 2408–2413. Corrigendum: *Angew. Chem. Int. Ed.* **2007**, *46*, 2970; see also Reference 2.

[2] Schemberg, J.; Schneider, K.; Fenske, D.; Müller, A. *ChemBioChem* **2008**, *9*, 595–602.

[3] (a) Pope, M. T.; Müller, A. *Angew. Chem. Int. Ed.* **1991**, *30*, 34–48. (b) Pauling, L. *J. Am. Chem. Soc.* **1929**, *51*, 2868–2880 (cited here although there is an error involved in the prediction of the structure of an important POM).

[4] (a) Müller, A.; Krickemeyer, E.; Bögge, H.; Schmidtmann, M.; Roy, S.; Berkle, A. *Angew. Chem. Int. Ed.* **2002**, *41*, 3604–3609. (b) Rehder, D.; Haupt, E. T. K.; Bögge, H.; Müller, A. *Chem. Asian J.* **2006**, *1*, 76–78; and references therein.

[5] (a) Müller, A.; Beckmann, E.; Bögge, H.; Schmidtmann, M.; Dress, A. *Angew. Chem. Int. Ed.* **2002**, *41*, 1162–1167. (b) Müller, A.; Botar, B.; Das, S. K.; Bögge, H.; Schmidtmann, M.; Merca, A. *Polyhedron* **2004**, *23*, 238–2385 (in that case, the cation transport was not published).

[6] (a) Hall, N. *Chem. Commun.* **2003**, 803–806. (b) Cronin, L.; *Angew. Chem. Int. Ed.* **2006**, *45*, 3576–3578.

[7] (a) Müller, A.; Roy, S. *J. Mater. Chem.* **2005**, *15*, 4673–4677. (b) Müller, A.; Gouzerh, P. *Chem. Soc. Rev.* **2012**, *41*, 7431–7463.

[8] Wörsdörfer, B.; Woycechowsky, K. J.; Hilvert, D. *Science* **2011**, *331*, 589–592.

[9] (a) Aureliano, M.; Crans, D. C. *J. Inorg. Biochem.* **2009**, *103*, 536–546. (b) Tiago, T.; Martel, P.; Guitiérrez-Merino, C.; Aureliano, M. *Biochim. Biophys. Acta* **2007**, *1774*, 474–480.

[10] Wittenkeller, L.; Abraha, A.; Ramasamy, R.; Mota de Freitas, D.; Theisen, L. A.; Crans, D. C. *J. Am. Chem. Soc.* **1991**, *113*, 7872–7881.

[11] Lobert, S.; Isern, N.; Henneington, B. S.; Correia, J. J. *Biochemistry* **1994**, *33*, 6244–6252.

[12] Rehder, D. *Bioinorganic Vanadium Chemistry*; Wiley: Chichester, 2008.

[13] Ueki, T.; Michibata, H. *Coord. Chem. Rev.* **2011**, *255*, 2249–2257.

[14] Yoshinaga, M.; Ueki, T.; Michibata, H. *Biochim. Biophys. Acta* **2007**, *1770*, 1423–1418.

[15] Ueki, T.; Furuno, N.; Michibata, H. *Biochim. Biophys. Acta* **2011**, *1810*, 457–464.

[16] (a) Fukui, K.; Ueki, T.; Ohya, H.; Michibata, H. *J. Am. Chem. Soc.* **2003**, *125*, 6352–6353. (b) Kawakami, N.; Ueki, T.; Amata, Y.; Kanamori, K.; Matsuo, K.; Gekko, K.; Michibata, H. *Biochim. Biophys. Acta* **2009**, *1794*, 647–679.

[17] Ueki, T.; Kawakami, N.; Toshishigi, M.; Matsuo, K.; Gekko, K.; Michibata, H. *Biochim. Biophys. Acta* **2009**, *1790*, 1327–1333.

[18] Pau, R. N.; Lawson, D.M. Molybdenum and Tungsten: Their Role in Biological Processes. In *Metal Ions in Biological Systems*; Sigel, A.; Sigel, H., Eds.; Dekker: New York, vol. *39*, 2002, pp 31–74.

[19] Kletzin, A.; Adams, M. W. W. *FEMS Microbiol. Rev.* **1996**, *18*, 5–63.

[20] Fenske, D.; Gnida, M.; Schneider, K.; Meyer-Klaucke, W.; Schemberg, J.; Henschel, V.; Meyer, A. K.; Knöchel, A.; Müller, A. *ChemBioChem* **2005**, *6*, 405–413. The Mo-storage system was first discovered in the aerobic nitrogen-fixing bacterium *A. vinelandii* [21] but owing to the ease of molybdate release, the published Mo content was too low. (But the storage protein also appears to be present in *Rhodopseudomonas palustris* and *Nitrobacter* species.) Its function is to supply the Mo-dependent nitrogenase system with Mo, in particular in an environment of Mo shortage.

[21] Pienkos, P. T.; Brill, W. J. *J. Bacteriol.* **1981**, *145*, 743–751.

[22] Pope, M. T. *Heteropoly and Isopoly Oxometalates*; Springer: New York, 1983.

[23] Kaim, W.; Schwederski, B. *Bioinorganic Chemistry: Inorganic Elements in the Chemistry of Life*, Wiley: Chichester, 1997.

[24] Lewis, A.; Moore, G. R.; Le Brun, N. E. *Dalton Trans.* **2005**, 3597–3610.

[25] For all aspects of this area see: Lehn, J.-M.; Atwood, J. L.; Davies, J. E. D.; MacNicol, D. D.; Vögtle, F., Eds. *Comprehensive Supramolecular Chemistry*; Elsevier/Pergamon: New York, vols. *1–11*, 1996. Bioorganic and bioinorganic systems are treated in volumes 4 and 5, respectively.

[26] Ramón-Maiques, S.; Marina, A.; Gil-Ortiz, F.; Fita, I.; Rubio, V. *Structure* **2002**, *10*, 329–342.

[27] Briozzo, P.; Evrin, C.; Meyer, P.; Assairi, L.; Joly, N.; Bärzu, O.; Gilles, A.-M. *J. Biol. Chem.* **2005**, *280*, 25533–25540.

[28] In preliminary biochemical studies, an essential function of ATP in Mo binding has been demonstrated [20]. However, systematic studies including the role of ATP in W binding have not been performed to date. These studies are required for postulating a reliable ATP-driven tungstate binding/release mechanism on a structural basis. The affinity between WSto and ATP/Mg^{2+} appears to be very high, as the protein structure contains ATP.

[29] (a) Müller, A.; Serain, C. *Acc. Chem. Res.* **2000**, *33*, 2–10. (b) Müller, A.; Shah, S. Q. N.; Bögge, H.; Schmidtmann, M. *Nature* **1999**, *397*, 48–50.

[30] Müller, A. *Nature* **1991**, *352*, 115.

[31] Müller, A.; Meyer, J.; Krickemeyer, E.; Beugholt, C.; Bögge, H.; Peters, F.; Schmidtmann, M.; Kögerler, P.; Koop, M. *J. Chem. Eur. J.* **1998**, *4*, 1000–1006.

[32] Lehn, J.-M. *Supramolecular Chemistry: Concepts and Perspectives*; Wiley: Weinheim, **1995**.

[33] Müller, A.; Reuter, H.; Dillinger, S. *Angew. Chem. Int. Ed. Engl.* **1995**, *34*, 2328–2361.

[34] Long, D.-L.; Cronin, L. *Chem. Eur. J.* **2006**, *12*, 3698–3706.

[35] Cronin, L. In *Comprehensive Coordination Chemistry II*; McCleverty, J. A.; Meyer, T. J., Eds.; Elsevier: Amsterdam, vol. *7*, 2004; pp 1–56.

[36] Umena, Y.; Kawakami, K.; Shen, J.-R.; Kamiya, N. *Nature* **2011**, *473*, 55–61.

[37] (a) Murray, J. W.; Maghlaoui, K.; Kargul, J.; Ishida, N.; Lai, T.-L.; Rutherford, A. W.; Sugiura, M.; Boussac, A.; Barber, J. *Energy Environ. Sci.* **2008**, *1*, 161–166. (b) Zaltsman, L.; Ananyev, G. M.; Bruntrager, E.; Dismukes, G. C. *Biochemistry* **1997**, *36*, 8914–8922.

[38] Sauer, K.; Yachandra, V. K. *Proc. Natl Acad. Sci. USA* **2002**, *99*, 8631–8636.

[39] Yano, J.; Kern, J.; Sauer, K.; Latimer, M. J.; Pushkar, Y.; Biesiadka, J.; Loll, B.; Saenger, W.; Messinger, J.; Zouni, A.; Yachandra, V. K. *Science* **2006**, *314*, 821–825.

[40] Sproviero, E. M.; Gascón, J.A.; McEvoy, J. P.; Brudvig, G. W.; Batista, V. S. *J. Am. Chem. Soc.* **2008**, *130*, 6728–6730.

[41] Barber, J.; Murray, J. W. *Philos. Trans. R. Soc. B* **2008**, *363*, 1129–1138.

[42] (a) Barber, J. *Inorg. Chem.* **2008**, *47*, 1700–1710. (b) Barber, J. *Phil. Trans. R. Soc. B* **2008**, *363*, 2665–2674.

[43] Guskov, A.; Kern, J.; Gabdulkhanov, A.; Broser, M.; Zouni, A.; Saenger, W. *Nat. Struct. Mol. Biol.* **2009**, *16*, 334–342.

[44] Armstrong, F. A. *Philos. Trans. R. Soc. B* **2008**, *363*, 1263–1270.

[45] Najafpour, M. M. *Orig. Life Evol. Biosph.* **2009**, *39*, 151–163.

[46] (a) Ford, G. C.; Harrison, P. O. M.; Rice, D. W.; Smith, J. M. A.; Treffry, A.; White, J. L.; Yariv, J. *Philos. Trans. R. Soc. Lond. B* **1984**, *304*, 551–565. (b) Crichton, R.R.; Declerq, J.-P. *Biochim. Biophys. Acta* **2010**, *1800*, 706–718.

[47] Chiancone, E.; Ceci, P. *Biochim. Biophys. Acta* **2010**, *1800*, 798–805.

[48] Haikarainen, T.; Papageorgiou, A. C. *Cell. Mol. Life Sci.* **2010**, *67*, 341–351.

[49] Tsou, C.-C.; Chiang-Ni, C.; Lin, Y.-S.; Chuang, W.-J.; Lin, M.-T.; Liu, C.-C.; Wu, J.-J. *Infect. Immun.* **2008**, *76*, 4038–4045.

[50] Zhao, G.; Ceci, P.; Ilari, A.; Giangiacomo, L.; Laue, T. M.; Chiancone, E.; Chasteen, N. D. *J. Biol. Chem.* **2002**, *27*, 27689–27696.

[51] Haikarainen, T.; Thanassoulas, A.; Stavros, P.; Nounesis, G.; Haatajy, S.; Papageorgiou, A. C. *J. Mol. Biol.* **2011**, *405*, 448–460.

[52] Prischi, F.; Konarev, P. V.; Iannuzzi, C.; Pastore, C.; Adinolfi, S.; Martin, S. R.; Svergun, D. I.; Pastore, A. *Nature Commun.* **2010**, *1*, 95.

[53] Stemmler, T. L.; Lesuisse, E.; Pain, D.; Dancis, A. *J. Biol. Chem.* **2010**, *285*, 26736–26743.

[54] Yoon, T.; Cowen, J. A. *J. Am. Chem. Soc.* **2003**, *125*, 6078–6084.

[55] Schagerlöf, U.; Elmlund, H.; Gakh, O.; Nordlund, G.; Hebert, H.; Lindahl, M.; Isaya, G.; Al-Karadaghi, S. *Biochemistry* **2008**, *47*, 4948–4954.

[56] Schüler, D. *FEMS Microbiol. Rev.* **2008**, *32*, 654–672.

[57] Yamamoto, D.; Taoka, A.; Uchihashi, T.; Sasaki, H.; Watanabe, H.; Ando, T.; Fukumori, Y. *Proc. Natl Acad. Sci. USA* **2010**, *107*, 9382–9387.

[58] Arakaki, A.; Webb, J.; Matsunaga, T. *J. Biol. Chem.* **2003**, *278*, 8745–8750.

[59] (a) Rao, C. N. R.; Raveau, B. *Transition Metal Oxides*, 2nd Edn.; Wiley-VCH: Weinheim, 1998. (b) Burgess, J. Water, Hydroxide and Oxide. In *Comprehensive Coordination Chemistry: The Synthesis, Reactions, Properties and Applications of Coordination Compounds*; Wilkinson, G.; Gillard, R. D.; McCleverty, J. A., Eds.; Pergamon: Oxford, vol. 2, 1987; pp 295–314.

[60] Leigh, G. J., Ed. *Nitrogen Fixation at the Millenium*; Elsevier: Amsterdam, 2002.

[61] (a) Masepohl, B.; Schneider, K.; Drepper, T.; Müller, A.; Klipp, W. Alternative Nitrogenases. In: *Nitrogen Fixation at the Millenium*; Leigh, G. J., Ed.; Elsevier: Amsterdam, 2002; pp 191–222. (b) Fay, A. W.; Blank, M. A.; Lee, C. C.; Hu, Y.; Hodgson, K. O.; Hedman, B.; Ribbe, M. W. *Angew. Chem. Int. Ed.* **2011**, *50*, 7787–7790.

[62] Smith, B. E.; Richards, R. L.; Newton, W. E., Eds. *Catalysts for Nitrogen Fixation: Nitrogenases, Relevant Chemical Models, and Commercial Processes*; Kluwer: Dordrecht, 2004.

[63] (a) Stiefel, E. I.; Matsumoto, K., Eds. *Transition Metal Sulfur Chemistry: Biological and Industrial Significance*, American Chemical Society: Washington, DC, 1996; (b) Müller, A.; Diemann, E. Sulfides and Metallothio Anions. In *Comprehensive Coordination Chemistry: The Synthesis, Reactions, Properties and Applications of Coordination Compounds*; Wilkinson, G.; Gillard, R. D.; McCleverty, J. A., Eds.; Pergamon: Oxford, vol. 2, 1987; pp 515–550, and 559–577, respectively (reviews refer to metal sulfide compounds and interdisciplinary aspects).

[64] Schneider, K.; Müller, A. Iron-Only Nitrogenase: Exceptional Catalytic, Structural and Spectroscopic Features. In: *Catalysts for Nitrogen Fixation: Nitrogenases, Relevant*

Chemical Models, and Commercial Processes; Smith, B. E.; Richards, R. L.; Newton, W. E., Eds.; Kluwer: Dordrecht, 2004; pp 281–307.

[65] Gollan, U.; Schneider, K.; Müller, A.; Schüddekopf, K.; Klipp, W. *Eur. J. Biochem.* **1993**, *215*, 25–35.

[66] (a) Taylor, A. B.; Stoj, C. S.; Ziegler, L.; Kosman, D. J.; Hart, P. J. *Proc. Nat. Acad. Sci. USA* **2005**, *43*, 15459–15464. (b) Fox, P. L. *BioMetals* **2003**, *16*, 9–40.

[67] Kowalewski, B.; Poppe, J.; Demmer, U.; Warkentin, E.; Dierks, T.; Ermler, U.; Schneider, K. *J. Am. Chem. Soc.* **2012**, *134*, 9768–9774.

PART II

DESIGN OF METALLOPROTEIN CAGES

3

DE NOVO DESIGN OF PROTEIN CAGES TO ACCOMMODATE METAL COFACTORS

Flavia Nastri, Rosa Bruni, Ornella Maglio, and Angela Lombardi

3.1 INTRODUCTION

One of the most ambitious and exciting purposes of chemists, biochemists, and biologists is to get the skill for designing novel enzymes able to catalyze any desired transformation [1–3]. Recently, outstanding papers appeared in the literature suggesting that this objective is not so far to be reached [4–6]. The first attempts to design proteins from scratch started 20 years ago [7]. Since then, *de novo* protein design has proven to be a powerful approach for addressing questions in protein stability and folding [8–13]. The features required for stabilizing secondary structure have been elucidated and the main factors that direct the collapse of the protein chain into a compact globular structure have been established. Thus, uniquely structured proteins have been designed and structurally characterized [4–6].

As soon as the principles and methods for designing well-defined protein structures became clear, many efforts have been devoted to design metalloproteins [14–19]. This goal is even more ambitious, since it aims to introduce functionality into an artificial protein through full control over the structure of both the protein scaffold and the metal site.

It is well known that, besides the coordinate bond, a plethora of interactions are responsible for finely tuning the metal site activity in metalloproteins [20]. The

Coordination Chemistry in Protein Cages: Principles, Design, and Applications, First Edition.
Edited by Takafumi Ueno and Yoshihito Watanabe.
© 2013 John Wiley & Sons, Inc. Published 2013 by John Wiley & Sons, Inc.

protein matrix shapes the primary coordination shell of the metal ion, dictating the composition, number, and geometry of the ligands. In addition, residues in the second shell, the immediate surroundings of the primary coordination sphere, influence a variety of structural and chemical facets, including hydrogen-bonding interactions to the ligands, pK_a values of the ligands, metal center oxidation state, and redox potential. Finally, neighboring side chains also exert steric and chemical controls over the ability of the metal ion(s) to bind or discriminate the substrates and to accommodate conformational changes. As a consequence, to construct an artificial metalloprotein several issues must be simultaneously satisfied: the design of a correct protein structure and the metal ion coordination requirements.

De novo-designed proteins have been extensively used as suitable scaffolds to accommodate in their interior metal-based prosthetic groups [8, 9, 11]. Several structural motifs, such as β-sheets, α/β-motif, and especially α-helices and helical bundles, can presently be designed with highest degree of confidence. In particular, the "rules" that control stability, oligomerization, and helix–helix orientation are now well in hand and a variety of *de novo*-designed α-helical coiled-coils and bundles, with native-like structures, have been reported [10, 11, 20]. Such structures, which may differ in the nature, number, and alignment of the helices, have successfully been used for the development of artificial metalloproteins [20].

The insertion of a metal-binding site into a *de novo*-designed protein requires to consider several demanding aspects: (1) the correct number of coordinating residues to properly position around the metal center, with the required side-chain conformation and (2) the secondary shell interactions, which are essential for modulating the properties of the metal site. A further aspect implies focusing on the stability/function tradeoff, which is essential for both structure and function [17, 21]. Enzymatic activity requires the presence of active site clefts replete with solvent-exposed hydrophobic groups and hydrogen-bonding groups, for proper binding of substrates. On the opposite, a folded conformation is stabilized by minimizing voids and maximizing the burial of hydrophobic groups. Therefore, to create cavities into a folded scaffold, without disrupting its well-defined tertiary structure, all the interactions, which constitute the main driving force for folding and which are necessary for substrate binding and activation, should accurately and simultaneously be optimized [20]. As a result, metalloprotein design is a complex problem that has been addressed through several strategies aimed at considering all the variables needed to specify structure and function. They span from the binary patterning of hydrophobic and hydrophilic residues, to the totally computational design methods, to combinatorial approach [22–31]. Because of the complexity of the design problem, a trial and error approach is often necessary for *de novo* designing a metalloprotein with a native-like structure. Through several cycles of design, synthesis, characterization, and redesign, it is possible to fine-tune the structural properties of the initial model and to tailor functional metal site into the interior [8, 32, 33].

This chapter, by reviewing the papers appeared in the literature in the last 10 years, illustrates the different strategies used for the development of *de novo*-designed protein cages housing mononuclear and dinuclear metal-binding sites, as well as hemes and metalloporphyrins.

We mainly focus on α-helical coiled-coil and bundle structures, which self-assemble with the metal cofactors. Fewer examples of artificial metalloproteins based on β-sheet structures have appeared in the literature and they are also described here.

3.2   *DE NOVO*-DESIGNED PROTEIN CAGES HOUSING MONONUCLEAR METAL COFACTORS

Metal ions/cofactors have been engineered into helical bundles by introducing metal-binding ligands at specific locations to mimic those in native proteins. Properly positioning the ligands that will coordinate the metal ion is critical for binding; however, the position of additional functional groups close to the coordination site may also be important.

Outstanding examples of the incorporation of mononuclear metal-binding sites into *de novo*-designed α-helical coiled-coil and bundle structures have appeared in the literature over the years.

In this respect, pioneering results are reported by Handel and DeGrado, who designed a four-helical bundle protein, $\alpha 4$, containing a 3-His carbonic anhydrase-like Zn^{2+}-binding site [34]. The designed protein was found to bind Zn^{2+} and NMR spectroscopy indicated that the $\alpha 4$-derived protein forms a three-coordinate site, consisting of three His side chains. The $\alpha 4$ fold was also used as a template for engineering a tetrahedral Cys_2His_2 Zn^{2+}-binding site, similar to the structural site of the His_2Cys_2 zinc finger proteins [35]. The direct contribution of each ligand to the coordination geometry and metal-binding affinity was also investigated through the synthesis of derivatives, in which each of the metal-binding ligands was mutated to alanine [36].

Remarkable examples have also been reported by Pecoraro and coworkers. Starting from the 1990s, Pecoraro's group elaborated the original design concept, by engineering metal-binding sites into *de novo*-designed coiled-coil scaffolds [37]. By using CoilSer and **TRI** families (see Table 3.1), the influence of the protein environment on metal ion coordination properties and specificity has been investigated in detail [6, 15, 38–50].

Starting from the CoilSer structure [51], Pecoraro and coworkers designed and structurally characterized, by X-ray diffraction, a parallel three-stranded coiled-coil peptide (CSL9C), able to bind As(III) in a trigonal thiolate coordination geometry (Fig. 3.1) [38]. This structure provides a suitable model of the active site of the ArsR repressor protein, whose structure has not been reported to date, contributing to a better understanding of the properties of the natural protein.

The primary structure of the **TRI** peptides consists of four repeats of the heptad sequence [(AcG($L_aK_bA_cL_dE_eE_fK_g$)$_4$G-NH$_2$)], with hydrophobic Leu residues placed at **a** and **d** positions, helix-inducing alanine residues at **c** position, and hydrophilic lysine and glutamate residues at the **b** and **f** positions, respectively, to promote water solubility [39]. The **TRI** peptides form amphipathic α-helices, which aggregate in aqueous solution, generating a coiled-coil with a hydrophobic interior (Fig. 3.2). A pH-dependent aggregation state was observed: two-stranded coiled-coils were

TABLE 3.1 Peptide Sequences of the TRI Family and Derivatives

Peptide		2 abcdefg	9 12 abcdefg	16 abcdefg	23 abcdefg	30 abcdefg
1. **TRI**	Ac-G	LKALEEK	LKALEEK	LKALEEK	LKALEEK	G-NH$_2$
2. **TRIL9C**	Ac-G	LKALEEK	**C**KALEEK	LKALEEK	LKALEEK	G-NH$_2$
3. **TRIL19C**	Ac-G	LKALEEK	LKALEEK	LKA**C**EEK	LKALEEK	G-NH$_2$
4. **TRIL9CL19C**	Ac-G	LKALEEK	**C**KALEEK	LKA**C**EEK	LKALEEK	G-NH$_2$
5. **TRIL12AL16C**	Ac-G	LKALEEK	LKA**A**EEK	**C**KALEEK	LKALEEK	G-NH$_2$
6. **TRIL16X**	Ac-G	LKALEEK	LKALEEK	**X**KALEEK	LKALEEK	G-NH$_2$
7. **TRIL12L**$_D$**L16C**	Ac-G	LKALEEK	LKA**L**$_D$EEK	**C**KALEEK	LKALEEK	G-NH$_2$
8. **GRAND**	Ac-G	LKALEEK	LKALEEK	LKALEEK	LKALEEK	LKALEEK G-NH$_2$
9. **GRAND** L16XL26AL30C	Ac-G	LKALEEK	LKALEEK	**X**KALEEK	LKA**A**EEK	**C**KALEEK G-NH$_2$
10. **GRAND**L12LD L16CL26AL30C	Ac-G	LKALEEK	LKA**L**$_D$EEK	**X**KALEEK	LKA**A**EEK	**C**KALEEK G-NH$_2$
11. **TRIL9CL23H**	Ac-G	LKALEEK	**C**KALEEK	LKALEEK	**H**KALEEK	G-NH$_2$
12. **CSL9XL23H**	Ac-E	WEALEKK	**X**AALESK	LQALEKK	HEALEHG	-NH$_2$

Sequence^a

^{a}X = Penicillamine. L$_D$ = D-leucine. Residues in bold indicate substitutions. Heptad positions of the residues are lettered.

FIGURE 3.1   Ribbon representation of the X-ray structure of the parallel three-stranded coiled-coil As(CSL9C)$_3$ (PDB code 2JGO). The structure of the metal-binding site is highlighted. The structure was generated with Visual Molecular Dynamics (VMD; http://www.ks. uiuc.edu/Research/vmd/).

preferred at low pH (<5), whereas three-stranded coiled-coils were observed above pH 6. This pH-dependent behavior may be related to the protonation state of the Glu side chains that occupy the helical interface **g** positions in the coiled-coil structure.

A mononuclear thiol-rich binding site was sculpted in the interior of these coiled-coils, by substituting Leu residues at either **a** or **d** positions with cysteine. The affinity of the peptides for a variety of heavy, and often toxic, metals, such as Hg(II), Cd(II), As(III), Pb(II), and Bi(III), was investigated.

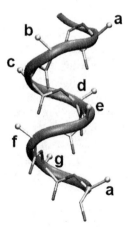

FIGURE 3.2   Amphipathic α-helix, placing hydrophobic residues (**a** and **d** positions) on one face and hydrophilic residues (**b** and **f** positions) on the other.

The binding affinity toward the different metal ions is strictly dependent upon the cysteine substitution site [40]. In particular, Hg(II) or Cd(II) binds at an **a** site (Cys substitution at position 9 or 16 in the peptide sequence) more strongly with respect to the **d** site (Cys substitution at position 12 or 19 in the peptide sequence). Thus, the placement of cysteines at different positions of the coiled-coil core yields similar, yet distinct, metal sites.

One interesting property of these *de novo*-designed proteins is their ability to stabilize unusual metal-coordination geometries. In particular, the aggregation preferences in different conditions were exploited to enforce uncommon coordination geometries on a metal ion.

Extensive studies were carried out on a series of peptides differing in length, in the number of hydrophobic interactions when forming the coiled-coil structures, and, consequently, in their self-association properties. These studies revealed that three-stranded systems, well structured in the *apo*-form, provide a trigonally symmetric framework, which forces metal ions, such as Hg(II) and Cd(II), to adopt in aqueous solutions unusual coordination geometries, as compared to small-molecule thiol ligands [15, 40, 41, 43]. For example, a trigonal Cys environment was found for Hg(II), in spite of the often preferred bis coordination. The most intriguing results were obtained with Cd(II), which adopted a mixture of trigonal-planar (CdS_3) and pseudotetrahedral (CdS_3O, O from an exogenous water molecule) coordinations [40]. The overall analysis carried out on Cd(II)-containing coiled-coil peptides clearly demonstrated that the steric hindrance of amino acids located in either the first or second coordination sphere plays a key role in controlling the coordination number and geometry of the Cd(II) ion [44–46]. Removal of the steric bulk above the metal-binding site provides a water pocket that led to a fully four-coordinate CdS_3O species (peptide **TRIL12AL16C**) [44].

Conversely, the pure CdS_3 geometry was achieved by increasing the steric constraints around the metal-binding site. To this end, two different strategies were used. The first strategy was aimed to modify the first coordination sphere, by replacing Cys with the nonnatural amino acid penicillamine (Pen), which possesses bulky methyl substituents on the C^β atom (peptide **TRIL16Pen**) [45]. In the second strategy the chirality of Leu, in the second coordination sphere, was modified from L to D (peptide **TRIL12L$_D$L16C**) [46]. As a consequence, the side chain of Leu was reoriented toward the C-terminus, providing steric hindrance in the metal-binding site.

Pecoraro and coworkers also succeeded in the design of a peptide containing multiple metal-binding sites. Taking advantage of the different cadmium affinities for the **a** and **d** sites, a **TRI** derivative with two Cys binding sites, **a** and **d**, separated by a layer of intervening Leu residues, was prepared (**TRIL9CL19C**) [47]. The binding of Cd(II) was followed by using ^{113}Cd-, ^{1}H-NMR, and circular dichroic spectroscopies. One of the most striking features of this **TRI** derivative is its ability to sequentially and selectively bind Cd(II) ions.

In fact, the two sites exhibit different binding affinities: the **a** site is exclusively occupied at first and binding of the **d** site occurs after the **a** site is filled. ^{113}Cd NMR spectroscopy revealed a similar coordination geometry for the two binding sites: a mixture of trigonal and tetrahedral geometry, with a trigonal:tetrahedral ratio of 45:55

and 55:45 for **a** and **d** sites, respectively. These similar ratios cannot explain the difference in the cadmium site affinity, which may result from different orientations of the Cys side chains in **a** and **d** sites.

Following these successful results, the same investigators attempted a more challenging obstacle: the design of a peptide capable of binding two Cd(II) equivalents in separate sites with different coordination geometries, one with a trigonal-planar geometry and the second one with a pseudotetrahedral geometry. This goal was achieved with **GRAND** peptide family (Table 3.1), which differs from the **TRI** family for an additional heptad repeat. Using the **GRAND** series, Pecoraro's group designed a "heterochromic" peptide, **GRAND**L16PenL26AL30C, capable of binding Cd(II) as four-coordinate CdS_3O at one site and as trigonal-planar CS_3 at a second site, as confirmed by the correlation of ^{113}Cd NMR and ^{111m}Cd perturbed-angular-correlation spectroscopy [48–50]. More interestingly, the two Cd(II) ions display different physical properties: the four-coordinate site binds Cd(II) at much lower pH value with respect to the three-coordinate site. Thus, the heterochromic peptide shows site-selective metal-ion recognition and influences the physical properties of the metal centers by controlling their coordination geometry.

Recently, DeGrado, Pecoraro, and coworkers successfully engineered metal-binding sites into the antiparallel three-helix bundle α_3D, a single-chain, well-characterized, and stable scaffold, well suited to tolerate mutations inside the sequence [52]. Based on the α_3D structure [53], the authors identified, by visual inspection, four potential sites along the bundle, able to host three Cys residues, one on each helix. Out of these four mutants, α_3DIV (Fig. 3.3), containing the Cys residues at an open end of the bundle, was selected as optimal. The peptide expressed in *Escherichia coli*, displayed structural stability in both the *apo*- and *holo*-forms. The folding behavior and the metal-binding properties of this new metalloprotein were studied in solution by CD, UV, and NMR. α_3DIV was capable of binding heavy metals, such as Cd(II), Hg(II), and Pb(II), with high affinity and the spectroscopic properties of the resulting complexes were similar to those observed for the homomeric three-stranded coiled-coil constructs.

It is evident, by the huge amount of data reported by Pecoraro and coworkers, that three-stranded coiled-coils are helpful scaffolds for understanding the biochemistry of different heavy metals, such as Cd(II), Hg(II), As(III), and Pb(II). Furthermore, the incorporation of metal-binding site within a single polypeptide, able to fold in a stable three-helix bundle, represents a step toward the construction of asymmetric metal-binding sites.

It is important to emphasize that the incorporation of metal-binding site into *de novo*-designed proteins is a useful approach not only to better understand the properties and the working mechanism of natural metalloproteins, but also to engineer artificial enzyme with specific chemical functions, well-defined structures, and/or interesting new physical/chemical properties. A successful example of an artificial metalloprotein, able to reproduce the catalytic properties of a natural enzyme, is the three-stranded coiled-coil protein (named 3SCC), shown in Figure 3.4 [6]. Based on the **TRI** family of peptides, Pecoraro and coworkers very recently have described, in an elegant contribution, the design and the structural characterization of a protein

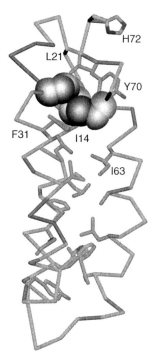

FIGURE 3.3   Molecular model of α_3DIV. Cys residues, located at the C-terminal end of the bundle, are shown as spheres. Hydrophobic residues F31, I14, I63, L21, and Y70, shown as sticks, form an hydrophobic "box" surrounding the Cys site. Reprinted with permission from Reference 52. Copyright 2011 Wiley.

(**TRI**L9CL23H) that approaches the catalytic performance of the natural enzyme carbonic anhydrase (CA). This synthetic metallohydrolase contains two different coordination environments with distinct functions: a catalytic metal site, ZnN_3O, as found in CA, and a separate HgS_3 site, introduced for structural stabilization. The Zn(II) active site lies at the C-terminus of the protein, tucked within the protein's hydrophobic core, while the Hg(II) binding site is located near the N-terminus and it helps to stabilize the structure at high pH. The artificial enzyme is capable of catalyzing the *p*-nitrophenyl acetate (*p*NPA) hydrolysis with an efficiency only ∼100-fold less than that of human CA and 550-fold better than comparable synthetic complexes. Similarly, the conversion of carbon dioxide and water in bicarbonate is only 500-fold less efficient than the natural enzyme. The Zn(II)His$_3$X site closely resembles that found in CA, as ascertained by X-ray crystallography carried out on CSL9PenL23H (see Table 3.1 for amino acid sequence), a peptide very similar to the **TRI** peptide family, but more amenable to crystallographic characterization [6]. The structural data indicate that the complex [Hg(II)]$_S$[Zn(II)(H$_2$O/OH$^-$)]$_N$(CSL9PenL23H)$_3$ contains two independent well-folded, parallel 3SCCs in the asymmetric unit. One contains Hg(II) bound to the sulfurs of Pen in a trigonal-planar structure and a four-coordinate Zn(II)

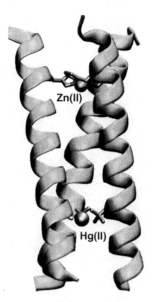

FIGURE 3.4   Ribbon representation of the crystal structure of $[Hg(II)]_S[Zn(II)(H_2O/OH^-)]_N(CSL9PenL23H)_3{}^{n+}$ (one of two trimers present in the asymmetric unit) at pH 8.5 (PDB code 3PBJ). Zn(II) and Hg(II) ions are represented as spheres. Side chains of coordinating residues, His and Pen, are represented in stick. The structure was generated with Visual Molecular Dynamics (VMD; http://www.ks.uiuc.edu/Research/vmd/).

bound to three His ligands and a chloride ion. The second 3SCC contains a T-shaped Hg(II) and a four-coordinate Zn(II) with three His residues and a water/hydroxyl ligand. This work represents an important step forward for enzyme design. Although this artificial enzyme is still two to three orders of magnitude away from the activity of the natural enzyme, its efficiency is impressive, also considering the lack of the second sphere structure of the natural counterpart. Thus, further improvements in the second-shell interactions, aimed to more closely mimic the structure of natural enzymes around the active site, will certainly help to approach the properties of the natural systems.

Using similar parallel coiled-coil scaffolds, Tanaka and coworkers examined the metal-binding properties of coiled-coil peptides, in order to rationalize the factors that impart metal ion stability and selectivity. In their early works, the authors designed an amphiphilic peptide (IZ) [YGG(IEKKIEA)$_4$], which folds into a native-like parallel triple-stranded coiled-coil. Starting from this peptide, the authors searched for proper mutations that could regulate the assembly of the peptide by metal ion coordination [54–56]. A metal-binding site was introduced into the hydrophobic core of the designed coiled-coil peptide, by replacing the hydrophobic isoleucines of the third heptad repeat with one or two His residues (IZ-3aH and IZ-3adH peptides) [54, 55]. These substitutions caused the *apo*-peptide to exist as a random coil structure due to the unfavorable hydrophobic burial of the polar histidines. Addition of

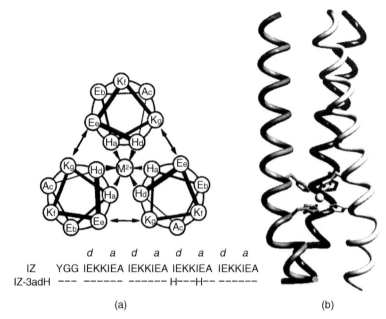

FIGURE 3.5 (a) Helical wheel representation of the third heptad of the IZ-3adH in the parallel orientation, viewed from the N- to the C-terminus. The sequences of the IZ and the IZ-3adH are also reported. (b) Molecular model of the Ni(II) complex of the IZ-3adH. Adapted with permission from Reference 55. Copyright 1998 American Chemical Society.

transition metal ions, such as Ni(II), Co(II), Zn(II), and Cu(II), promotes a metal-induced peptide self-assembly into a triple-stranded-α-helical structure [55]. Interestingly, the two peptides show different coordination geometry and metal-binding affinities. IZ-3adH, which holds two His residues at the **a** and **d** positions of the third heptad repeat, shows high affinity for Ni(II), suggesting a six-coordinate octahedral geometry of the metal-binding site (Fig. 3.5). On the other hand, the IZ-3aH peptide, which has only one His residue, is capable of binding Zn(II) and Cu(II), but not Ni(II). Furthermore, EPR analysis of the Cu(II) complex indicated a tetragonal coordination geometry, with the coordination positions occupied by three His nitrogens and one exogenous water.

In a subsequent work, the same authors modified the IZ peptide sequence, by introducing a "soft" metal-binding site in its hydrophobic core (IZ-AC), in order to investigate the metal ion selectivity [57]. Isoleucines were substituted with Cys and Ala residues, resulting in the destabilization of the coiled-coil structure. As observed for the previously studied systems, "soft" metal ions, such as Cd(II), Cu(I), and Hg(II), induce the unstructured peptide to fold into a three-stranded α-helical bundle, as monitored by circular dichroism spectroscopy. In contrast, "harder" metal ions, such as Ni(II), Co(II), and Zn(II), do not induce peptide self-assembly, thus indicating metal ion selectivity. Furthermore, the IZ-AC peptide not only folds into a three-stranded α-helical bundle upon metal coordination, but forces Cd(II) to adopt an unusual trigonal coordination geometry.

The next step was the design of heterometal-binding sites. In particular, Cu(II) and Ni(II) binding sites were introduced into the hydrophobic core of a three-stranded helical bundle [58]. The peptide (IZ(5)-2a3adH) was obtained by inserting, into a sequence consisting of five heptad repeats, three His residues placed in spatial vicinity of each other. The metal-binding properties of this new derivative were analyzed by CD, NMR, and ESR spectroscopies. All the data demonstrated that IZ(5)-2a3adH is capable to simultaneously bind Ni(II) and Cu(II) and that the metal ion coordination induces an α-helical structure. Ni(II) binds to four His residues at the **3a** and **3d** position, with a square-planar coordination geometry, while the Cu(II) ion engages three nitrogens of the His residues, at the **2a** position, and an oxygen donor, possibly derived from a water molecule. Interestingly, binding of the first metal ion induces the folding of the peptide into the triple-stranded coiled-coil, thereby promoting the second metal ion binding.

Recently, Tanaka's group focused on the *de novo design* of type 1 blue copper protein [59]. Using a previously designed four-helical coiled-coil protein [60], a Cu(II)-binding site with distorted geometry was engineered (AM2C). Two His residues and one Cys residue were located in the core of the protein, to allow the Cu(II) coordination. The AM2C folding, metal-binding properties, and coordination geometry were ascertained by CD, UV-vis, EPR, and X-ray absorption fine structure (EXAFS) spectroscopies. All the data revealed that the protein adopts the expected structure, even in the *apo*-form, and encompasses a type 1 copper complex. The Cu(II) ion is coordinated by three endogenous protein ligands (two His and one Cys), with a distorted trigonal geometry, and by an exogenous ligand (Cl⁻ ion) (Fig. 3.6). Interestingly, the electrochemical properties of the blue copper site in AM2C are similar to azurin, both in the redox potential value and in the reversibility of the Cu(II/I) redox

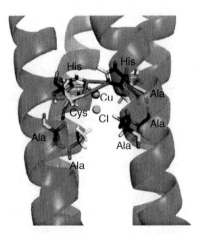

FIGURE 3.6   Active site of the minimized model structure of the AM2C-Cu^{2+} complex. Reprinted with permission from Reference 59. Copyright 2010 American Chemical Society.

process, which does not cause structural changes. In summary, the successful *de novo* design of such artificial blue copper protein confirms that α-helical coiled-coil proteins are suitable scaffold for accommodating metal-binding sites, with different and elaborate geometries.

The research activity of Ogawa and coworkers further expanded the repertoire of coiled-coil motifs utilized as protein cages for housing metal cofactors. In an effort to incorporate redox activity into designed metalloproteins, Ogawa's group designed a coiled-coil polypeptide, C16C19-GGY [Ac-K(IEALEGK)$_2$ (CEACEGK)(IEALEGK)GGY-NH$_2$], in which Cys residues were placed at the **a** and **d** positions of the third heptad repeat (position 16 and 19 in the peptide sequence) [61, 62]. The design was based on the expectation that the formation of a two-stranded coiled-coil would arrange four Cys in a tetrahedral metal-binding site similar to that found in rubredoxin, a naturally occurring electron-transfer protein. The resulting peptide undergoes a significant conformational change from monomeric random coil to a metal-bridged coiled-coil upon binding to a variety of soft metal ions, such as Cd(II), Hg(II), Ag(I), Au(I), and Cu(I) [62]. Interestingly, the inorganic cofactors are not only able to induce peptide self-assembly but also to dictate the nature of the oligomerization state. In particular, Cd(II) and Hg(II) adducts exist as a two-stranded coiled-coil containing a single metal ion, the binding of Au(I) produces peptide hexamers, and Ag(I) and Cu(I) promote the assembly into a four-stranded coiled-coil, encapsulating a tetranuclear metal ion thiolate cluster [63, 64]. Significantly, "harder" transition metal ions, such as Fe(II), Co(II), Ni(II), Zn(II), or Pb(II), do not cause peptide conformational change, thus demonstrating metal-binding specificity.

According to Pecoraro's studies, these results indicate that the C16C19-GGY peptide do not have a strong preference for a particular coiled-coil geometry; in contrast, the coordination properties of the different metal ions strongly influence the structure of the resulting metalloprotein.

The Cu(I)-binding properties of C16C19-GGY were studied in more detail [64]. It was found that the addition of one equivalent of Cu(I) to the monomeric peptide results in the cooperative formation of a metalloprotein, containing a multinuclear Cu(I)-thiolate cluster within the hydrophobic core of the self-assembled four-helix bundle. The multinuclear Cu(I) site is responsible for an intense room temperature long-lived (microsecond) luminescence, centered at 600 nm, a behavior also reported for natural proteins, such as metallothionein, ACE1 transcription factor, and copper-responsive repressor CopY, all containing polynuclear Cu(I) clusters, buried within the protein. The luminescence is quenched by the addition of either ferricyanide, oxygen, or urea to respectively indicate that the emitting species is associated with the reduced Cu(I) state, has significant triplet character, and is quenched upon exposure to bulk solvent. EXAFS data suggested a model for the copper cluster in which only four of the eight available Cys residues coordinate the metal ions: the four copper ions are bridged by Cys sulfur atoms and have a terminal N/O ligand. Furthermore, spectroscopic titrations showed that subsequent additions of Cu(I) result in the occupation of a second, lower affinity, metal-binding site in the designed metalloprotein. This occupancy does not significantly affect the conformation of the synthetic metalloprotein but quenches the emissive state of the polynuclear Cu(I) center.

This work clearly demonstrates how the *de novo* design of a synthetic metallo-protein can lead to the development of a novel entity, possessing unique physical properties.

Recently, Nanda and coworkers computationally designed a four-helix bundle protein, CCIS1 (Coiled Coil Iron Sulfur protein 1), capable of encapsulating in its hydrophobic core a Fe_4S_4 cluster [65]. This design is particularly significant because most of the known natural FeS-coordinating proteins are rich in beta structures and they generally have cysteine residues, coordinating the cluster, in exposed loop regions. The protein, expressed and purified from *E. coli*, is folded and readily self-assembled with Fe_4S_4 clusters *in vitro*, as confirmed by a wide range of analytical and spectroscopic measurements. This result is particularly encouraging because the design strategy utilized can be extended into a multi-FeS cluster protein, by simply duplicating and translating the binding site along the coiled-coil axis. This opens new possibilities for constructing functional analogs of the natural multi-FeS redox chains.

Tackling the challenge of developing synthetic artificial endonucleases and eluci-dating principles that govern metal binding and interactions with DNA, Franklin and coworkers, in a novel approach, designed a series of "chimeric proteins," consist-ing of helix-turn-helix (HTH) motifs [66]. The design approach used, *the chimeric design or modular turn substitution*, consists in exchanging superimposable loops from different proteins, to make a new mutant capable of performing the functions of both parents. In the beginning, chimeras were built by incorporating a Ca(II)-binding EF-hand loop of calmodulin into the HTH motif of the *engrailed* or antennape-dia homeodomain, a DNA-binding motif. Although functionally different, the local topology of the HTH and EF-hand motifs are similar, and thus the turn of one motif could be modularly replaced with the turn of the other. The designed systems are able to bind different metal ions: the HTH motif was modified by turn-sequence substi-tution, thus allowing binding of Ca(II) and lanthanide ions, while still retaining the parental structures [67]. Notably, the trivalent lanthanide chimeras are catalytically competent, able to hydrolyze phosphate esters including DNA, and to bind and cleave DNA, with sequence preferences. Further work was aimed to design transition metal-binding sites into *engrailed* homeodomain. In one successful design, the octarepeat Cu(II) loop (PHGGGWGQ) from the prion protein (PrP) was incorporated into the HTH motif of *engrailed* homeodomain [68].

While DNA-binding and cleavage assays were not performed with this chimera, it showed exceptional specificity for Cu(II). In particular, the new chimeric system spectroscopically mimics the PrP octarepeat. Copper(II) binding was demonstrated by Trp fluorescence titrations, EPR spectroscopy, and ESI mass spectrometry. The peptide was found to bind one equivalent of copper; similar to the PrP octarepeat itself, the EPR spectrum was consistent with a type 2 Cu(II) site, in a N_3O or N_4 coor-dination environment, characterized by two weak axial ligands (water molecules). Also resembling the PrP, circular dichroism studies showed that the chimeric protein, predominantly disordered in the *apo*-form, experiences a slight structural enhance-ment upon copper binding. Thus, the HTH/PrP chimeric system represents a useful model to understand the structural flexibility and reactivity of a single prion site, isolated in a new context. The results by Franklin and coworkers demonstrate how a robust fold can be redesigned to incorporate catalytic activity into a minimalist recognition motif. The designed HTH/EF-hand chimeras represent the first examples of a small peptidic artificial nuclease with sequence discrimination and demonstrate that the HTH motif is a robust scaffold, useful to build novel metallopeptide constructs.

In contrast to the design of α-helical proteins, the *de novo* design of functional metalloprotein containing β-structure proteins is less well developed, even though such β-motifs frequently occur in natural metalloproteins. In this context, the *de novo* design of a redox-active rubredoxin mimic, RM1, by Dutton, DeGrado, and coworkers, is a rare example of a structural and functional artificial metallo-β-sheet protein [69]. The design of the protein started out from the active site structure of rubredoxin, the simplest iron–sulfur proteins. RM1 incorporates the two rubredoxin β-hairpins, that make up the metal-binding site, which are fused together through a highly stable hairpin motif, the tryptophan zipper (Trpzip) [70]. The design also included a connection that incorporates a third strand in each β-sheet to enhance the hydrophobic core. The identities of the remaining amino acids were determined by recursive calculations with the computational design algorithm SCADS (statistical computationally assisted designed strategy) [71], which provided site-dependent side-chain probabilities. The resulting molecule was significantly shorter (40 vs. 54 amino acids) and had a different topology from that of the natural protein. UV-vis spectroscopy demonstrated that RM1 is capable of binding transition metal ions in a tetrahedral tetrathiolate geometry and in a peptide/metal-ion stoichiometry consistent with the design. CD spectroscopy and analytical ultracentrifugation indicated that RM1 is monomeric and folds in the presence and absence of metal ions. More interestingly, investigation of redox properties indicated that RM1 reversibly cycles between the Fe(II) and Fe(III) oxidation states: it shows activity for 16 cycles under aerobic conditions, compared to two to three times for previously reported rubredoxin mimics based on thioredoxin scaffold [72].

Often the metal ion-binding sites of metalloproteins are formed between two or more elements of secondary structure. A careful examination of the three-dimensional structure of natural proteins can be used to identify geometric relationships between these secondary structural elements, thus facilitating the design of metalloprotein mimics. Our laboratory has approached the challenge of constructing a rubredoxin model using the retrostructural analysis approach [73].

By a detailed analysis of the crystal structure of the natural iron–sulfur proteins, we found a Cys-Xxx-Xxx-Cys antiparallel β-hairpin motif occurring in a tandem twofold repeat in the iron-binding sites of rubredoxins. These sites could be described to 0.78 Å rms deviation by using a C_2-symmetric arrangement of the identical idealized β-hairpins. This finding was the basis for the design of a minimal dimeric model. We successfully selected the minimum set of constituents necessary for an accurate reconstruction of the active site structure and, using symmetry considerations, we designed a β-hairpin peptide, named METP (miniaturized electron-transfer protein), as rubredoxin model (Fig. 3.7). METP is made up of an undecapeptide sequence, with two properly spaced Cys residues that dimerizes (via a C_2 axis) in the presence of tetrahedrally coordinating metal ions, giving rise to a compact sandwiched structure. The characterization by several spectroscopic techniques showed that METP retains

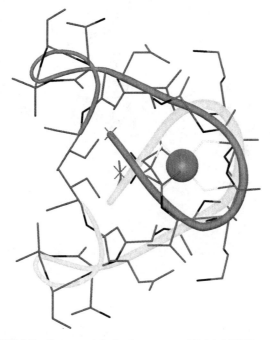

FIGURE 3.7   Energy minimized structure of Fe(II)-METP complex.

the main features of the parent protein. It binds Co(II), Zn(II), Fe(II), and Fe(III), strictly resembling the rubredoxin active site spectroscopic features. Further, the structural characterization in solution by NMR of the Zn(II)-METP complex confirmed the success of the used design approach, in the construction of metalloprotein models. By changing the nature of the coordinating residues, it may allow to develop models for several tetrahedral metal-binding sites, such as the Cys_2His_2 or His_3Cys zinc-binding motifs and the copper-binding domain of blue copper proteins.

3.3   *DE NOVO*-DESIGNED PROTEIN CAGES HOUSING DINUCLEAR METAL COFACTORS

Dinuclear metal cofactors are widely distributed in biology. Examples of naturally occurring dinuclear metal cofactors involve the di-iron, di-manganese, di-copper, di-nickel, and di-zinc clusters, which are found in important classes of enzymes, such as monoxygenases, desaturases, oxidases, hydroxylases, and hydrolases [74]. In numerous cases, the dinuclear metal site is housed into a four-helix bundle fold. Our group, in collaboration with DeGrado's group at the University of Pennsylvania, used a retrostructural analysis of the natural di-iron oxo proteins, in order to dissect the main factors which contribute to define the structure and geometry of the metal-binding site in this class of proteins [75, 76]. Starting from this analysis,

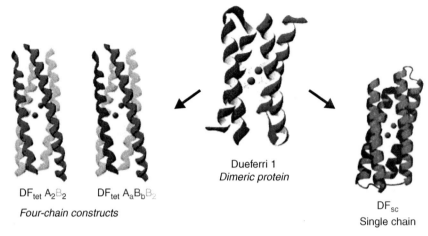

DF$_{tet}$ A$_2$B$_2$ DF$_{tet}$ A$_a$B$_b$B$_2$

Four-chain constructs

Dueferri 1
Dimeric protein

DF$_{sc}$
Single chain

FIGURE 3.8 Different DF protein constructs, as models of di-iron proteins: four-chain constructs (DFtet), dimeric protein (DF1), and single chain (DF$_{sc}$).

a *de novo* design approach was followed for the development of the DF (Due Ferri) family of artificial proteins, which are models of di-iron and di-manganese metalloproteins (Fig. 3.8).

Design started from the analysis of several members of the di-iron metalloproteins, and in particular was focused on ferritin [77], bacterioferritin [78], rubrerythrin [79, 80], ribonucleotide reductase R2 subunit (R2) [81], D9 ACP desaturase [82], and the catalytic subunit of methane monooxygenase [83]. Even though these proteins show less than 5% sequence identity, their active sites are housed within a very simple pseudo-222-symmetric four-helix bundle. To build the prototype scaffold housing the dinuclear metal center, we followed the symmetry concept, trying to reproduce in a symmetric model the quasi-symmetrical structure of a metalloprotein [84]. The use of C_2 symmetry has been shown to be particularly advantageous, because it simplifies the design, reduces the size of the molecules to be synthesized, and simplifies their structural characterization. The first developed model, DF1, is made up of two 48-residue helix–loop–helix (α2) motifs, able to specifically self-assemble into an antiparallel four-helix bundle [13,84,85]. Table 3.2 reports the amino acid sequence of the α2 motif (entry 1) together with the intended helical secondary structure. To provide a Glu4His2 liganding environment for the di-iron center, each α2 subunit contains a Glu$_a$-Xxx$_b$-Xxx$_c$-His$_d$ sequence in helix 2, which donates a His side chain (His^{39}/His$^{39'}$) as well as a bridging Glu carboxylate (Glu^{36}/Glu$^{36'}$) to the site. A second Glu carboxylate ligand on helix 1 (Glu^{10}/Glu$^{10'}$) provides a fourth protein ligand per metal ion. Liganding side chains were placed in the appropriate rotamers to allow interaction with a di-iron center. The final positions of the helices were dictated by three different requirements: (1) the geometry of the liganding site was restrained to bind di-iron with two bridging Glu carboxylates, two nonbridging Glu side chains, and the Nδ of two His side chains; (2) the helical packing angles and

TABLE 3.2 Peptide Sequences of DF Family

Peptide	g	abcdefg	abcdefg	abcdefg	abcdefg abcdefg	abcdefg	abcdefg	abcdef
1. DF1	DY		LRELLKL	ELQLIKQ	YREALEYV--KL----PV	LAKILED	EEKHIEW	LETILG
2. DF3	DY		LRELLKG	ELQGIKQ	YREALEYT--HN----PV	LAKILED	EEKHIEW	LETILG
3. DF2	MDY		LRELYKL	EQQAMKL	YREASERV--GD----PV	LAKILED	EEKHIEW	LETING
4. DF2t	MDY		LRELYKL	EQQAMKL	YREASEK--ARN-PEKKSV	LQKILED	EEKHIEW	LETING
5. DF1sc(1-28)	DE		LRELLKA	EQQAIKI	YKEVLKKA-KE----GD			
DF1sc(29-59)	EQELARL		IQEIVKA	EKQAVKV	YKEAAE----KA----RN			
DF1sc(60-88)	PEKRQV		IDKILED	EEKHIEW	LKAASK-----QGN			
DF1sc(89-114)	AEQFASL		VQQILQD	EQRHVEE	IEKKN			
6. DFtet: A	K	LKELKSK	LKELLKL	ELQAIKQ	YKELKAE----LKEL			
Aa	E	LKELKSE	LKELLKL	ELQAIKQ	FKELKAE----LKEL			
Ab	K	LKKLKSR	LKKLLKL	ELQAIHQ	YKKLKAR----LKKL			
B	E	LEELESE	LKKILKL	EERHIEW	LEKLEAK----LKEL			

[a]The liganding residues are in bold. Helical regions are underlined, and the heptad positions of the residues are lettered.

distances were constrained to match those typically observed in the active sites of di-iron proteins; and (3) precise twofold symmetry between the two pairs of helices was enforced. Each metal ion is five-coordinate and a sixth vacant site lies on adjoining faces of the two metal ions, providing a potential site for binding of exogenous ligands. Satisfaction of side-chain packing requirements was assured by positioning hydrophobic side chains to fill the core, as dictated by the steric environment of the backbone structure. In addition, Glu and Lys residues were chosen for helix-favoring, solvent-accessible sites, to provide water solubility and to drive the assembly into the desired antiparallel topology. Finally, an idealized γ-α_L-β interhelical loop was included between the two pairs of helices.

The DF1 sequence was also carefully engineered to include second-shell inter-actions, which are crucial for defining structural and functional properties of metal-binding sites. Thus, in DF1 a Tyr residue at position 17 donates a second-shell hydrogen bond to the nonbridging $Glu^{10'}$ of the other monomer in each of the α_2 subunits (the same interaction exists between the symmetrically related pair, $Tyr^{17'}$ and Glu^{10}). Similarly, an Asp residue at position $35'$ forms a hydrogen bond with the imidazole Nε of the His ligand in the neighboring helix of the dimer. This Asp is further involved in a salt bridge interaction with a Lys at position 38. This hydrogen-bond network consisting of Lys/Asp/His in DF1 is similar to that observed in the active sites of natural proteins, such as methane monooxygenase, where the lysine residue is replaced by an arginine to form an Arg/Asp/His cluster. DF1 was able to bind Zn(II), Co(II), or Fe(II) ions, and detailed information of its overall structure was obtained from the analysis of the crystal structure of the di-Zn(II) and of the di-Mn(II) form of the protein [84, 85], as well as the NMR structure of the *apo*-form [21]. The overall structure is very similar to the intended designed model, consisting of an antiparallel pair of helical hairpins with the desired topology, and the structure of the active site is also very close to that found in the diferrous and dimanganous forms of naturally occurring proteins.

The C_2 symmetric DF1 protein set the stage for the development of the other analogs. In particular, an iterative cycle of redesign and structural characterization was subsequently applied in order to improve the DF1 properties and to obtain artificial proteins with specific catalytic activities and functions. Although DF1 behaves as a native-like protein irrespective of its ligation state, it had several limitations: it was designed for maximal stability, therefore the interior of the protein was efficiently packed with a large number of hydrophobic side chains, resulting in a high degree of conformational stability. This behavior accounted for the absence of functionality, since enzymatic functions usually require a certain level of conformational freedom.

In the first round of the redesign process, which leads to the DF1 and DF2 subsets (see Table 3.2), several changes in the sequence, as well in the loop structure, were done, in order to redefine the accessibility of the active site, too much constrained in the extremely stable prototype DF1. In particular, substitution of Leu^{13} and $Leu^{13'}$ in DF1 with smaller side-chain amino acids, such as Ala and Gly, afforded a cavity large enough to allow the access of small molecules to the metal center. The L13A-DF1 [86] and L13G-DF1 [87] variants, as well as the DF2 subset [88–91], bind exogenous ligands, such as phenol and acetate, and display ferroxidase activity.

However, optimization of the catalytic activity required further changes in the second-shell ligands, in order to provide the protein with stability and, at the same time, with the flexibility required for function.

This redesign strategy would necessarily require the preparation and purification of significant quantities of hundreds to thousands of variants, to be screened for binding of small molecules, and reactivity toward a variety of substrates. To advance this goal, four-helix bundles based on heterotetrameric systems, named DF_{tet} [92], were developed, with the aim of bridging the gap between "rational design" and "combinatorial" approaches (Table 3.2). DF_{tet} constructs can be mixed and matched to allow easy generation of combinatorial diversity. First, an A_2B_2 heterotetramer was generated [92]. By combining n variants of A and m of B, such a system would generate $n \times m$ variants (e.g., 10 variants each of A and B would provide 100 combinations). Next, an $A_aA_bB_2$ heterotetramer was designed, in which the two A chains were distinct [93]. Ten variants of each of these three chains would now generate 1000 different peptides. The A_2B_2 heterotetramer (designated DF_{tet}-A_2B_2) was designed to be an antiparallel coiled-coil (Fig. 3.8). The helices were extended relative to those of DF1 (33 residues in DF_{tet} vs. 24 residues in DF1) to increase the tetramer stability by increasing the size of the hydrophobic core. Next, side chains were placed onto the A and B helices, resulting in an A_2B_2 heterotetramer with C_2 symmetry. The residues within a 12-residue region of the binding site are essentially identical to the corresponding residues of DF2.

Residues at the remaining **a** and **d** positions were modeled as leucines because this side chain effectively filled the interior volume of the bundle. The nature of the residues at the remaining **e**, **g**, **b**, and **c** positions were chosen to specifically stabilize only one of the possible topologies for an antiparallel A_2B_2 heterotetramer. For this purpose, a simplified energy function, which considers only the charge of the residues, was used. The target function selected for the sequence with the largest energy gap between the desired structure and an alternatively folded structure, rather than to simply minimize the energy of the desired conformation. This represented the first example in which "negative design" against alternatively folded structures had been explicitly coded and experimentally tested.

The sequence of DF_{tet}-A_2B_2 was further modified to create DF_{tet}-$A_aA_bB_2$ [94]. In this design, one helix–helix interface was redesigned to provide high specificity for the desired three-component assembly. In particular, Glu, Lys, and Arg residues were introduced such that they would be able to interact favorably in the desired three-component hetero-oligomer. The resulting proteins were found to assemble into a helical tetramer with considerable thermodynamic stability and capable of binding Co(II), Zn(II), and Fe(II) in the expected stoichiometry. CD spectroscopy, size exclusion chromatography, and analytical ultracentrifugation showed that the desired tetramer formed when the three components were combined in the proper stoichiometry (one A_a and A_b per two moles of B).

A number of asymmetrical variants of DF_{tet}-$A_aA_bB_2$ and symmetrical variants of DF_{tet}-A_2B_2 were prepared and evaluated. The variants of DF_{tet}-A_2B_2 are designated as G_4-DF_{tet} (in which Leu^{15} and Ala^{19} of both A chains were substituted with Gly), L_2G_2-DF_{tet} (in which Leu^{15} was retained and Ala^{19} was changed to Gly), A_2G_2-DF_{tet}, and G_2A_2-DF_{tet} (which have Gly or Ala at the indicated positions). All the variants were screened for their ability to react with Fe(II) and O_2, and the variant with the fewest steric restrictions, namely G_4-DF_{tet}, exhibited the most rapid rate of oxidation and formation of the oxo species with no detectable intermediates. The same variant showed the greatest binding ability toward phenol, which binds to the di-ferric site. Finally, the two-electron oxidation of 4-aminophenol to the corresponding quinone monoimine, catalyzed by the proteins in atmospheric O_2, was investigated. Again, G_4-DF_{tet} showed a $\approx$ 1000-fold rate enhancement relative to the background reaction when the initial rates of the reaction in the presence and absence of the protein were compared. The G_4-DF_{tet} catalyzed this reaction for at least 100 turnovers. Changing either of the Gly residues at position 19 or 15 to Ala gave a protein whose rate was decreased between 2.5- and 5-fold. All the results on DF_{tet} variants provide deep insights into the active site requirements for function. They clearly demonstrate that the catalytic efficiency is sensitive to the side-chain hindrance around the metal center, thus highlighting the specificity of the design. Further, the combinatorial approach was proven very promising for identifying potential candidates for defined applications.

Even though combinatorial studies on the DF_{tet} subset allows to select G_4-DF_{tet} as the best candidate for function, structural data were not available; therefore, it was not possible to finely correlate active site structure and activity. In order to acquire structural information, it was considered useful to transfer the key mutations, identified with the DF_{tet} analysis, either into a single-chain protein or into the well-characterized DF1 framework.

First, a 114-residue protein (DF_{sc}, Fig. 3.8 and Table 3.2) was designed, using the helical backbone of the DF_{tet} peptides as template [95]. The identities of 26 residues were predetermined, including the primary and secondary active site ligands, residues involved in active site accessibility, and the turn sequences (which were selected from database analyses). The remaining 88 amino acids were determined using a side-chain repacking algorithm written by Saven and coworkers termed SCADS [71]. This algorithm is based upon a recently developed statistical theory of protein sequences. Rather than sampling sequences, the theory directly provides the site-specific amino acid probabilities, which are then used to guide sequence design. The resulting sequence (DF_{sc}) expresses well in *E. coli* and is highly soluble (at least 2.5 mM). DF_{sc} stoichiometrically binds a variety of divalent metal ions, including Zn(II), Co(II), Fe(II), and Mn(II), with micromolar affinities. ^{15}N-HSQC NMR spectra of both the *apo-* and Zn(II)-proteins reveal narrow line-widths and excellent dispersion in the amide and aliphatic regions; further, the NMR structure is in good agreement with the designed structure.

To gain advantages from the symmetric nature of the dimeric derivatives, which simplifies synthesis and structural data interpretation, we also inserted the G_4-DF_{tet}-deriving mutations into DF1. Gly residues were introduced at both positions 9 and 13 of both α_2 monomer, with the aim to open up the active site and create a cavity large enough to accommodate exogenous ligands. However, the introduction of helix-destabilizing Gly residues and the loss of the hydrophobic driving force strongly destabilize the fold of the protein. Indeed, a single mutation of Leu$^{13,13'}$

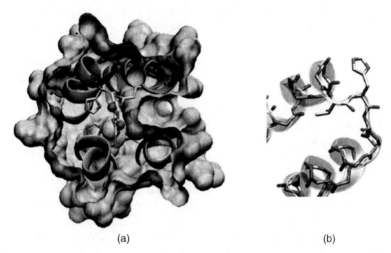

(a)   (b)

FIGURE 3.9   (a) Surface representation of the di-Zn(II)-DF3 NMR structure (PDB code 2KIK). An arrow highlights the active site access channel. (b) Details of the loop structure. The figures were generated with Visual Molecular Dynamics (VMD; http://www.ks.uiuc.edu/ Research/vmd/).

to Gly destabilizes DF1 by 10.8 kcal/mol per dimer, precluding the introduction of the second glycine residue [13]. To compensate for the thermodynamic cost of carving an active-site access channel onto the protein, it was necessary to optimize the interhelical loop conformation. To increase the conformational stability of the DF scaffold, we modified the sequence of the interhelical turn, which adopts a "Rose-like" α_R-α_L-β conformation [89]. The original Val^{24}-Lys^{25}-Leu^{26} of DF1 was changed to Thr^{24}-His^{25}-Asn^{26}. In models, His^{25} appeared capable of forming stabilizing hydrogen-bonded C-capping interactions of helix 1, while Asn^{26} could form N-capping interactions either with helix 2 or with the carbonyl group of Thr^{24}, depending on its rotamer. The newly designed protein DF3 (Fig. 3.9, Table 3.2) showed improved water solubility (up to 3 mM) and active site accessibility, while retaining the unique native-like structure, as assessed by NMR structural characterization [5]. Most importantly, DF3, with its well-defined active site, displays ferroxidase and oxidase activity. In fact, it catalyzes the two-electron oxidation of 3,5-ditert-butyl-catechol to the corresponding 3,5-ditert-butyl-quinone. The catalytic efficiency toward structurally related substrates illustrates the selectivity of the active site.

Taken together, the outcome from DF investigation demonstrates that the *de novo* design of four-helix bundles as cages for dinuclear metal-binding sites can afford artificial proteins holding not only the coordination and spectroscopic features of the natural counterparts, but notably their functionality.

The *de novo* design approach has also been used for the construction of peptide scaffolds housing metal-binding sites for distinct metal cofactors. A remarkable example was reported by Holm and coworkers, who designed a series of helix–loop–helix peptides to cage a [Ni-X-Fe$_4$S$_4$]-bridged structure [96, 97]. This metal-binding site was proposed for the A-cluster of the nickel-containing enzyme carbon monoxide dehydrogenase (CODH). Four helix–loop–helix 63mer peptides were designed and synthesized in order to assess the utility of peptides as scaffolds for the stabilization of complex metal-bridged assembly, consisting of two discrete fragments that are juxtaposed wholly or in part by one or more covalent bridges. Two α-helices were connected through a loop containing the ferredoxin consensus sequence Cys-Ile-Ala-Cys-Gly-Ala-Cys; a Cys residue was selected as the X bridging group and was positioned toward the *N*-terminus, separated by one residue from the first Cys of the consensus sequence. Three other binding residues were incorporated in appropriate positions to constitute a binding site for Ni(II), with a distorted square-planar geometry. One of the peptides was designed with an N$_3$S (His$_3$Cys) site, and each of the other three with N$_2$S$_2$ (His$_2$Cys$_2$) sites. The peptides were found to be able to bind one [Fe$_4$S$_4$] cluster per 63mer. In addition, XAS and EXAFS data provided strong evidence that the nickel binds in the desired site in two of the metallopeptides [97].

3.4   *DE NOVO*-DESIGNED PROTEIN CAGES HOUSING HEME COFACTOR

A high number of protein cages for heme cofactors are found in Nature. The architecture and composition of the protein cages for heme have been selected by Nature in order to allow a single prosthetic group to serve numerous and diverse chemical functions [74, 98]. In fact, the heme group embedded into the protein matrix is able to promote a variety of functions, such as dioxygen storage and transport, electron transfer, hydroxylation and oxidation of organic substrates, and hydrogen peroxide disproportion [100]. Variations in the shape, volume, and chemical composition of the binding site, in the mode of heme-binding and in the number and nature of heme-protein interactions, are the main factors which contribute to the functional specificity of the heme [99, 100].

Many efforts have been devoted to the design of protein incorporating the heme cofactor [101, 102]. The design strategy for the development of these systems is very challenging because it requires two sets of factors to be taken into account simultaneously: (1) the construction of an artificial protein that adopts the unique desired folding and (2) the engineering into the interior of the designed protein of a proper cavity able to accommodate the large hydrophobic heme cofactor.

The use of α-helical peptides able to cage the heme group has been and still remains the focus of the majority of the works in this research area, because the α-helix motif is a recurring structural motif that surrounds the heme cofactor in numerous natural hemeproteins. Over the years, a large number of helical hemeprotein models have appeared in the literature. They differ in structural complexity and can be grouped into two classes, according to the nature of the interactions between the heme and the helical peptides: (1) models in which peptide chains are covalently linked to the heme and (2) models in which peptide chains incorporate one or more heme groups, by noncovalent self-assembling around the heme. Excellent examples of covalent heme-peptide assemblies are the peptide-sandwiched mesoheme, reported by Benson and coworkers [103, 104], and the peptide-sandwiched deuteroheme (Mimochromes), developed by us [101, 104–107]. Along our research activity on Mimochromes, recently reviewed [20], we have succeeded with the design of peroxidase-like activity into the last born Mimochrome VI, which exhibits catalytic performance similar to the natural counterpart horse radish peroxidase [108].

This section will mainly focus on noncovalent heme-peptide adducts and in particular on the use of four-helix bundles [109] as protein cages for mono-heme and multi-heme binding. It should be highlighted that the assembly of α-helices into the four-helix bundle motif is an ubiquitous and functionally important architecture found in several hemeproteins, such as cytochrome c' and cytochrome b_{562} [110–112].

DeGrado, Dutton, and coworkers first described *de novo*-designed four-helix bundle proteins that spontaneously fold in aqueous solution and assemble with one or more hemes [113, 114]. These systems bind heme with a bis-His axial coordination, affording a symmetrical coordination site, which simplifies the design and the structural and functional characterization. In the first pioneering work [113], Choma, DeGrado, and coworkers utilized one member of a four-helix bundle family, designed through an incremental process [115], to incorporate a heme group. In particular, the prototypes VAVH$_{25}$(S-S) and retro(S-S) heme-binding proteins were originated from the α_2B sequence [115], a helix–loop–helix peptide that dimerizes to form a four-helix bundle (Table 3.3, Fig. 3.10). Each monomer of the α_2 dimer is made up of two amphiphilic helical sequences, comprising the heptad repeat (Leu$_a$-Glu$_b$-Glu$_c$-Leu$_d$-Leu$_e$-Lys$_f$-Lys$_g$). The sequence linking the helices contains a single Pro residue to aid in helix termination, as well as two Arg residues to promote a reversal in the overall peptide chain direction. The arrangement of the Leu residues was based on the helical packing observed in naturally occurring four-helix bundle proteins. Glutamic acid and lysine were chosen for the hydrophilic, helical residues, and their side-chain conformations were modeled to form favorable ion pairs along one face of the helix. In addition, the helix has a number of negatively charged residues at its NH$_2$-terminus and positively charged residues at its COOH-terminus to stabilize helix formation by favorably interacting with the helical dipole. Several substitutions were made into the α_2B sequence in order to create a hydrophobic cavity to accommodate a bis-His-ligated heme (Fig. 3.10). The centrally located Leu at position 25 of each α_2 monomer was replaced by a His to bind a single heme; to achieve good

TABLE 3.3   Peptide Sequences of Heme-Binding Four-Helix Bundles

Peptide		Sequence[a]	
α_2	Ac-C	GELEELLKKLKELLKGPRR	GELEELLKKLKELLKG-NH$_2$
VAVH25	Ac-C	GELEELLKK**A**KELLKGPRR	GE**VEEH**LKK**V**KELLKG-NH$_2$
RETRO	Ac-C	GKLLEK**V**KKL**HEE**VEGRRP	GKLLEK**A**KKLLEELEG-NH$_2$

[a]Residues in bold indicate substitutions.

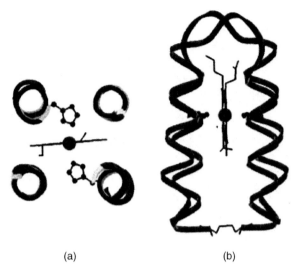

(a)   (b)

FIGURE 3.10   Molecular model of the retro(S-S) peptide with bound heme. (a) View looking down the core of the bundle at the heme-binding site. (b) Lateral view of the protein model, with the loop positioned at the top and the disulfide positioned at the bottom of the bundle. Adapted with permission from Reference 101. Copyright 2001 American Chemical Society.

packing between the heme and the cavity, positions 22 and 29 were changed to Val and position 11 was changed to Ala. Finally, in order to ensure an unambiguous dimeric aggregation state, the peptides were N-terminally linked through a disulfide bridge, to get the $VAVH_{25}$(S-S) model. Retro(S-S) is a variant of $VAVH_{25}$(S-S) with a reversed peptide sequence. The design successfully resulted in the construction of four-helix bundles able to bind the heme with a 1:1 stoichiometry, with a bis-His ligation. Interestingly, a low-spin ferric heme predominated in retro(S-S), which binds the heme more tightly than $VAVH_{25}$(S-S). On the opposite, a mixture of low- and high-spin ferric hemes was present in $VAVH_{25}$(S-S), consistent with less strong coordination by the iron axial ligands. The negative reduction midpoint potentials, -170 and -220 mV for VAVH25(S-S) and retro(SS), respectively, suggested that the heme is more accessible to solvent than originally anticipated. All the data indicated the correctness of the design and set the beginning of a very prolific production of a variety of heme-containing four-helix bundle.

Dutton and coworkers reached impressive results in the design of protein structures, which assemble arrays of cofactors and reproduce native-like functions. They used the terms *maquettes*, French word for scale model of something, to refer to these molecular models [2]. The maquettes are based on a sequence with high α-helix-forming propensities, with residues arranged in the heptad repeat pattern, typical of left-handed coiled-coils (Table 3.4, entry 1). Such a sequence spontaneously assembles into a four-helix bundle scaffold, with glutamate and lysine exposed to the solvent and leucine buried into the interior.

TABLE 3.4 Peptide Sequences of the Maquettes Obtained from an Iterative Design Strategy

Peptide		Sequence[a]						Scaffold
		abcdefg	abcdefg	abcdefg	abcdefg			
1. α-helix-forming peptide		ELLKL	LEELLKK	LEELLKL	LEELLKK		L	
2. H10H24	CGGG	**ELWKL**	**HEELLKK**	**FEELLKL**	**HEERLKK**		**L**	(α-SS-α)$_2$
3. H10A24	CGGG	**ELWKL**	**HEELLKK**	**FEELLKL**	**AEERLKK**		**L**	(α-SS-α)$_2$
4. H10A24-L6I,L13F	CGGG	**EIWKL**	**HEEFLKK**	**FEELLKL**	**HEERLKK**		**L**	(α-SS-α)$_2$
5. L31M	CGGG	**EIWKL**	**HEEFLKK**	**FEELLKL**	**HEERLKK**		**M**	(α-SS-α)$_2$
6. HP1	CGGG	**EIWKQ**	**HEEALKK**	**FEEALKQ**	**FEE-LKK**		**L**	(α-SS-α)$_2$
7. HP7	G	**EIWKQ**	**HEDALQK**	**FEEALNQ**	**FED-LKQ**	GGSGCGSG	**L**	(α-loop-α)$_2$
	G	**EIWKQ**	**HEDALQK**	**FEEALNQ**	**FED-LKQ**		**L**	

[a]Residues in bold indicate substitutions. Heptad positions of the residues are lettered.
Crosslinking cysteines are underlined.

The prototype hemeprotein maquette (α-SS-α)$_2$ [114], later referred also as H10H24 (Table 3.4, entry 2), was designed, in collaboration with the DeGrado group, by combining the general features of the minimalist α_2 artificial protein with the B and D helices of cytochrome bc_1 [116, 117]. The molecule is made up of a 62-residue parallel α_2 system, in which two helices of 31 residues are cross-linked by a disulfide bridge. In order to insert a heme-binding site into the four-helix bundle, Leu at positions 10 and 24 was replaced by His. Further, as observed in the cytochrome *b* subunit of the natural cytochrome bc_1 complex, Phe and Arg replaced Leu at positions 17 and 27. The self-assembly of the α_2 dimers generates an artificial protein housing four bis-His heme-binding sites per four-helix bundle (Fig. 3.11a). The spectroscopic data were fully consistent with the working model since [H10H24]$_2$ binds four equivalents of ferric heme with sub-nanomolar to micromolar dissociation constants. Further, the electrochemical properties of the artificial proteins, including heme–heme redox interactions, closely resemble those of native proteins and are fully consistent with the working model for their structures. However, both the *apo*-peptides and the heme complexes showed dynamic behavior that precluded structure determination.

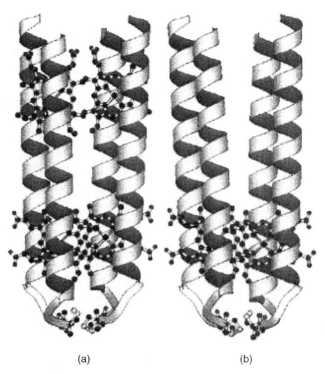

(a)   (b)

FIGURE 3.11 Molecular models of maquettes: (a) cytochrome *c* maquette and (b) maquette variant containing only one couple of heme at the 10,10' positions. Adapted with permission from Reference 101. Copyright 2001 American Chemical Society.

Next, starting from the H10H24 prototype maquette, Dutton and coworkers developed a variety of variants with the aim of providing a framework to analyze the key factors which regulates hemeprotein reduction midpoint potential [118–120]. His to Ala substitutions, either at position 10 or 24 of the sequence, gave molecules with different heme/four-helix bundle stoichiometry and allowed independent determination of the heme redox potential (Fig. 3.11b). In particular, the H10A24 maquette scaffold (Table 3.4, entry 3) was used to determine the proton-binding/release mechanism coupled to the heme redox reaction and the effects of several factors (such as heme peripheral substitution, electrostatic interactions with charged amino acids in heme vicinity, as well as with other heme cofactors) in the modulation of the heme redox potential [118–120]. The complete electrochemical characterization of the H10A24 maquette allowed Dutton and coworkers to underline the basic principles for designing artificial redox/proton-coupled proteins [118], summarized as follows: (1) the designed scaffold should provide an highly hydrophobic environment for tight binding of the heme cofactor; (2) to attain pH redox-dependent behaviors, the scaffold must retain the fold upon side-chain protonation changes; (3) the burial of protonatable groups into the hydrophobic protein interior could modulate the pK values of acidic side chains, thus allowing redox-coupled proton exchange; (4) modulation of heme redox potential can be achieved by charge compensation: negatively charged residues favor ferric heme, thus determining negative redox potential values; (5) since charge compensation effects are distributed between many groups, the design of hemeprotein with pH-independent redox behaviors can be accomplished by removing all charged groups in the proximity of the heme. Using the factors alone and in combination, the heme reduction potential in the maquette could be modulated by up to 435 mV, nearly half the range observed for natural hemeproteins.

In parallel with the analysis of the redox properties of these artificial proteins, a huge amount of work was dedicated to design structurally defined maquette variants. In order to attribute structural specificity to the molten globule H10H24 prototype maquette, an iterative redesign strategy was followed [121, 122]. The strategy was aimed at refining the packing interaction into the hydrophobic core of the four-helix bundle (**a** and **d** positions), by single substitutions of Leu residues with β-branched and/or aromatic residues. Since the **a** positions are occupied by key residues for heme binding, that is, His[10] and His[24], and one **d** position is occupied by an arginine, Arg[27], necessary for heme redox potential modulation, substitutions were restricted at Leu[6], Leu[13], and Leu[20] (Table 3.4, entry 4). The H10H24-L6I,L13F variant showed a uniquely ordered structure with native-like behaviors, as assessed by NMR and chemical denaturation experiments [123, 124]. Further, the crystal structure in the *apo*-form of the derivative H10H24-L6I,L13F,L31M variant, later referred to as L31M (Table 3.4, entry 5), was determined [125]. One of the main results obtained from the iterative design process and the structural characterization was that both H10H24-L6I,L13F and L31M variants are well specific in the *apo*-state, but become molten globules upon heme binding. All the information obtained from the analysis of the maquette variants was used for the goal of developing a structurally defined heme-binding maquette, housing the functionality of proton/coupled redox behaviors [126]. Starting from the X-ray structure of L31M, and using the knowledge on natural proteins and all the data collected on the H10H24 variant series, a new molecule was designed, referred to as HP1 (Table 3.4, entry 6). Redesign of L31M started from the evidences that a structural reorganization of the maquettes must occur, in order to satisfy the heme coordination into the four-helix bundle core. A search in the Protein Data Bank (PDB) for the conformational statistics of the His residues in naturally occurring bis-His hemeproteins revealed that a rotation $> 50°$ along the helical axes was needed in order to accommodate the heme group inside the bundle and coordinate it with the typical histidine rotamers. Such rotation causes Glu at positions 11 and 18 to become buried with respect to their relative position in the *apo*-form, and in turn Leu at positions 9 and 23 to become solvent exposed, thus creating a mismatch in the alignment and packing of the helices and producing a nonnative-like hemeprotein. The strategy used for the design of HP1 was therefore aimed at revising the hydrophobic/hydrophilic pattern of the initial four-helix bundle motif, through the modification of 6 of the 31 residues in each helix. In particular, Leu[9] and Leu[23] were converted to Gln, and Phe[13] and Leu[20] at the interfacial position were substituted with Ala. The Glu[11] and Glu[18] were not varied in order to keep their effect in the modulation of the redox potential. The final design afforded a protein able to bind two hemes with a well-defined structure and the conserved functionality of proton-coupled redox activity, as observed for the H10A24 maquette. HP1 binds two hemes per four-helix bundle with K_d values < 20 nM. The UV-vis spectra were consistent with a bis-His coordination in a low-spin state for both hemes. Interestingly, the *apo*-form was unstructured, thus demonstrating that the structural specificity in these new maquettes is strictly related to heme binding.

One of the most remarkable properties of the maquettes is that they can accommodate allosteric regulation of protein conformational switches. Structural characterization, in solution and in the solid state, of H10H24-L6I,L13F and L31M (Table 3.4, entry 4 and 5) revealed that the helical interface, which lies between the helices not constrained by heme binding, has an unusually low degree of surface complementarity and a high degree of interhelix motion. As a consequence, the four-helix bundle in the maquettes can adopt either an *anti* topology, in which the helices are antiparallel, or a *syn* topology with the helices of different subunits parallel (Fig. 3.12).

Dutton and coworkers demonstrated that a conformational switch of the two subunits between the two different *syn* or *anti* topologies can occur upon a specific driving stimulus, such as heme binding/release or heme redox changes [127]. Proof of principle of this phenomenon was demonstrated on the H10A24 and the related H10S24 maquettes, by using an optical probe (pyrene or coproporphyrin), positioned as external pendant near the loop region, inserted to unambiguously assign topology. Depending on the topologies, bound heme groups can either interact strongly or weakly. In the *syn* topology, the two heme groups, coordinated by the histidines at the $10,10'$ positions of each α-SS-α subunit, are proximal, and thus capable of participating in charge interactions. In the *anti* topology, the hemes are distal and electrostatic interactions are expected to be much weaker. Charge repulsion between ferric hemes drives the conformational switch between the *syn* and *anti* topologies, which in turn determines the structural changes in the overall symmetry of the bundle and alterations in surface patterns of the amino acids. This work

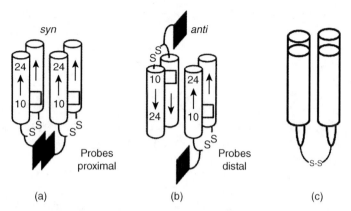

FIGURE 3.12   Possible topologies of the four-helix bundle maquettes: (a) *syn* and (b) *anti*. Reprinted with permission from Reference 127. Copyright 2001 American Chemical Society. (c) Candelabra motif scaffold.

proves that these simple protein scaffolds can also readily accommodate independent allosteric and cooperative regulation of the switch, modulated, for example, by ionic strength [127].

All the results obtained from the iterative redesign strategy and structural characterization of the maquettes were used for the design of a di-heme-containing four-helix bundle, with bis-His ligated hemes, which showed oxygen transport properties [128, 129]. This new maquette (Table 3.4, entry 7) houses the allostery and cooperativity of natural proteins and binds O_2 with affinities and exchanges timescales matching those of natural globins. Iterative redesign strategy suggested that to allow maquette O_2 binding at room temperature and inhibit heme oxidation exclusion of water from the protein interior was needed. This goal was accomplished by constraining helical motion through loop redesign. Using the knowledge on the conformational switch maquette behaviors, the loops were redesigned to link the helices across the most mobile interface and were connected by a disulfide linkage to afford a scaffold termed "candelabra" motif (see Fig. 3.12c). In order to simplify design and characterization, only two heme-binding sites per four-helix bundle were kept in this new molecules, by placing histidine at positions 7 and 42 of the α-loop α monomer sequence (see Table 3.4, entry 7). The exclusion of water from the protein interior was demonstrated measuring hydrogen/deuterium (H/D) exchange by NMR. By binding one heme group per bundle, the authors demonstrated the formation of an oxy-ferrous complex, with UV-visible spectral behaviors similar to those of native neuroglobin, and an half-time of 50 ms. Binding of CO determines one His displacement, with the formation of a stable carboxy-ferrous heme with a 400 ms half-time. Interestingly, O_2 binding is 10-fold preferred with respect to CO, thus suggesting that the distal histidine may stabilize the bound oxygen.

In conclusion, the design and characterization of heme maquettes demonstrates that it is possible to construct designed hemeprotein cages with properties remarkably similar to those found in natural proteins. As outlined by the authors, in these simple maquettes, functionality can be assured by a delicate balance between helical motion and helical strain. In particular, glutamate burial upon heme binding in the maquette scaffold induced distal histidine strain and allowed small ligands to bind at the active site.

With the aim of expanding the application of the maquettes scaffold in the development of artificial proteins, Gibney and coworkers shortened the prototype $(\alpha$-SS-$\alpha)_2$ maquette scaffold to design a five-coordinated hemeprotein [130, 131]. First, the maquette was truncated by seven amino acids to afford the sequence (Ac-CGGGEIWLK·HEEFIKK·FEERIKKL-CONH$_2$), named [Δ7-H10I14I21]$_2$, which contains a single His per helix at position 10 (**a** position of the heptad). The thermodynamics of the heme-peptide binding in both the ferrous and ferric state was fully evaluated and revealed insight into the relative roles of ligand preference, steric hindrance, and heme–heme electrostatic repulsion in hemeprotein design. Finally, the same truncated maquette scaffold was used to analyze different binding motifs, such as bis-pyridyl and mono-His-heme coordination. In particular, substitution of one His heme ligand at position 10 with 1-methyl-L-histidine (H1m) in the maquette sequence [Δ7-H10I14I21]$_2$ afforded a five-coordinate high-spin ferrous hemeprotein similar to deoxymyoglobin [130]. Further, by using this simplified maquette scaffold, Gibney et al. demonstrated the feasibility of using nonnatural amino acid side chains to ligate the heme iron and the effects derived from the ligand modifications on the heme affinity and redox properties [131].

As outlined above, the work by Dutton and coworkers on the heme-maquette design was guided by the principles of binary patterning in all the steps. The artificial proteins were designed by using an iterative redesign approach, in which the systematical variation of hydrophobic/hydrophilic side chains allowed the refinement of the prototype structure to afford molecules with desired function and/or defined tertiary structure.

Several alternative strategies have also been followed in the development of four-helix bundle cages for housing heme cofactors. The development of numerous computational methods for backbone and side-chain optimization set the stage for the computational *de novo* design of heme-binding four-helix bundle scaffold. To this aim, DeGrado and coworkers applied the parameterization methods, previously used to develop antiparallel four-helical bundles as models for di-iron proteins (see Section 3.3), to the design of hemeproteins [132]. The starting point for the design was the mathematical parameterization of the backbone coordinates of the transmembrane di-heme four-helix bundle in cytochrome bc_1 [28]. The backbone geometries of the helical bundle in cytochrome bc_1 can be well described using a simple D_2-symmetrical model, and the usual nomenclature for coiled-coils, with residues of the heptad repeats designated **a** through **g**.

The design started by generating the backbone coordinates through symmetry operations from a single idealized monomer, which was fit to the coordinates of cytochrome bc_1. A 25-residue peptide sequence was designed, in which, similar to cytochrome bc_1, a His residue was placed at **d** position in the N-terminal heptad and a Gly residue was placed at **a** position in the C-terminal heptad (Fig. 3.13c).

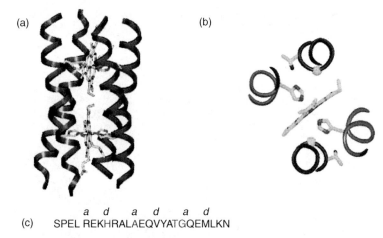

(a)   (b)

a   d   a   d   a   d
(c)   SPEL REKHRALAEQVYATGQEMLKN

FIGURE 3.13 Model structure of a D_2-symmetric artificial di-hemeprotein: (a) D_2-symmetric four-helix bundle, (b) heme-binding site, and (c) peptide sequence with key residues (His, Ala, Thr, and Gly) highlighted in bold. Reprinted with permission from Reference 132. Copyright 2004 American Chemical Society.

The tetramerization of the peptide creates two heme-binding sites, with the heme iron positioned on the central axis of the coiled-coil (Fig. 3.13a). The heme group is axially coordinated by the Nε of two His side chains contained in helices 1 and 3 (Fig. 3.13b), and the orientation of each His residue is fixed by hydrogen bonding between its Nδ and a Thr residue on the neighboring helices 2 and 4 (Fig. 3.13b). This second-shell interaction and the His to heme iron coordination specifically define the geometry of the bundle. The remaining core residues were selected in order to optimize the hydrophobic packing and to destabilize alternative structures. Binding of the heme to the peptide caused the formation of a well-folded tetramer, with UV-visible and EPR spectra typical of bis-His ligated *b*-type cytochromes, in low-spin state. Titration of the peptide with heme supported a 4:2 stoichiometry for the heme/peptide complex, and NMR spectra demonstrated that the protein exists in solution in a unique conformation. All these results demonstrated that the D_2-symmetrical model used is well suited to the analysis and design of tubular proteins that contain binding sites within their central cores, as already demonstrated for the di-iron site (see Section 3.3).

DeGrado and coworkers also used the SCADS computational algorithm [71, 133] to design hemeprotein assemblies containing nonbiological cofactors, such as the iron diphenylporphyrin cofactor [134]. Design started with a homo-tetrameric four-helix bundle, based on a 34-residue sequence (see Table 3.5 and Fig. 3.14a). In order to create a scaffold that selectively binds the target cofactor, the sequence was designed by several steps, using a Monte Carlo-simulated annealing protocol that considered the following constraints: (1) a metal–metal distance between 17 and 19 Å; (2) optimal His Nε to Fe bonding interactions; (3) second-shell hydrogen bonds between His Nδ

TABLE 3.5   Peptide Sequences of PA$_{TET}$ and PA$_{SC}$

		Sequence[a]							
Peptide		a	d	a	d	a	d	a	d	a
PA$_{TET}$ (1-34)	S	LEEALQE	AQQTAQE	AQQALQK	HQQAFQK	FQKYG
PA$_{SC}$ (1-31)	S	PEEAMQE	AQQTARE	AEQAMQK	HRQAYDKGD
PA$_{SCT}$ (32-60)		QQK	ALQTAKE	FQQAMQK	HKQY	MNPQAISE
PA$_{SC}$ (61-84)		S	VQKTARY	FEQAMQK	HRQAYDKGD
PA$_{SC}$ (85-108)		QQK	ALQTAKE	AQQAMQK	HSQALRG

[a]Keystone residues are bolded. Helical regions are underlined and the heptad positions of the residues are lettered.

and Thr Oγ; (4) minimal steric clashes; and (5) maintenance of D_2 symmetry. The remaining residues were defined by computation using SCADS. The spectroscopic characterization of the model protein in the presence of the cofactor demonstrated the correctness of the design. In addition, the value of the midpoint redox potential (103 mV vs. NHE) suggested the cofactor to be inserted in a hydrophobic protein microenvironment, as hypothesized.

Subsequently, a monomeric 108-residue single-chain version of the hemeprotein assembly was designed (see Table 3.5 and Fig. 3.14b) [135]. Design strategy

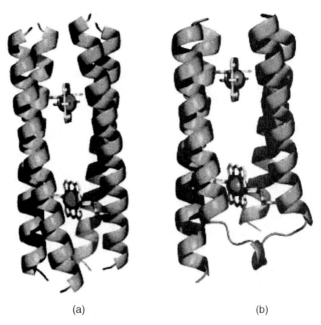

(a)   (b)

FIGURE 3.14   Molecular model of porphyrin assemblers: (a) PA$_{TET}$ and (b) PA$_{SC}$. Reprinted with permission from Reference 135. Copyright 2007 American Chemical Society.

proceeded as for the homo-tetrameric bundle by the insertion of "keystone" residues involved in the iron-cofactor coordination (His) and second-shell interactions (Thr), whereas all the remaining residues were defined by computation using SCADS. In order to connect the monomeric peptide sequence of tetrameric porphyrin assembler (PA$_{TET}$) into a single-chain (PA$_{SC}$), interhelical loops were introduced into the structure by grafting known interhelical backbone structures onto the PA$_{TET}$ model. The helices were shortened to obtain a compact (108-residue) single-chain peptide, made up of three 24-residue helices and one 28-residue helix. The helices were connected through loops of natural protein structures taken from a protein database with reduced redundancy, PDB Select [136]. A program called STITCH was developed to identify loops that would be most suitable to connect a given pair of unconnected helices. PA$_{SC}$ was expressed from a synthetic gene (10–15 mg/L pure protein) in *E. coli* and characterized in solution by NMR and UV-vis spectroscopy. All the data showed that it bound with high specificity to the desired cofactors, suggesting that a uniquely structured protein and well-defined site were indeed been generated.

An alternative strategy for the design of four-helix bundle with controlled topology and housing the heme cofactor is the chemical synthesis of template-assembled synthetic proteins (TASP), first introduced by Mutter and coworkers [137]. In this approach, four α-helices are covalently connected to a cyclic decapeptide template to afford a self-assembled four-helix bundle, housing a specific cofactor. This approach offers the advantage that the engineering of antiparallel four-helix bundle is easily accomplished through the covalent attachment of different peptide chains onto the template. Haehnel and coworkers used this approach for the development of modular organized proteins (named MOP), designed to provide a suitable scaffold for the incorporation of a variety of cofactors [138–140]. The first molecule, MOP1, able to bind hemes, was engineered using four α-helical peptides covalently connected to a cyclic decapeptide template [138]. A bis-heme-binding protein was constructed on the basis of the cytochrome *b* subunit of the mitochondrial cytochrome bc_1 complex. Two helices, H1 and H2, were designed in an attempt to reproduce the main features of the cytochrome *b* A-D helices: H1 constitutes the heme-binding helices and were covalently linked to the template in a parallel orientation. They are antiparallel to a parallel H2 helix couple, which plays the role of shielding the heme-binding pocket against the solvent. A suitable chemo-selective synthetic strategy was developed in order to control the directionality of the helical segments. The spectroscopic properties of the bound heme revealed the successful assembly of the model hemeprotein, in that they resembled those of the natural protein. Characterization by mass spectrometry and circular dichroism supported the hypothesized structure. The free energy of folding shows a stabilizing effect by the two heme groups, which have respective redox midpoint potentials of -106 and -170 mV.

Further improvement in the design of MOPs resulted in the development of artificial hemeproteins with enzymatic activity, supported on cellulose membranes [141]. A library of mono-His-ligated hemeproteins was prepared using the peptide sequences derived from MOP1. Amino acid substitutions were made in order to search for the best candidates in terms of oxygenase activity. A total of 352 synthetic hemeproteins were screened on the cellulose support, by monitoring the kinetics of verdoheme formation. This screening allowed the selection of four proteins for a full characterization in solution, which revealed that the heme oxygenase activity was strictly related to the heme coordination and redox potential.

The same strategy was also used by Haehnel and coworkers for the construction of different metal-binding sites housed within the four-helix bundle scaffold [142, 143]. In particular, mononuclear copper-binding sites were inserted into a four-helix bundle scaffold, covalently linked onto a cellulose support. In order to produce either a type 1 or type 2 copper center, several His and Cys residues were inserted at different positions in the helix. The cellulose-linked TASP approach allowed to screen a high number of variants. In detail, 96 sequences were screened onto the cellulose for copper binding, and one-third among them was found to coordinate copper. Three sequences out of the 96 were fully characterized in solution. They showed to be highly stable and well packed with copper-binding sites with features intermediate between those of type 1 and type 2 copper centers. The refinement of the design, by engineering the second coordination sphere of the copper ion, afforded a set of 180 proteins, which were screened for metal ion binding. Characterization in solution revealed that substitutions in the second coordination sphere gave artificial proteins with different coordination environments. In conclusion, these works highlighted the feasibility of the TASP approach in the construction of artificial proteins based on the four-helix bundle scaffold and revealed that the properties of the cellulose-bound proteins are also retained in solution. In this respect, this strategy is very promising in that it could be, in principle, applied to the construction of different scaffolds, housing a variety of metal-binding sites.

The screening of a library of variants was also used by Hecht and coworkers for the design of heme-binding proteins based on the four-helix bundle scaffold [144–146]. By using the binary patterning approach, two libraries of sequences were tested. In the first round of the design, out of 30 binary code sequences tested, 15 were found to bind heme with a broad range of affinities and with spectroscopic features resembling those of natural cytochromes [144]. Subsequently, several binary code proteins exhibited peroxidase activity at rates rivaling natural peroxidases [145] or carbon monoxide binding [146]. Even though the CO-binding affinities are similar to that of myoglobin, with dissociation constants in the nanomolar range, resonance Raman studies indicated that none of the proteins contribute an H-bond donor or a lone pair for interaction with the CO, as often observed in the CO adducts of natural hemeproteins. Probably, the lack of explicit design makes these binary code proteins unable to make specific directional interactions with bound ligands.

More recently, the designed proteins were tested for their ability to be immobilized on a gold electrode. The novel proteins, when immobilized onto the electrode surface, hold their peroxidase-like activity [147]. Interestingly, the artificial proteins create a specific environment for the heme, since different protein sequences induce different activities. In addition, the trends of the activity observed for the protein in solution are in agreement with those observed for the immobilized proteins, thus revealing that the immobilization does not modify the artificial hemeprotein features. Finally, these results open up the possibility of application of these novel hemeproteins in the construction of biosensors.

3.5   SUMMARY AND PERSPECTIVES

The artificial metalloproteins reviewed in this chapter clearly show that *de novo* protein design provides an attractive approach for the construction of models, which appeared valuable to probe the features required for the function of complex metalloproteins. It should be emphasized that the metal-binding sites in proteins are not isolated entities, but rather function within the context of much larger tertiary and quaternary structures. As a consequence, the availability of an appropriate protein scaffold greatly assists in the successful design of synthetic metalloproteins. In this respect, *de novo* protein design is nowadays able to provide highly stable and simple scaffolds, which contain the essential elements that are believed essential for protein activity. These minimal models may easily allow to examine how the sequence and structure of the protein affect the specificity and affinity for metal ion binding, the physical and spectroscopic properties of the metal cofactor, and, ultimately, its function.

Several different strategies, ranging from "rational design" to "computational" and "combinatorial" approaches have proven to be useful in obtaining structural and functional metalloprotein models. Also, the use of different structural topologies and assemblies may result helpful for functional and structural studies. This is quite evident in four-helix bundles, which can be obtained by a single chain, by two noncovalently associated helix–loop–helix motifs, or by four distinct helices that come together by noncovalent self-assembly. Each class provides different advantages. For example, the symmetric nature of the dimeric derivatives simplifies the interpretation of data whereas the helices in the four-chain constructs can be combinatorially assembled to generate a large number of helical bundle proteins, which can be screened for activity.

Thus, the combination of quite diverse approaches to design appears very productive. As already demonstrated by several examples, particularly challenging design problems may require the marriage of rational computer-based design methods with combinatorial selection methods. In this context, the computational procedure may allow to identify the best possible sequence library, rather than the best possible sequence.

From these considerations, and from the impressive progress made in the field, it appears that in the near future *de novo*-designed proteins will be utilized not only to cage natural-occurring metal cofactors, but also to create novel metal-binding sites, which can add new functionality to proteins.

ACKNOWLEDGMENTS

The authors wish to thank Prof. Vincenzo Pavone for stimulating discussions and continuous support and Dr. Claudia Vicari for helping in the elaboration of drawing. Special thanks to all coworkers at the University of Napoli and at the University of Pennsylvania, who have contributed to the results on DF models, described in this chapter.

REFERENCES

[1] Bolon, D. N.; Voigt, C. A.; Mayo, S. L. *Curr. Opin. Chem. Biol.* **2002**, *6*, 125–129.

[2] Koder, R. L.; Dutton, P. L. *Dalton Trans.* **2006**, 3045–3051.

[3] Ueno, T.; Abe, S.; Yokoi, N.; Watanabe, Y. *Coord. Chem. Rev.* **2007**, *251*, 2717–2731.

[4] Röthlisberger, D.; Khersonsky, O.; Wollacott, A. M.; Jiang, L.; DeChancie, J.; Betker, J.; Gallaher, J. L.; Althoff, E. A.; Zanghellini, A.; Dym, O.; Albeck, S.; Houk, K. N.; Tawfik, D. S.; Baker, D. *Nature* **2008**, *453*, 190–195.

[5] Faiella, M.; Andreozzi, C.; Torres Martin de Rosales, R.; Pavone, V.; Maglio, O.; Nastri, F.; DeGrado, W. F.; Lombardi, A. *Nat. Chem. Biol.* **2009**, *5*, 882–884.

[6] Zastrow, M. L.; Peacock, A. F. A.; Stuckey, J. A.; Pecoraro, V. L. *Nat. Chem.* **2012**, *4*, 118–123.

[7] Bryson, J. W.; Betz, S. F.; Lu, H. S.; Suich, D. J.; Zhou, H. H.; O'Neil, K. T.; DeGrado, W. F. *Science* **1995**, *270*, 935–941.

[8] Kohn, W. D.; Hodge, R. S. *Trends Biotechnol.* **1998**, *16*, 379–389.

[9] Baltzer, L. *Curr. Opin. Struct. Biol.* **1998**, *8*, 466–470.

[10] Baltzer, L.; Nilsson, H.; Nilsson, J. *Chem. Rev.* **2001**, *101*, 3153–3163.

[11] DeGrado, W. F.; Summa, C. M.; Pavone, V.; Nastri, F.; Lombardi, A. *Annu. Rev. Biochem.* **1999**, *68*, 779–819.

[12] Razeghifard, R.; Wallace, B. B.; Pace, R. J.; Wydrzynski, T. *Curr. Protein Pept. Sci.* **2007**, *8*, 3–18.

[13] Hill, B.; Raleigh, D. P.; Lombardi, A.; DeGrado, W. F. *Acc. Chem. Res.* **2000**, *33*, 745–754.

[14] Kennedy, M. L.; Gibney, B. R. *Curr. Opin. Struct. Biol.* **2001**, *11*, 485–490.

[15] Ghosh, D.; Pecoraro, V. L. *Curr. Opin. Chem. Biol.* **2005**, *9*, 97–103.

[16] Lu, Y.; Berry, S. M.; Pfisterand, T. D. *Chem. Rev.* **2001**, *101*, 3047–3080.

[17] Xing, G.; DeRose, V. J. *Curr. Opin. Chem. Biol.* **2001**, *5*, 196–200

[18] Rosati, F.; Roelfes, G. *ChemCatChem* **2010**, *2*, 916–927.

[19] Barker, P. D. *Curr. Opin. Struct. Biol.* **2003**, *13*, 490–499.

[20] Maglio, O.; Nastri, F.; Lombardi, A. Structural and functional aspects of metal binding sites in natural and designed metalloproteins. In: *Ionic Interactions in Natural and Synthetic Macromolecules*; Ciferri, A.; Perico, A., Eds.; John Wiley & Sons Inc.: Hoboken, NJ, **2012**; p. 361.

[21] Maglio, O.; Nastri, F.; Pavone, V.; Lombardi, A.; DeGrado, W. F. *Proc. Natl Acad. Sci. USA* **2003**, *100*, 3772–3777.

[22] Hecht, M. H.; Das, A.; Go, A.; Bradley, L. H.; Wei, Y. *Protein Sci.* **2004**, *13*, 1711–1723.

[23] Bradley, L. H.; Thumfort, P. P.; Hecht, M. H. *Methods Mol. Biol.* **2006**, *340*, 53–69.

[24] Bradley, L. H.; Wei, Y.; Thumfort, P. P.; Wurth, C.; Hecht, M. H. *Methods Mol. Biol.* **2007**, *352*, 155–166.

[25] Gordon, D. B.; Hom, G. K.; Mayo, S. L.; Pierce, N. A. *J. Comput. Chem.* **2003**, *24*, 232–243.

[26] Larson, S. M.; England, J. L.; Desjarlais, J. R.; Pande, V. S. *Protein Sci.* **2002**, *11*, 2804–2813.

[27] Bolon, D. N.; Mayo, S. L. *Proc. Natl Acad. Sci. USA* **2001**, *98*, 14274–14279.

[28] North, B.; Summa, C. M.; Ghirlanda, G.; DeGrado, W. F. *J. Mol. Biol.* **2001**, *311*, 1081–1090.

[29] Voigt, C. A.; Mayo, S. L.; Arnold, F. H.; Wang, Z. G. *Proc. Natl Acad. Sci. USA* **2001**, *98*, 3778–3783.

[30] Lazar, G. A.; Johnson, E. C.; Desjarlais, J. R.; Handel, T. M. *Protein Sci.* **1999**, *8*, 2598–2610.

[31] Desjarlais, J. R.; Handel, T. M. *J. Mol. Biol.* **1999**, *290*, 305–318.

[32] Holm, R. H.; Solomon, E. I. *Chem. Rev.* **2004**, *104*, 347–348.

[33] Saven, J. G. *Curr. Opin. Colloid Interface Sci.* **2010**, *15*, 13–17.

[34] Handel, T. M.; Williams, S. A.; Menyhard, D.; DeGrado, W. F. *J. Am. Chem. Soc.* **1990**, *112*, 6710–6711.

[35] Regan, L.; Clarke, N. D. *Biochemistry* **1990**, *29*, 10878–10883.

[36] Klemba, M.; Regan, L. *Biochemistry* **1995**, *34*, 10094–10100.

[37] Peacock, F. A. A.; Iranzo, O.; Pecoraro, V. L. *Dalton Trans.* **2009**, 2271–2444.

[38] Touw, D. S.; Nordman, C. E.; Stuckey, J. A.; Pecoraro, V. L. *Proc. Natl Acad. Sci. USA* **2007**, *104*, 11969–11974.

[39] Dieckmann, G. R.; McRorie, D. K.; Lear, J. D.; Sharp, K. A.; DeGrado, W. F.; Pecoraro, V. L. *J. Mol. Biol.* **1998**, *280*, 897–912.

[40] Matzapetakis, M.; Farrer, B. T.; Weng, T.-C.; Hemmingsen, L.; Penner-Hahn, J. E.; Pecoraro, V. L. *J. Am. Chem. Soc.* **2002**, *124*, 8042–8054.

[41] Farrer, B. T.; McClure, C. P.; Penner-Hahn, J. E.; Pecoraro, V. L. *Inorg. Chem.* **2000**, *39*, 5422–5423.

[42] Farrer, B. T.; Pecoraro, V. L. *Proc. Natl Acad. Sci. USA* **2003**, *100*, 3760–3765.

[43] Dieckmann, G. R.; McRorie, D. K.; Tierney, D. L.; Utschig, L. M.; Singer, C. P.; O'Halloran, T. V.; Penner-Hahn, J. E.; DeGrado, W. F.; Pecoraro, V. L. *J. Am. Chem. Soc.* **1997**, *119*, 6195–6196.

[44] Lee, K.-H.; Matzapetakis, M.; Mitra, S.; Marsh, E. N. G.; Pecoraro, V. L. *J. Am. Chem. Soc.* **2004**, *126*, 9178–9179.

[45] Lee, K.-H.; Cabello, C.; Hemmingsen, L.; Marsh, E. N. G.; Pecoraro, V. L. *Angew. Chem. Int. Ed.* **2006**, *45*, 2864–2868.

[46] Peacock, A. F. A.; Hemmingsen, L.; Pecoraro, V. L. *Proc. Natl Acad. Sci. USA* **2008**, *105*, 16566–16571.

[47] Matzapetakis, M.; Pecoraro, V. L. *J. Am. Chem. Soc.* **2005**, *127*, 18229–18233.

[48] Iranzo, O.; Cabello, C.; Pecoraro, V. L. *Angew. Chem. Int. Ed.* **2007**, *46*, 6688–6691.

[49] Iranzo, O.; Jakusch, T.; Lee, K.-H.; Hemmingsen, L.; Pecoraro, V. L. *Chem. Eur. J.* **2009**, *15*, 3761.

[50] Iranzo, O.; Chakraborty, S.; Hemmingsen, L.; Pecoraro, V. L. *J. Am. Chem. Soc.* **2011**, *133*, 239–251.

[51] Lovejoy, B.; Choe, S.; Cascio, D.; McRorie, D.; DeGrado, W. F.; Eisenberg, D. *Science* **1993**, *259*, 1288–1293.

[52] Chakraborty, S.; Kravitz, J. Y.; Thulstrup, P. W.; Hemmingsen, L.; DeGrado, W. F.; Pecoraro, V. L. *Angew. Chem. Int. Ed.* **2011**, *50*, 2049–2053.

[53] Walsh, S. T. R.; Cheng, H.; Bryson, J. W.; Roder, H.; DeGrado, W. F. *Proc. Natl Acad. Sci. USA* **1999**, *96*, 5486–5491.

[54] Suzuki, K.; Hiroaki, H.; Kohda, D.; Tanaka, T. *Protein Eng.* **1998**, *11*, 1051–1055.

[55] Suzuki, K.; Hiroaki, H.; Kohda, D.; Nakamura, H.; Tanaka, T. *J. Am. Chem. Soc.* **1998**, *120*, 13008–13015.

[56] Kiyokawa, T.; Kanaori, K.; Tajima, K.; Koike, M.; Mizuno, T.; Oku, J.-I.; Tanaka, T. *J. Peptide Res.* **2004**, *63*, 347–353.

[57] Li, X.; Kazuo, S.; Kanaori, K.; Tajima, K.; Kashiwada, A.; Hiroaki, H.; Kohda, D.; Tanaka, T. *Protein Sci.* **2000**, *9*, 1327–1333.

[58] Tanaka, T.; Mizuno, T.; Fukui, S.; Hiroaki, H.; Oku, J.; Kanaori, K.; Tajima, K.; Shirakawa, M. *J. Am. Chem. Soc.* **2004**, *126*, 14023–14028.

[59] Shiga, D.; Nakane, D.; Inomata, T.; Funahashi, Y.; Masuda, H.; Kikuchi, A.; Oda, M.; Noda, M.; Uchiyama, S.; Fukui, K.; Kanaori, K.; Tajima, K.; Takano, Y.; Nakamura, H.; Tanaka, T. *J. Am. Chem. Soc.* **2010**, *132*, 18191–18198.

[60] Shiga, D.; Nakane, D.; Inomata, T.; Masuda, H.; Oda, M.; Noda, M.; Uchiyama, S.; Fukui, K.; Takano, Y.; Nakamura, H.; Mizuno, T.; Tanaka, T. *Biopolymers* **2009**, *91*, 907–916.

[61] Kharenko, O. A.; Ogawa, M. Y. *J. Inorg. Biochem.* **2004**, *98*, 1971–1974.

[62] Ogawa, M. Y.; Fan, J.; Fedorova, A.; Hong, J.; Kharenko, O. A.; Kornilova, A. Y.; Lasey, R. C.; Xie, F. *J. Braz. Chem. Soc.* **2006**, *17*, 1516–1521.

[63] Kharenko, O. A.; Kennedy, D. C.; Demeler, B.; Maroney, M. J.; Ogawa, M. Y. *J. Am. Chem. Soc.* **2005**, *127*, 7678–7679.

[64] Xie, F.; Sutherland, D. E. K.; Stillman, M. J.; Ogawa, M. Y. *J. Inorg. Biochem.* **2010**, *104*, 261–267.

[65] Grzyb, J.; Xu, F.; Weiner, L.; Reijerse, E. J.; Lubitz, W.; Nanda, V.; Noy, D. *Biochim. Biophys. Acta* **2010**, *1797*, 406–413.

[66] Harris, K. L.; Lim, S.; Franklin, S. J. *Inorg. Chem.* **2006**, *45*, 10002–10012.

[67] Welch, J. T.; Kearney, W. R.; Franklin, S. J. *Proc. Natl Acad. Sci. USA.* **2003**, *100*, 3725–3730.

[68] Shields, S. B.; Franklin, S. J. *Biochemistry* **2004**, *43*, 16086–16091.

[69] Nanda, V.; Rosenblatt, M. M.; Osyczka, A.; Kono, H.; Getahun, Z.; Dutton, P. L.; Saven, J. G.; DeGrado, W. F. *J. Am. Chem. Soc.* **2005**, *127*, 5804–5805.

[70] Cochran, A. G.; Skelton, N. J.; Starovasnik, M. A. *Proc. Natl Acad. Sci. USA* **2001**, *98*, 5578–5583.

[71] Saven, J. G. *Curr. Opin. Struct. Biol.* **2002**, *12*, 453–458.

[72] Benson, D. E.; Wisz, M. S.; Liu, W.; Hellinga, H. W. *Biochemistry* **1998**, *37*, 7070–7076.

[73] Lombardi, A.; Marasco, D.; Maglio, O.; Di Costanzo, L.; Nastri, F.; Pavone, V. *Proc. Natl Acad. Sci. USA* **2000**, *97*, 11922–11927.

[74] Lippard, S. J.; Berg, J. M. *Principles of Bioinorganic Chemistry*; University Science Books: Mill Valley, CA, **1994**.

[75] Maglio, O.; Nastri, F.; Torres Martin de Rosales, R.; Faiella, M.; Pavone, V.; DeGrado, W. F.; Lombardi, A. *C. R. Chimie* **2007**, *10*, 703–720.

[76] Calhoun, J. R.; Nastri, F.; Maglio, O.; Lombardi, A.; DeGrado, W. F. *Biopolymers* **2005**, *80*, 264–278.

[77] Lawson, D. M.; Artymiuk, P. J.; Yewdall, S. J.; Smith, J. M. A.; Livingston, J. C.; Treffry, A.; Luzzago, A.; Levi, S.; Arosio, P.; Cesareni, G.; Thomas, C. D.; Shaw, W. V.; Harrison, P. M. *Nature* **1991**, *349*, 541–544.

[78] Frolow, F.; Kalb (Gilboa), A. J.; Yariv, J. *Nat. Struct. Biol.* **1994**, *1*, 453–460.

[79] deMare, F.; Kurtz, D. M., Jr., Nordlund, P. *Nat. Struct. Biol.* **1996**, *3*, 539–546.

[80] Sieker, L. C.; Holmes, M.; Le Trong, I.; Turley, S.; Santarsiero, B. D.; Liu, M. Y.; LeGall, J.; Stenkamp, R. E. *Nat. Struct. Biol.* **1999**, *6*, 308–309.

[81] Andersson, M. E.; Högbom, M.; Rinaldo-Matthis, A.; Andersson, K. K.; Sjöberg, B.-M.; Nordlund, P. *J. Am. Chem. Soc.* **1999**, *121*, 2346–2352.

[82] Lindqvist, Y.; Huang, W.; Schneider, G.; Shanklin, J. *EMBO J.* **1996**, *15*, 4081–4092.

[83] Rosenzweig, A. C.; Brandstetter, H.; Whittington, D. A.; Nordlund, P.; Lippard, S. J.; Frederick, C. A. *Proteins* **1997**, *29*, 141–152.

[84] Lombardi, A.; Summa, C.; Geremia, S.; Randaccio, L.; Pavone, V.; DeGrado, W. F. *Proc. Natl Acad. Sci. USA* **2000**, *97*, 6298–6305.

[85] Geremia, S.; Di Costanzo, L.; Randaccio, L.; Engel, D. E.; Lombardi, A.; Nastri, F.; DeGrado, W. F. *J. Am. Chem. Soc.* **2005**, *127*, 17266–17276.

[86] Di Costanzo, L.; Wade, H.; Geremia, S.; Randaccio, L.; Pavone, V.; DeGrado, W. F.; Lombardi, A. *J. Am. Chem. Soc.* **2001**, *123*, 12749–12757.

[87] DeGrado, W. F.; Di Costanzo, L.; Geremia, S.; Lombardi, A.; Pavone, V.; Randaccio, L. *Angew. Chem. Int. Ed. Engl.* **2003**, *42*, 417–420.

[88] Pasternak, A.; Kaplan, S.; Lear, J. D.; DeGrado, W. F. *Protein Sci.* **2001**, *10*, 958–969.

[89] Lahr, S. J.; Engel, D. E.; Stayrook, S. E.; Maglio, O.; North, B.; Geremia, S.; Lombardi, A.; DeGrado, W. F. *J. Mol. Biol.* **2005**, *346*, 1441–1454.

[90] Maglio, O.; Nastri, F.; Calhoun, J. R.; Lahr, S.; Wade, H.; Pavone, V.; DeGrado, W. F.; Lombardi, A. *J. Biol. Inorg. Chem.* **2005**, *10*, 539–549.

[91] Wade, H.; Stayrook, S. E.; DeGrado, W. F. *Angew. Chem. Int. Ed. Engl.* **2006**, *45*, 4951–4954.

[92] Summa, C. M.; Rosenblatt, M. M.; Hong, J. K.; Lear, J. D.; DeGrado, W. F. *J. Mol. Biol.* **2002**, *321*, 923–938.

[93] Marsh, E. N.; DeGrado, W. F. *Proc. Natl Acad. Sci. USA* **2002**, *99*, 5150–5154.

[94] Kaplan, J.; DeGrado, W. F. *Proc. Natl Acad. Sci. USA* **2004**, *101*, 11566–11570.

[95] Calhoun, J. R.; Kono, H.; Lahr, S.; Wang, W.; DeGrado, W. F.; Saven, J. G. *J. Mol. Biol.* **2003**, *334*, 1101–1115.

[96] Laplaza, C. E.; Holm, R. H. *J. Am. Chem. Soc.* **2001**, *123*, 10255–10264.

[97] Musgrave, K. B.; Laplaza, C. E.; Holm, R. H.; Hedman, B.; Hodgson, K. O. *J. Am. Chem. Soc.* **2002**, *124*, 3083–3092.

[98] Holm, R. H.; Kennepohl, P.; Solomon, E. I. *Chem. Rev.* **1996**, *96*, 2239–2314.

[99] Dolphin, D., Ed. *The Porphyrins*, Vol. 7; Academic Press: New York, **1979**.

[100] Smith, L. J.; Kahraman, A.; Thornton, J. M. *Proteins* **2010**, *78*, 2349–2368.

[101] Lombardi, A.; Nastri, F.; Pavone, V. *Chem. Rev.* **2001**, *101*, 3165–3189.

[102] Reedy, C. J.; Gibney, B. R. *Chem. Rev.* **2004**, *104*, 617–650.

[103] Kennedy, M. L.; Silchenko, S.; Houndonougbo, N.; Gibney, B. R.; Dutton, P. L.; Rodgers, K. R.; Benson, D. R. *J. Am. Chem. Soc.* **2001**, *123*, 4635–4636.

[104] Benson, D. R.; Hart, B. R.; Zhu, X.; Doughty, M. B. *J. Am. Chem. Soc.* **1995**, *117*, 8502–8510.

[105] Nastri, F.; Lombardi, A.; D'Andrea, L. D.; Sanseverino, M.; Maglio, O.; Pavone, V. *Biopolymers* **1998**, *47*, 5–22.

[106] Lombardi, A.; Nastri, F.; Marasco, D.; Maglio, O.; De Sanctis, G.; Sinibaldi, F.; Santucci, R.; Coletta, M.; Pavone, V. *Chem. Eur. J.* **2003**, *9*, 5643–5654.

[107] Di Costanzo, L.; Geremia, S.; Randaccio, L.; Nastri, F.; Maglio, O.; Lombardi, A.; Pavone, V. *J. Biol. Inorg. Chem.* **2004**, *9*, 1017–1027.

[108] Nastri, F.; Lista, L.; Ringhieri, P.; Vitale, R.; Faiella, M.; Andreozzi, C.; Travascio, P.; Maglio, O.; Lombardi, A.; Pavone, V. *Chem. Eur. J.* **2011**, *17*, 4444–4453.

[109] Kamtekar, S.; Hecht, M. *FASEB J.* **1995**, *9*, 1013–1022.

[110] Weber, P. C.; Bartsch, R. G.; Cusanovich, M. A.; Hamlin, R. C.; Howard, A.; Jordan, S. R.; Kamen, M. D.; Meyer, T. E.; Weatherford, D. W.; Xuong, N. G.; Salemme, F. R. *Nature* **1980**, *286*, 302–304.

[111] Xavier, A. V.; Czerwinski, E. W.; Bethge, P. H.; Mathews, F. S. *Nature* **1978**, *275*, 245–247.

[112] Lederer, F.; Glatigny, A.; Bethge, P. H.; Bellamy, H. D.; Mathews, F. S. *J. Mol. Biol.* **1981**, *148*, 427–448.

[113] Choma, C. T.; Lear, J. D.; Nelson, M. J.; Dutton, P. L.; Robertson, D. E.; DeGrado, W. F. *J. Am. Chem. Soc.* **1994**, *116*, 856–865.

[114] Robertson, D. E.; Farid, R. S.; Moser, C. C.; Urbauer, J. L.; Mulholland, S. E.; Pidikiti, R.; Lear, J. D.; Wand, A. J.; DeGrado, W. F.; Dutton, P. L. *Nature* **1994**, *368*, 425–432.

[115] Ho, S. W.; DeGrado, W. F. *J. Am. Chem. Soc.* **1987**, *109*, 6751–6758.

[116] Yun, C. H.; Crofts, A. R.; Gennis, R. B. *Biochemistry* **1991**, *30*, 6747–6754.

[117] Xia, D.; Yu, C. A.; Kim, H.; Xia, J. Z.; Kachurin, A. M.; Zhang, L.; Yu, L.; Deisenhofer, J. *Science* **1997**, *277*, 60–66.

[118] Shifman, J. M.; Moser, C. C.; Kalsbeck, W. A.; Bocian, D. F.; Dutton, P. L. *Biochemistry* **1998**, *37*, 16815–16827.

[119] Shifman, J. M.; Gibney, B. R.; Sharp, R. E.; Dutton, P. L. *Biochemistry* **2000**, *39*, 14813–14821.

[120] Gibney, B. R.; Isogay, Y.; Rabanal, F.; Reddy, K. S.; Grosset, A. M.; Moser, C. C.; Dutton, P. L. *Biochemistry* **2000**, *39*, 11041–11049.

[121] Gibney, B. R.; Rabanal, F.; Skalicky, J. J.; Wand, A. J.; Dutton, P. L. *J. Am. Chem. Soc.* **1999**, *121*, 4952–4960.

[122] Gibney, B. R.; Huang, S. S.; Skalicky, J. J.; Fuentes, E. J.; Wand, A. J.; Dutton, P. L. *Biochemistry* **2001**, *40*, 10550–10561.

[123] Skalicky, J. J.; Bieber, R. J.; Gibney, B. R.; Rabanal, F.; Dutton, P. L.; Wand, A. J. *J. Biomol. NMR* **1998**, *11*, 227–228.

[124] Skalicky, J. J.; Gibney, B. R.; Rabanal, F.; Bieber-Urbauer, R. J.; Dutton, P. L.; Wand, A. J. *J. Am. Chem. Soc.* **1999**, *121*, 4941–4951.

[125] Huang, S. S.; Gibney, B. R.; Stayrook, S. E.; Dutton, P. L.; Lewis, M. *J. Mol. Biol.* **2003**, *326*, 1219–1225.

[126] Huang, S. S.; Koder, R. L.; Lewis, M.; Wand, A. J.; Dutton, P. L. *Proc. Natl Acad. Sci. USA* **2004**, *101*, 5536–5541.

[127] Grosset, A. M.; Gibney, B. R.; Rabanal, F.; Moser, C. C.; Dutton, P. L. *Biochemistry* **2001**, *40*, 5474–5487.

[128] Koder, R. L.; Valentine, K. G.; Cerda, J.; Noy, D.; Smith, K. M.; Wand, A. J.; Dutton, P. L. *J. Am. Chem. Soc.* **2006**, *128*, 14450–14451.

[129] Koder, R. L.; Anderson, J. L. R.; Solomon, L. A.; Reddy, K. S.; Moser, C. C.; Dutton, P. L. *Nature* **2009**, *458*, 305–309.

[130] Zhuang, J.; Amoroso, J. H.; Kinloch, R.; Dawson, J. H.; Baldwin, M. J.; Gibney, B. R. *Inorg. Chem.* **2004**, *43*, 8218–8220.

[131] Privett, H. K.; Reedy, C. J.; Kennedy, M. L.; Gibney, B. R. *J. Am. Chem. Soc.* **2002**, *124*, 6828–6829.

[132] Ghirlanda, G.; Osyczka, A.; Liu, W.; Antolovich, M.; Smith, K. P.; Dutton, P. L.; Wand, A. J.; DeGrado, W. F. *J. Am. Chem. Soc.* **2004**, *126*, 8141–8147.

[133] Kono, H.; Saven, J. G. *J. Mol. Biol.* **2001**, *306*, 607–627.

[134] Cochran, F. V.; Wu, S. P.; Wang, W.; Nanda, V.; Saven, J. G.; Therien, M. J.; DeGrado, W. F. *J. Am. Chem. Soc.* **2005**, *127*, 1346–1347.

[135] Bender, G. M.; Lehmann, A.; Zou, H.; Cheng, H.; Fry, H. C.; Engel, D.; Therien, M. J.; Blasie, J. K.; Roder, H.; Saven, J. G.; DeGrado, W. F. *J. Am. Chem. Soc.* **2007**, *129*, 10732–10740.

[136] Hobohm, U.; Sander, C. *Protein Sci.* **1994**, *3*, 522–524.

[137] Mutter, M.; Vuilleumier, S. *Angew. Chem. Int. Ed. Engl.* **1989**, *28*, 535–554.

[138] Rau, H. K.; Haehnel, W. *J. Am. Chem. Soc.* **1998**, *120*, 468–476.

[139] Rau, H. K.; DeJonge, N.; Haehnel, W. *Angew. Chem. Int. Ed.* **2000**, *39*, 250–253.

[140] Rau, H. K.; DeJonge, N.; Haehnel, W. *Proc. Natl Acad. Sci. USA* **1998**, *95*, 11526–11531.

[141] Monien, B. H.; Drepper, F.; Sommerhalter, M.; Lubitz, W.; Haehnel, W. *J. Mol. Biol.* **2007**, *371*, 739–753.

[142] Schnepf, R.; Hörth, P.; Bill, E.; Wieghardt, K.; Hildebrandt, P.; Haehnel, W. *J. Am. Chem. Soc.* **2001**, *123*, 2186–2195.

[143] Schnepf, R.; Haehnel, W.; Wieghardt, K.; Hildebrandt, P. *J. Am. Chem. Soc.* **2004**, *126*, 14389–14399.

[144] Rojas, N. R. L.; Kamtekar, S.; Simons, C. T.; Mclean, J. E.; Vogel, K. M.; Spiro, T. G.; Farid, R. S.; Hecht, M. H. *Protein Sci.* **1997**, *6*, 2512–2524.

[145] Moffet, D. A.; Certain, L. K.; Smith, A. J.; Kessel, A. J.; Beckwith, K. A.; Hecht, M. H. *J. Am. Chem. Soc.* **2000**, *122*, 7612–7613.

[146] Moffet, D. A.; Case, M. A.; House, J. C.; Vogel, K.; Williams, R. D.; Spiro, T. G.; McLendon, G. L.; Hecht, M. H. *J. Am. Chem. Soc.* **2001**, *123*, 2109–2115.

[147] Das, A.; Hecht, M. H. *J. Inorg. Biochem.* **2007**, *101*, 1820–1826.

4

GENERATION OF FUNCTIONALIZED BIOMOLECULES USING HEMOPROTEIN MATRICES WITH SMALL PROTEIN CAVITIES FOR INCORPORATION OF COFACTORS

TAKASHI HAYASHI

4.1 INTRODUCTION

Hemoproteins are among the most versatile of metalloproteins. These proteins have an iron porphyrin located within the interior of the protein. Naturally occurring iron porphyrins are generally referred to as "heme." The structure of heme comprises four pyrrole rings in a relatively planar arrangement due to the requirement for extensive electronic delocalization. In some cases, protein matrices distort the heme structure from planarity to induce unique heme reactivities. In nature, several types of heme cofactors with different peripheral substituents have been identified [1]. Heme b (protoporphyrin IX iron complex) (**1**) is the simplest and most commonly seen member of the hemoprotein family (Fig. 4.1). One of the most important characteristics of heme b is that the prosthetic group is noncovalently bound within the heme pocket via multiple interactions including the coordination between iron and a protein-based ligand, hydrophobic contacts, and electrostatic interactions. Examples of hemoproteins containing heme b are myoglobin, hemoglobin, various peroxidases, and cytochrome P450s, among others. The physicochemical properties and/or enzymatic activities of these proteins are mainly generated by precise interactions between heme and

Coordination Chemistry in Protein Cages: Principles, Design, and Applications, First Edition.
Edited by Takafumi Ueno and Yoshihito Watanabe.
© 2013 John Wiley & Sons, Inc. Published 2013 by John Wiley & Sons, Inc.

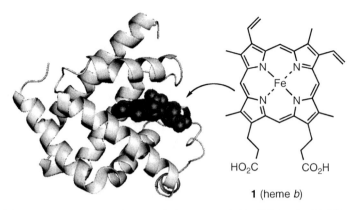

1 (heme *b*)

FIGURE 4.1   Crystal structure of sperm whale metmyoglobin (PDB code: 1A6K) and molecular structure of heme *b*.

the protein matrix, including axial ligation. This suggests that the heme pocket that is constructed by specific assembly of amino acid residues represents molecular architecture with a specific function.

Some of the heme *b*-containing hemoproteins can be converted into the apo-form under acidic conditions. Although the apo-forms of hemoproteins are in partially defolded states, the addition of heme *b* into an apoprotein solution triggers refolding to generate the corresponding reconstituted proteins. Furthermore, it is well known that heme *b* can be replaced with several heme derivatives such as deuteroheme-IX, mesoheme-IX, and hematoheme-IX in myoglobin and peroxidases [2]. Comparisons of physicochemical properties between native proteins and proteins reconstituted with various heme derivatives have provided important insights into our understanding of the structural and electronic effects on the heme peripheral groups including vinyl, methyl, and propionate groups, which interact with the amino acid residues of the heme pocket. From the data, it is inferred that the heme–protein interaction is relatively sensitive with regard to its effects in modulating protein function. We, therefore, have a chance to modify the function of a protein by engineering the protein structure.

To engineer hemoproteins for different functions, at least three different methods can be used: (i) modification of a reactive residue on the protein surface by appropriate chemical reagents, (ii) selective replacement of amino acid residues by site-directed mutagenesis, and (iii) substitution of native heme with an artificial metal complex. The first and second methods are common and relatively conventional whereas the third method has not been as widely used. However, heme substitution has great potential to dramatically change hemoprotein properties because heme itself acts as a reaction center in protein matrices. Moreover, the heme pocket of the apo-form of heme proteins tends to be relatively specialized. Insertion of an artificially designed prosthetic group into the cavities of hemoproteins has the potential to yield functional biomaterials [3, 4].

Recent efforts to insert various metal complexes into a protein cavity have included insertions into the cavities of proteins other than hemoproteins such as biotin-binding site of streptavidin [5]. These efforts have begun to show promise in generating new biocatalysts. This challenging work is expected to contribute to the development of the next generation of advances in bioinorganic chemistry, coordination chemistry, and protein engineering. This chapter focuses on the heme pocket as an attractive protein cavity for development of new functions and describes several representative examples of reconstituted hemoproteins with artificial prosthetic groups, which show remarkable gas-binding properties and enzymatic activities.

4.2   HEMOPROTEIN RECONSTITUTION WITH AN ARTIFICIAL METAL COMPLEX

The precursor of a reconstituted hemoprotein (known as apoprotein) can usually be obtained using a method developed by Teale [6]. The native heme is removed from the heme pocket by lowering the pH of a solution of the protein to pH 3 or lower, and the heme is separated from the protein in an extraction using 2-butanone. The heme moiety moves to the organic layer immediately and the apoprotein remains in aqueous solution. After conventional purification by column chromatography, the colorless apoprotein is generally obtained in a good yield. If the heme–heme pocket interaction is relatively strong (e.g., horseradish peroxidase (HRP)), acidic conditions of pH $= 1$–2 are required to remove the heme from the heme pocket. In the case of cytochrome P450$_{cam}$, it takes 1–2 days to obtain the apo-form, because the heme is buried deep within the protein matrix. On the other hand, the apo-form of myoglobin can be obtained immediately. The acid-acetone method, in which the apo-form is obtained as precipitated protein, is another way to remove native heme from the heme pocket [7].

The insertion of an artificial metal complex into the heme pocket is carried out simply by addition of the complex into an apoprotein solution at 4°C [8]. Although the reaction time and the yield of the reconstituted protein depend on the structures of protein and the artificial group, the heme–heme pocket interaction of myoglobin normally reaches equilibrium within 30 min. Figure 4.2 provides

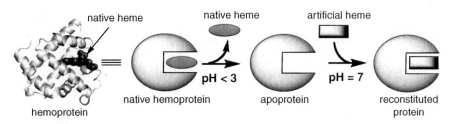

FIGURE 4.2   Representative scheme for reconstitution of a heme protein.

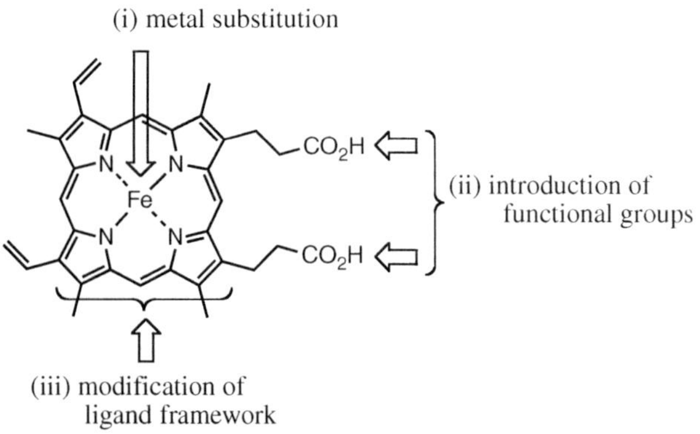

(i) metal substitution

(ii) introduction of functional groups

(iii) modification of ligand framework

FIGURE 4.3 Three methods for molecular design of an artificial cofactor.

a representative scheme of reconstitution of a hemoprotein with an artificial heme derivative.

The molecular design of an artificial prosthetic group is quite important in order to obtain suitable stability and functionality in the heme pocket. As shown in Figure 4.3, there are at least three methods that can be used to design an artificial prosthetic group: (i) metal substitution within the porphyrin framework, (ii) introduction of functional groups at the heme peripheral side chains, and (iii) modification of the ligand framework. Reconstitution with Zn, Co, or Cu porphyrin complexes has a long history that spanned 40 years [9]. In contrast, attempts to regulate the functions of hemoproteins function using methods (ii) and (iii) represent relatively new efforts. In this chapter, modification of hemoproteins with an artificial cofactor prepared by method (ii) or (iii) is mainly described.

4.3 MODULATION OF THE O_2 AFFINITY OF MYOGLOBIN

Myoglobin, the best-known oxygen storage protein in vertebrates, has one molecule of heme b bound in the heme pocket via histidine coordination at the proximal site. Dioxygen is bound in the distal site of the heme pocket which possesses several essential residues to support the O_2-binding event. The imidazole ring of the distal histidine, His64, directly interacts with the bound O_2 via a hydrogen-bonding interaction. After the crystal structures of the oxy form and the deoxy form of myoglobin from sperm whale were published, a lot of structural and mechanistic studies focusing on O_2, CO, and NO binding to myoglobin were reported [10, 11]. For example, it is well known that the O_2 affinities of most myoglobins and hemoglobins in vertebrates are within 10^5–10^6 M^{-1} at 20°C, and CO affinities are higher by approximately 10–50-fold. In contrast, several unique hemoproteins have been found to exhibit extremely high O_2 affinity in microorganisms and plants. These hemoproteins may play a role in the removal of O_2 from the tissue. For example, the O_2 affinity of hemoglobin of the

Ascaris nematode is 3.8×10^8 M^{-1}. This hemoglobin has high O_2 selectivity (M' value (K_{CO}/K_{O_2}) = 0.02–0.03), because there is a special hydrogen-bonding network in the distal site [12]. From this finding, several groups have focused on efforts to modify the distal site of sperm whale myoglobin by site-directed mutagenesis to enhance high O_2 affinity [13]. However, satisfactory results have not yet been obtained because it may not be possible to develop such a systematic hydrogen-bonding network in the myoglobin mutants.

Over the past half-century, many chemists and biochemists have studied O_2 and CO-binding events in a series of deoxymyoglobins and hemoglobins in attempts to gain an understanding of the relationship between the heme structure and function. These studies have given us important insights into the role of the peripheral substituents on the process of binding of small molecules. Therefore, replacement of native heme with an appropriate iron complex may be another way to modulate O_2 affinity in myoglobin as shown in Figure 4.4.

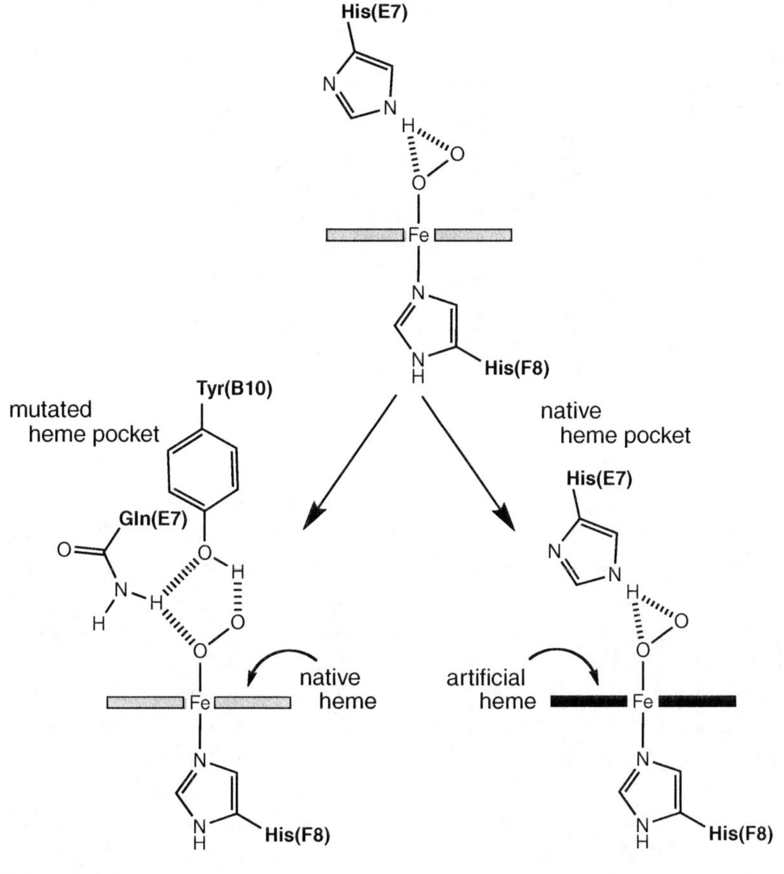

FIGURE 4.4   Schematic representation of oxygenated cofactors in protein matrices.

Recently, Neya et al. prepared α-azamesoporphyrin XIII (**2**) and β,δ-diazamesoporphyrin III (**3**), and inserted these complexes into apomyoglobin to yield the reconstituted myoglobins, rMb(**2**) and rMb(**3**), respectively. The deoxy forms of these myoglobins revealed high O_2 affinities with binding constants of $5.5 \times 10^7 \, M^{-1}$ and $3.05 \times 10^7 \, M^{-1}$, respectively. These values are 40–100-fold higher than that of native myoglobin. It was proposed that enhancement of the O_2 affinity using these azaporphyrins is derived from the strong equatorial ligand field of the heme iron in a narrower coordination cavity [14, 15].

Another artificial prosthetic group candidate for enhancement of O_2 affinity is a porphycene iron complex which represents a constitutional isomer of iron porphyrin, in which two bipyrrole moieties are linked via ethylene bridges with the D_{2h} symmetry [16]. The decrease of the symmetry from the D_{4h}-porphyrin framework lowers the LUMO level of the porphycene ring, and the Lewis acidity of the coordinated iron atom causes an increase in O_2 affinity. According to these characteristic properties, Hayashi et al. designed and prepared iron porphycene **4** with two propionate side chains as a new prosthetic group and then inserted it into apomyoglobin [17]. The crystal structure of rMb(**4**) with 2.25 Å resolution indicates that the porphycene complex **4** is located in the normal heme pocket with axial histidine coordination (Fig. 4.5) [18]. In addition, an acid titration of rMb(**4**) suggests that the His93 imidazole is tightly bound to the iron atom in the porphycene framework as a result of the

(a)

(b)

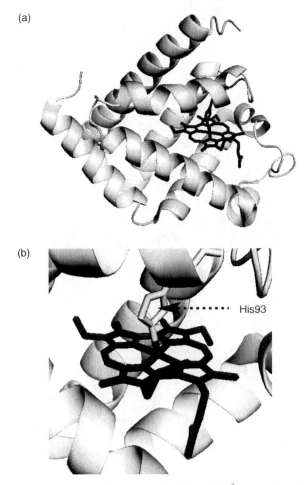

His93

FIGURE 4.5   Crystal structure of met-rMb(**4**) with a 2.25 Å resolution. Water, imidazole as a buffer, and sulfonate molecules are omitted to clarify. (a) Overall view of rMb(**4**). (b) Enlarged view of the heme pocket.

strong Lewis acidity of the iron atom, because the p$K_{1/2}$ value (which represents a pH value corresponding to 50% dissociation of a prosthetic group from the heme pocket) is 3.1. This is 1.4 pH units below the p$K_{1/2}$ value of native myoglobin. Table 4.1 shows the kinetic parameters for O$_2$ and CO binding to native deoxymyoglobin and several reconstituted deoxymyoglobins determined by laser-flash photolysis and stopped flow kinetics. Notably, the O$_2$ affinity of deoxy-rMb(**4**) is much higher (2300-fold) than the O$_2$ affinity of native sperm whale deoxymyoglobin. The O$_2$-binding constant determined from the ratio between k_{on}(O$_2$) and k_{off}(O$_2$) is 1.6×10^9 M^{-1} for rMb(**4**), in which the association of O$_2$ with the deoxy protein is accelerated, and the dissociation of O$_2$ is remarkably decelerated relative to the measurements of the

TABLE 4.1 Kinetic Parameters for Binding of O_2 and CO to Sperm Whale Myoglobins and *Ascaris* Hemoglobin[a]

	$k_{on}(O_2)$ ($\mu M^{-1} s^{-1}$)	$k_{off}(O_2)$ (s^{-1})	$K_a(O_2)$ (M^{-1})	k_{auto} (h^{-1})[b]	$k_{on}(CO)$ ($\mu M^{-1} s^{-1}$)	$k_{off}(CO)$ (s^{-1})	$K_a(CO)$ (M^{-1})	M'
WT-Mb[c]	17	28	6.1×10^5	0.10	0.51	0.050	1.0×10^7	16
Mb[H64A][c,f]	90	5700	1.6×10^4	87	3.9	0.13	3.0×10^7	1900
rMb(**4**)[c]	91	0.057	1.6×10^9	0.024	11	0.070	1.6×10^8	0.10
rMb[H64A](**4**)[c,f]	290	5.9	4.9×10^7	22	110	0.086	1.3×10^9	27
WT-Mb(Co)[d,g]	40	2800	1.4×10^4	—	—	—	—	—
rMb(**5**)[d]	95	82	1.2×10^6	—	—	—	—	—
Ascaris Hb[e]	1.5	0.004	3.8×10^8	—	0.17–0.21	0.018	0.94–1.2×10^7	0.02–0.032

[a] In 100 mM phosphate buffer, pH 7.0, at 25°C.
[b] Autoxidation was monitored at 37°C.
[c] Reference 19.
[d] Reference 20.
[e] Reference 12.
[f] H64A mutant.
[g] Wild-type myoglobin reconstituted with protoporphyrin IX cobalt complex.

native protein [19]. Although the enhancement of O_2 affinity seems to be derived from several factors, the formation of a stable Fe–O_2 σ bond with energetically favorable overlap of the d_z^2-orbital with a π^*-orbital of O_2 appears to be essential. The cobaltous porphycene **5** is also capable of stabilizing O_2 in the myoglobin matrix by 100-fold compared to the measurements of myoglobin reconstituted with the protoporphyrin IX cobalt complex [20]. Furthermore, the O_2 affinity of the H64A mutant reconstituted with **4**, rMb[H64A](**4**), where His64 is replaced with Ala, is still higher than that of native myoglobin, although it is known that the bound O_2 species is destabilized by the lack of hydrogen bonding with the imidazole moiety of His64 in the myoglobin distal pocket. In contrast, the process of autoxidation that generates the metform of rMb[H64A](**4**) from the oxygenated species is significantly accelerated relative to the autoxidation of the native myoglobin. This suggests that hydrogen bonding clearly depresseses the autoxidation process. Moreover, the CO affinity of rMb(**4**) is not remarkably enhanced relative to the O_2 affinity. As a result, the M' value ($=K_{CO}/K_{O_2}$) is 0.1, indicating that the O_2/CO discrimination is dramatically changed by the substitution of native heme with **4** in spite of the absence of amino acid replacements. These findings clearly indicate that the replacement of native heme with an appropriate metal complex is an effective strategy for regulating the function of myoglobin.

4.4 CONVERSION OF MYOGLOBIN INTO PEROXIDASE

HRP, one of the best-known heme-containing peroxidases with high catalytic activity toward H_2O_2-dependent phenol oxidation, has a single heme as a cofactor. This heme is ligated by a proximal histidine and also has a distal histidine in the heme pocket. Myoglobin also has the same heme b and two characteristic histidines, His93 and His64. The peroxidase activity of native myoglobin is quite low because myoglobin does not have an appropriate H_2O_2 activation system and a substrate-binding site near the heme pocket. However, the structural similarities of myoglobin and HRP have encouraged us to convert myoglobin into a peroxidase by chemical modification [21, 22].

4.4.1 Construction of a Substrate-Binding Site Near the Heme Pocket

As mentioned above, there is no substrate-binding site in myoglobin because, by nature, myoglobin is a simple O_2 storage protein and not an enzyme. Taking these characteristics into account, several research groups focused on constructing a substrate-binding site near the heme pocket. This was done by modifying the heme-propionate groups. Two heme-propionate side chains attached at the 6- and 7-positions of the heme framework are located outside of the heme pocket. Thus, one can easily imagine that an appropriate modification of the terminus of the propionate side chain(s) would assist a hydrophobic substrate in associating with the heme moiety.

Hayashi et al. first introduced benzene moieties into each of the heme-propionate side chains and evaluated the peroxidase activity of the corresponding reconstituted myoglobins. The benzene moieties of the modified protoheme IX **6** became localized near the heme pocket thus forming a hydrophobic surface supporting the access of aryl substrates such as phenol derivatives (Fig. 4.6) [23]. In fact, titration of guaiacol, (2-methoxyphenol), into a solution of myoglobin

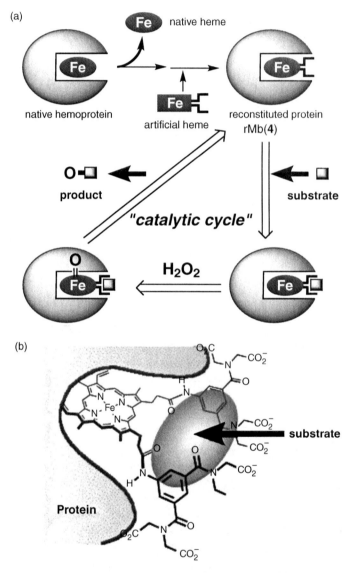

FIGURE 4.6   (a) Catalytic cycle of substrate oxidation catalyzed by reconstituted myoglobin with the artificial substrate-binding site. (b) Plausible structure of the heme pocket in rMb(**6**).

reconstituted with **6**, rMb(**6**), provides clear UV-vis spectral changes of the characteristic heme Soret band with isosbestic points. This indicates that guaiacol is specifically bound in the myoglobin heme pocket with a dissociation constant of 80 μM. These remarkable spectral changes are not detected upon the addition of guaiacol into native myoglobin. From Michaelis–Menten kinetics, the k_{cat}/K_M value for H_2O_2-dependent guaiacol oxidation catalyzed by rMb(**6**) is 13-fold larger than the corresponding value observed for native myoglobin. In addition, the peroxygenase activities of rMb(**6**) toward thioanisole sulfoxidation and styrene epoxidation were also found to be higher than the corresponding values observed for native myoglobin [24].

Next, **6** was inserted into the heme pocket of a mutant with the distal histidine replaced by aspartic acid to yield a hybrid protein rMb[H64D](**6**) [25]. This was pursued because Watanabe et al. had already reported that the H64D mutant activates H_2O_2 smoothly and then accelerates the process of oxidation of small substrates. Although the k_{cat} value for guaiacol oxidation is not enhanced by rMb[H64D](**6**), the catalytic efficiency represented by k_{cat}/K_M is remarkably increased by 430-fold relative to the native protein (Table 4.2). Moreover, to enhance the k_{cat} value of the catalytic oxidation, modified hemes **7** and **8** were prepared in which one of the propionate side chains is not substituted [26]. The reconstituted proteins are designated rMb[H64D](**7**) and rMb[H64D](**8**), respectively. From the evaluation of the activities of these proteins, it was found that preservation of one of the propionate side chains is essential for maintaining and/or increasing the k_{cat} value. As a result, the peroxidase activity of rMb[H64D](**8**) toward guaiacol oxidation is significantly increased with the k_{cat} value of 24 s^{-1} and the k_{cat}/K_M value is similar to that observed for native enzyme HRP [27]. In addition, it is interesting that only rMb[H64D](**8**) was found to catalyze the hydroxylation of ethylbenzene to yield 2-phenylethanol. The initial turnover number of this reaction is 0.3 min^{-1} at 25°C, pH 7.0. In the case of hydroxylation of a C–H bond, a substrate is generally required to be located near the Compound I active site as shown in Fig. 4.7. The findings described above may reflect the synergetic effects derived from the enhanced reactivity of Compound I of

TABLE 4.2   Kinetic Parameters of H_2O_2-dependent Guaiacol Oxidation Catalyzed by Sperm Whale Myoglobins[a]

	k_{cat} (s^{-1})	K_m (mM)	k_{cat}/K_M (s^{-1} M^{-1})
WT-Mb	2.8	54	53
rMb(**6**)	6.2	3.4	1800
Mb[H64D]	9.0	1.8	5100
rMb[H64D](**6**)	1.2	0.052	23 000
rMb[H64D](**7**)	13	0.63	21 000
rMb[H64D](**8**)	24	0.29	85 000
HRP[b]	—	—	72 000

[a] In 20 mM sodium malonate buffer, pH 6.0, at 25°C, [protein] = 4.0 μM, [H_2O_2] = 100 mM.
[b] Reference 27; pH 6.0, [H_2O_2] = 1.0 mM.

rMb[H64D](**8**) and efficient substrate binding provided by the benzene moieties of the modified heme-propionate side chains.

Casella et al. have also demonstrated modification of one of the heme-propionate side chains with an amino acid to enhance the peroxidase activity of myoglobin [28]. The histidine-linked heme **9** was found to enhance H_2O_2-dependent peroxidase activity toward several phenol derivatives. The k_{cat}/K_M value for rMb(**9**) toward the oxidation of tyramine (4-(2-aminoethyl)phenol) oxidation was determined to be 158 M^{-1}, which is clearly higher than that of native myoglobin by 6.5-fold ($k_{cat}/K_M = 24$ M^{-1} s^{-1} at 25°C, pH 6). In addition, to further increase the catalytic efficiency, a hybrid myoglobin rMb[T67R/S92D](**9**), where **9** was inserted into the apo-form of the double mutant, was prepared [29]. Kinetic studies indicated that both k_{cat} and K_M were improved and that the catalytic efficiency was clearly enhanced by 7.5-fold relative to native myoglobin.

The results obtained with these two reconstituted proteins suggest that chemical modification of heme cofactors with optimization of the heme pocket structure should be an effective strategy for generating a new high-performance heme enzyme.

4.4.2 Replacement of Native Heme with Iron Porphyrinoid in Myoglobin

In the catalytic cycle of HRP, the most important step is O–O bond cleavage of the hydroperoxide complex, Fe–OOH, generated from the reaction of H_2O_2 with ferric heme. In nature, the reactive intermediate known as compound I (oxoferryl porphyrin π-cation radical) is generated by heterolysis of the O–O bond. This is induced by the general base/acid behavior of the distal amino acid residues and strong coordination of the proximal histidine (Fig. 4.7) [30].

When comparing the ligation of the His93–heme iron of myoglobin with the His170–heme iron ligation of HRP, one must realize that the strength of the coordination bond of HRP is greater than that of myoglobin because the imidazole of the proximal ligand moiety of HRP is partially converted to imidazolate due to partial ionization of the imidazole NH of His170 as a result of an interaction with Asp247 [31]. It has been proposed that the strong coordination from the proximal His to the heme-iron activates H_2O_2 and smoothly leads to the production of the compound I species via heterolytic cleavage of the O–O bond. Therefore, the porphycene iron complex is an attractive candidate to use as a cofactor to enhance the ligation between His and the heme-iron (see Section 4.3). In fact, the reaction of cumene hydroperoxide with ferric rMb(**4**) provides cumyl alcohol and acetophenone with a ratio of 42:1 [18]. This supports the proposal that heterolysis of the O–O bond to yield compound I is dominant in rMb(**4**). Next, an experiment which monitored catalytic guaiacol oxidation upon addition of H_2O_2 indicated that the initial rate obtained by rMb(**4**) is 11-fold faster at pH 7.0 than the initial rate observed for native myoglobin. Furthermore, rMb(**4**) was also found to accelerate the rate of styrene epoxidation and thioanisole sulfoxidation by fivefold relative to the native protein.

Recently, Hayashi et al. also reconstituted apomyoglobin with iron corrole **10** and evaluated the peroxidase activity of the reconstituted protein rMb(**10**) [32]. The corrole ligand consists of a tetrapyrrole macrocycle which lacks one meso carbon relative to the porphyrin framework and has a trianionic character which may stabilize a high-valent state of a metal ion bound within the core. The corrole complex **10** was found to be bound in the myoglobin heme pocket with a $pK_{1/2}$ value of 4.8. This value is slightly higher than that of the native myoglobin by 0.3 units. Although the strength of the ligation between His64 and corrole-iron is not high, the initial rate of H_2O_2-dependent guaiacol oxidation is remarkably enhanced by rMb(**10**)

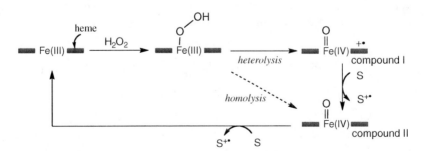

FIGURE 4.7 Catalytic cycle of HRP.

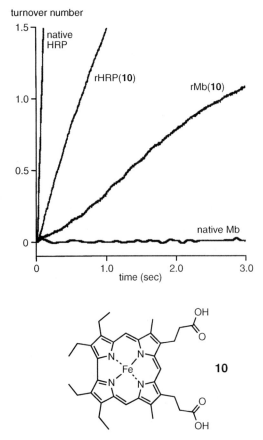

FIGURE 4.8   Molecular structure of corrole iron complex **10** and the time course of guaiacol oxidation catalyzed by native myoglobin, rMb(**10**), native HRP, and rHRP(**10**) at 20°C, pH 7.0: [protein] = 0.1 μM, [guaiacol] = 1 μM, [H$_2$O$_2$] = 1 mM.

(Fig. 4.8). Although the reason for the enhancement of the oxidation reaction has not yet been clarified, it is expected that the corrole framework stabilizes the oxidized ferryl intermediates, compound I and II, and then suppresses any uncoupling processes such as catalase activity.

4.4.3   Other Systems Used in Enhancement of Peroxidase Activity of Myoglobin

Niemeyer et al. introduced DNA oligomers at the terminus of one or both of the heme-propionate side chain(s) to yield covalent a DNA–heme adduct, **11** or **12**, respectively. Each DNA-modified heme cofactor was then inserted into apomyoglobin to obtain the DNA-conjugated myoglobins, rMb(**11**) or rMb(**12**), respectively [33, 34]. The peroxidase activities of the reconstituted myoglobins toward H$_2$O$_2$-dependent oxidation of Amplex Red (10-acetyl-3,7-dihydroxyphenoxazine) and ABTS (2,2′-azinobis(3-ethylbenzothiazoline-6-sulfonic acid ammonium salt)) were found to be clearly enhanced relative to those of native myoglobin. Interestingly, the peroxidase activity was found to be dependent on the oligonucleotide sequences (12–24-mers). Particularly, the introduction of the single-stranded oligonucleotides into both of the heme-propionate side chains in myoglobin was found to accelerate the peroxidase activity relative to the reconstituted myoglobin with a single-stranded oligonucleotide at only one of the propionate side chains. In addition, the reconstituted myoglobins were immobilized onto the streptavidin-coated microplate wells functionalized with complementary biotinylated oligomers via specific Watson–Crick base pair interactions. The conjugates clearly retain peroxidase activity.

In an additional experiment, a ruthenium-bipyridyl complex, $Ru(bpy)_3^{2+}$ moiety was attached to the terminus of one of the heme-propionate side chains in the DNA-linked heme [35]. The reconstituted protein with the ruthenium complex-linked heme **13** was successfully immobilized via hybridization to microbeads containing complementary pairs of oligonucleotides. The hybrid materials exhibited much higher peroxidase activity than the activity observed for native myoglobin toward the oxidation of Amplex Red to resorufin upon irradiation with visible light instead of H_2O_2. The hybrid system of rMb(**13**) acts as a light-triggered photocatalyst in the presence of $[Co(NH_3)_5Cl]^{2+}$ ion as a sacrificial electron acceptor.

4.5 MODULATION OF PEROXIDASE ACTIVITY OF HRP

It is known that the native heme cofactor in HRP is also removable and that reconstituted HRP can be obtained upon insertion of an appropriate metal complex into the apoprotein by conventional methods. In addition, HRP has a suitable heme pocket for producing compound I and II, because these reactive intermediates are detectable at room temperature. There have been several previous reports describing the peroxidase activities of reconstituted HRP with modified heme cofactors which have improved our understanding of the role of the peripheral groups on the enzyme function. However, there had been no reports of remarkable enhancement of peroxidase activity in reconstituted HRP until recently.

Hayashi et al. prepared and characterized HRP reconstituted with iron porphycene **4**, rHRP(**4**) in an attempt to produce a modified peroxidase with an activity superior to that of the native enzyme [36]. One of the characteristics of rHRP(**4**) is that the EPR spectrum of rHRP(**4**) exhibits low-spin signals ($S = 1/2$) at 5 K, although the native HRP enzyme has high-spin character ($S = 5/2$). This finding suggested that strong coordination of the proximal histidine to the iron atom could be achieved by replacement of the porphyrin with porphycene. The addition of H_2O_2 into a solution of rHRP(**4**) produced a characteristic broad band at 820 nm which was detected by transient UV-vis spectroscopic technique. The near-infrared band is consistent with that of the porphycene π-cation radical species generated by an electrochemical method. Upon addition of $[Fe(CN)_6]^{4-}$, this band disappears within 100 ms. These findings indicate that the transient species can be assigned as an oxoferryl porphycene π-cation radical intermediate. The peroxidase activity of rHRP(**4**) toward guaiacol oxidation was found to be similar to that of native HRP whereas the sulfoxidation of thioanisole catalyzed by rHRP(**4**) is dramatically accelerated by 12-fold relative to that of the native protein (Fig. 4.9). Detailed kinetic studies on the catalytic oxidation indicated that the rate of formation of the compound I-like species for rHRP(**4**) is slightly slower whereas each rate constant of the two substrate-oxidation steps which occur in the reaction with guaiacol are significantly larger than those of the native protein. This result suggests that the reactivities of the oxoferryl species of iron porphycene are higher than the reactivities observed for the native heme cofactor in HRP. This, therefore, represents a remarkable example of a reconstituted enzyme exceeding the natural activity of native HRP. Hayashi et al. also reported on

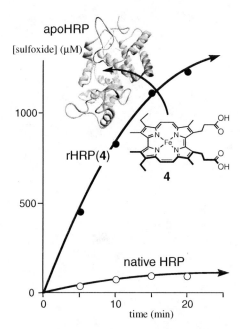

FIGURE 4.9   Thioanisole sulfoxidation catalyzed by native HRP and rHRP(**4**) at pH 7.0 and 20°C.

the peroxidase activity of **10** in the HRP heme pocket (Fig. 4.8) [32]. The kinetic investigation unexpectedly indicated that catalytic oxidation of guaiacol by rHRP(**10**) is slower although the peroxidase activity of reconstituted myoglobin rMb(**10**) is dramatically increased. One of the reasons could be that rMb(**10**) and rHRP(**10**) have different resting states with different oxidation states of the iron center. The former is ferric whereas the latter is the ferryl species, as indicated by EPR and UV-vis studies.

4.6   MYOGLOBIN RECONSTITUTED WITH A SCHIFF BASE METAL COMPLEX

Over the last decade, metal porphyrinoids as well as several Schiff base metal complexes have been evaluated as artificial cofactors for myoglobin because it is well known that Schiff base metal complexes such as manganese salen (*N,N'*-ethylenebis(salicylimine) manganese complex) catalyzes various substrate oxidations. Watanabe et al. were the first research group to insert Mn(III) or Cr(III) salophen (*N,N'*-bis(salicylidene)-1,2-phenilenediamine metal complex) into apomyoglobin [37]. The reconstituted protein was then characterized by mass spectroscopy. In addition, after Ala71 was replaced with Gly to reduce the steric interaction between the Schiff base framework and the heme pocket at the Ala71 methyl group, it was successfully demonstrated in the crystal structure of A71G mutant myoglobin reconstituted with 3,3′-dimethylsalophen iron complex, chromium complex, and manganese complex (**14**), that the salophen complexes are located in the normal heme pocket with M–His93 coordination [38]. Furthermore, myoglobin reconstituted with 3,3′-dipropylsalophen manganese complex rMb[A71G](**14**) exhibits peroxygenase activity toward H_2O_2-dependent thioanisole sulfoxidation (Fig. 4.10). In addition, the activities of rMb[A71G](**15**) and rMb[A71G](**16**) reconstituted with salen manganese complexes, **15** and **16**, are higher by 21-fold and 31-fold, respectively, relative to those observed for complexes without the protein matrix [39]. Moreover, the oxomanganese complexes of rMb[A71G](**15**) and rMb[A71G](**16**) generated by H_2O_2 undergo enantioselective sulfoxidation of thioanisole to yield the

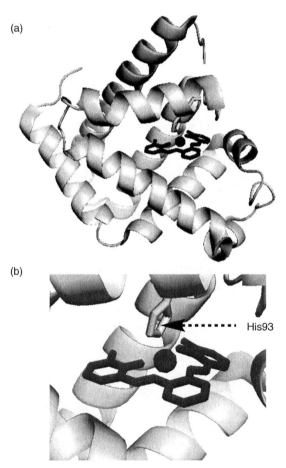

FIGURE 4.10   Crystal structure of A71G-myoglobin reconstituted with manganese salophen **14**, rMb[A71G](**14**), with 1.45 Å resolution. (a) Overall view of rMb[A71G](**14**). (b) Enlarged view of the heme pocket.

corresponding *R*-sulfoxide with the 17%ee and 27%ee values, respectively. According to the calculated structures of the salen cofactors in the protein matrix, the position of the metal center of the salen ligand depends on the bulkiness of the 3,3′-substituent groups and their interactions with the heme pocket. It is particularly notable that the H64D/A71G mutant produces *S*- and *R*-sulfoxides when reconstituted with rMb[H64D/A71G](**15**) and rMb[H64D/A71G](**16**), respectively. Therefore, Watanabe et al. pointed out that the rate and enantioselectivity of the thioanisole sulfoxidation should be regulated by the precise positions of the oxometal species and bound substrate in the distal site of the heme pocket.

M = Mn³⁺ (**14**)

R = CH₃ (**15**)
R = *n*-Pr (**16**)

In contrast, Lu et al. attached a manganese salen moiety **17** onto the myoglobin heme pocket via covalent bonding at the 72 and 103 positions after replacement of these residues with cysteine (Fig. 4.11) [40]. This rMb[L72C/Y103C](**17′**) protein accelerates thioanisole sulfoxidation by 163-fold compared to **17** without the protein matrix and the *S*-configuration product was obtained with a 51%ee.

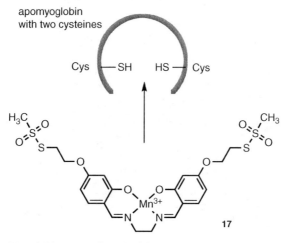

FIGURE 4.11   Myoglobin reconstituted with salen manganese complex **17** via covalent linkage.

Furthermore, rMb[T39C/H64X/L72C](**17′**) was prepared with the distal histidine replaced by phenylalanine or arginine, and pH-dependent sulfoxidation and ABTS oxidation reactions were monitored. The H64R mutant exhibits a relatively high turnover number for thioanisole sulfoxidation with 47.7%ee compared with that observed for H64F mutant at pH 8.0 [41]. This suggests that the hydrogen-bonding network between the distal residue and H_2O_2 is important for activation of the manganese-bound H_2O_2. This has also been seen in the native enzyme.

4.7   A REDUCTASE MODEL USING RECONSTITUTED MYOGLOBIN

4.7.1   Hydrogenation Catalyzed by Cobalt Myoglobin

It is known that a cobalt(II) protoporphyrin IX complex (**18**) can be easily incorporated into apomyoglobin using the conventional method. If the cobalt complex could be reduced to the Co(I) species, a cobalt hydride species would be available in the heme pocket. Willner et al. introduced an eosin chromophore (Eo) onto the surface of rMb(**18**) to serve as a photosensitizer [42, 43]. This was accomplished via covalent bonding to obtain the eosin-attached cobalt myoglobin, rMb(**18**)-Eo (Fig. 4.12). Reduction of Co(II) to Co(I) via electron transfer from photoexcited eosin moiety in rMb(**18**)-Eo was then attempted. The obtained Co(I) species can be converted into an intermediate with a Co–H bond which serves as a strong nucleophile. In fact, it was found that acetylene could be catalytically hydrogenated to ethylene within the myoglobin matrix with a quantum yield of 1×10^{-2} in the presence of Na_2EDTA as a sacrificial electron donor. Although H_2 evolution may also occur in this system, the hydrogenation of acetylene was found to be dominant. Electrochemical reduction of rMb(**18**)-Eo was also demonstrated and the hydrogenation from acetylene dicarboxylic acid to maleic acid was monitored.

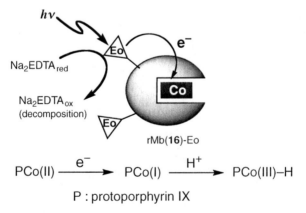

FIGURE 4.12   Reconstituted cobalt myoglobin with eosin moieties (Eo) on the protein surface.

4.7.2   A Model of Hydrogenase Using the Heme Pocket of Cytochrome *c*

Hydrogenase catalyzes a reaction that leads to the evolution of H_2. This metalloenzyme is attractive not only for the study of enzymatic chemistry but also from the viewpoint of the construction of a fuel cell system. In nature, there are several members of the hydrogenase family, and one of them is known to have a unique diiron core with a dithiolate ligand in the active site. As the structure of the hydrogenase known as [FeFe]-H_2ase gradually becomes apparent, efforts to construct enzyme models that include a diiron carbonyl cluster have become the subject of increasing attention. However, the number of appropriate functional models is quite limited because almost all models are only soluble in organic solvents and unstable in water. Recently, to overcome these problems, Hayashi et al. have focused on the heme pocket of cytochrome *c*, because the interior of the protein has two cysteine residues, Cys14 and Cys17, which seem to be suitable for use as bridging ligands to support the diiron core, although these residues are linked with heme *c* in the heme pocket (Fig. 4.13) [44]. After the removal of native heme from cytochrome *c* by a conventional method using Ag_2SO_4, the addition of $Fe_2(CO)_9$ into a solution of apocytochrome *c* formed the $(\mu\text{-}S\text{-}Cys)_2Fe_2(CO)_6$ complex in the protein matrix. Although a crystal structure has not been obtained, the UV-vis and IR spectra were found to be consistent with those obtained by the previously reported model complex. This new type of artificial

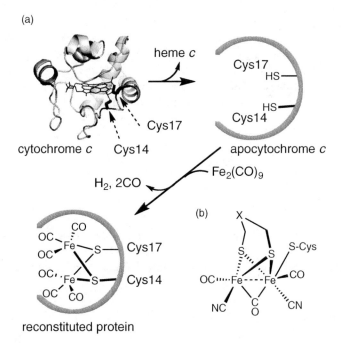

FIGURE 4.13   (a) Construction of diiron cluster $(\mu\text{-}S\text{-}Cys)_2Fe_2(CO)_6$ in the cytochrome *c* protein cavity after the removal of heme *c*. (b) Native diiron core in [FeFe]-H_2ase.

protein containing a diiron core was found to be soluble in water and works well as a catalyst for H_2 evolution in the presence of $[Ru(bpy)_3]^{2+}$ and ascorbate as a photosensitizer and sacrificial reagent, respectively. Interestingly, H_2 evolution was not observed for a diiron complex containing an oligopeptide, YKCAQCH, as a ligand which includes two cysteines in the sequence. Therefore, these results indicate that cytochrome c will provide a suitable protein matrix for obtaining a diiron core with catalytic activity.

4.8   SUMMARY AND PERSPECTIVES

One of the landmarks in hemoprotein chemistry is considered to be the clarification of the 3D structures of myoglobin and hemoglobin which occurred in the 1960s. Over the past five decades, several three-dimensional structures of hemoproteins with unique and specific heme-binding sites have been solved by X-ray crystal structure analysis and NMR spectroscopy, and they became available in the PDB database. These structural data have given us important insights into the function of proteins at the molecular level and indicate that the heme pockets contribute significantly to the modulation of the function of a series of hemoproteins. Therefore, one can easily envisage that the modification of a cofactor and/or heme pocket would produce changes in the inherent functions of hemoproteins [45]. In fact, to engineer various functionalities into hemoproteins, one of the most effective methods will be the incorporation of an appropriate artificial metal complex into apoproteins after the removal of native heme. The present survey has attempted to highlight these chemical strategies that have attracted great attention by researchers engaged in modification of hemoproteins. Although the construction of functionalized hemoproteins by replacement of native heme with an artificial metal complex has just begun, it is likely that very attractive and/or unexpected features may be obtained from the interactions between an artificial cofactor and an appropriate heme protein cavity. Moreover, these methods will provide new strategies for generation of functionalized biomaterials.

ACKNOWLEDGMENTS

The preparation of this chapter was supported by Grants-in-Aid for Scientific Research from MEXT and JSPS. I thank Drs. Takashi Matsuo, Hideaki Sato, and Akira Onoda for their important contributions to the hemoprotein modification works.

REFERENCES

[1] Turano, P.; Lu, Y. In *Handbook on Metalloproteins*; Bertini, I., Sigel, A., Sigel, H., Eds; Marcel Dekker: New York, 2001; pp 269–356.

[2] Hayashi, T. In *Handbook of Porphyrin Science*; Kadish, K. M., Smith, K. M., Guilard, R., Eds; World Scientific: Singapore, 2010; Vol. 5, pp 1–69.

[3] Hayashi, T.; Hisaeda, Y. *Acc. Chem. Res.* **2002**, *35*, 35–43.

[4] Fruk, L.; Kuo, C. H.; Torres, E.; Niemeyer, C. M. *Angew. Chem. Int. Ed.* **2009**, *48*, 1550–1574.

[5] Kohlar, V.; Wilson, Y. M.; Lo, C.; Sardo, A.; Ward, T. R. *Curr. Opin. Biotech.* **2010**, *21*, 744–752.

[6] Teale, F. W. *Biochim. Biophys. Acta.* **1959**, *35*, 543–543.

[7] Yonetani, T. *J. Biol. Chem.* **1967**, *242*, 5008–5013.

[8] Ascoli, F.; Rosaria, M.; Fanelli, R.; Antonini, E. *Methods Enzymol.* **1981**, *76*, 72–87.

[9] Sano, S. In *The Porphyrins*; Dolphin, D., Ed.; Academic Press: New York, 1979; Vol. 7, pp 377–402.

[10] Phillips, G. N., Jr. In *Handbook of Metalloproteins*; Messerschmidt, A., Huber, R., Poulos, T., Wieghardt, E., Eds; John Wiley: Chichester, 2001; Vol. 1, pp 5–15.

[11] Springer, B. A.; Sligar, S. G.; Olson, J. S.; Phillips, G. N., Jr. *Chem. Rev.* **1994**, *94*, 699–714.

[12] Goldberg, D. E. *Chem. Rev.* **1999**, *99*, 3371–3378.

[13] Travaglini-Allocatelli, C.; Cutruzzolà, F.; Brancaccio, A.; Vallone, B.; Brunori, M. *FEBS Lett.* **1994**, *352*, 63–66.

[14] Neya, S.; Kaku, T.; Funasaki, N.; Shiro, Y.; Iizuka, T.; Imai, K.; Hori, H. *J. Biol. Chem.* **1995**, *270*, 13118–13123.

[15] Neya, S.; Hori, H.; Imai, K.; Kawamura-Konishi, Y.; Suzuki, H.; Shiro, Y.; Iizuka, T.; Funasaki, N. *J. Biochem.* **1997**, *121*, 654–660.

[16] Sessler, J. L.; Gebauer, A.; Vogel, E. In *Porphyrin Handbook*; Kadish, K. M., Smith, K. M., Guilard, R., Eds; Academic Press: San Diego, 2000; Vol. 2, pp 1–54.

[17] Hayashi, T.; Dejima, H.; Matsuo, T.; Sato, H.; Murata, D.; Hisaeda, Y. *J. Am. Chem. Soc.* **2002**, *124*, 11226–11227.

[18] Hayashi, T.; Murata, D.; Makino, M.; Sugimoto, H.; Matsuo, T.; Sato, H.; Shiro, Y.; Hisaeda, Y. *Inorg. Chem.* **2006**, *45*, 10530–10536.

[19] Matsuo, T.; Dejima, H.; Hirota, S.; Murata, D.; Sato, H.; Ikegami, T.; Hori, H.; Hisaeda, Y.; Hayashi, T. *J. Am. Chem. Soc.* **2004**, *126*, 16007–16017.

[20] Matsuo, T.; Tsuruta, T.; Maehara, K.; Sato, H.; Hisaeda, Y.; Hayashi, T. *Inorg. Chem.* **2005**, *44*, 93391–9396.

[21] Ozaki, S.; Roach, M. P.; Matsui, T.; Watanabe, Y. *Acc. Chem. Res.* **2001**, *34*, 818–825.

[22] Watanabe, Y.; Hayashi, T. In *Progress in Inorganic Chemistry*; Karlin, K. D., Ed.; Wiley-Interscience: Hoboken, 2005; Vol. 54, pp 449–493.

[23] Hayashi, T.; Hitomi, Y.; Ando, T.; Mizutani, T.; Hisaeda, Y.; Kitagawa, S.; Ogoshi, H. *J. Am. Chem. Soc.* **1999**, *121*, 7747–7750.

[24] Hayashi, T.; Matsuda, T.; Hisaeda, Y. *Chem. Lett.* **2003**, *32*, 496–497.

[25] Sato, H.; Hayashi, T.; Ando, T.; Hisaeda, Y.; Ueno, T.; Watanabe, Y. *J. Am. Chem. Soc.* **2004**, *126*, 436–437.

[26] Matsuo, T.; Fukumoto, K.; Watanabe, T.; Hayashi, T. *Chem. Asain J.* **2011**, *6*, 2491–2499.

[27] Savenkova, M. I.; Kuo, J. M.; Oritiz de Montellano, P. R. *Biochemistry* **1998**, *37*, 10828–10836.

[28] Monzani, E.; Alzuet, G.; Casella, L.; Redaelli, C.; Bassani, C.; Sanangelantoni, A. M.; Gullotti, M.; De Gioia, L.; Santagostini, L.; Chillemi, F. *Biochem* **2000**, *39*, 9571–9582.

[29] Roncone, R.; Monzani, E.; Murtas, M.; Battaini, G.; Pennati, A.; Sanangelantoni, A. M.; Zuccotti, S.; Bolognesi, M.; Casella, L. *Biochem. J.* **2004**, *377*, 717–724.

[30] Gajhede, M. In *Handbook of Metalloproteins*; Messerschmidt, A., Huber, R., Poulos, T., Wieghardt, E., Eds; John Wiley: Chichester, 2001; Vol. 1, pp 195–210.

[31] Poulos, T. L.; Kraut, J. *J. Biol. Chem.* **1980**, *255*, 8199–8205.

[32] Matsuo, T.; Hayashi, A.; Abe, M.; Matsuda, T.; Hisaeda, Y.; Hayashi, T. *J. Am. Chem. Soc.* **2009**, *131*, 15124–15125.

[33] Fruk, L.; Müller, J.; Niemeyer, C. M. *Chem. Eur. J.* **2006**, *12*, 7488–7457.

[34] Glettenberg, M.; Miemeyer, C. M. *Bioconjug. Chem.* **2009**, *20*, 969–975.

[35] Kuo, C. H.; Fruk, L.; Niemeyer, C. M. *Chem. Asian J.* **2009**, *4*, 1064–1069.

[36] Matsuo, T.; Murata, D.; Hisaeda, Y.; Hori, H.; Hayashi, T. *J. Am. Chem. Soc.* **2007**, *129*, 12906–12907.

[37] Ohashi, M.; Koshiyama, T.; Ueno, T.; Yanase, M.; Fujii, H.; Watanabe, Y. *Angew. Chem. Int. Ed.* **2003**, *42*, 1005–1008.

[38] Ueno, T.; Ohashi, M.; Kono, M.; Kondo, K.; Suzuki, A.; Yamane, T.; Watanabe, Y. *Inorg. Chem.* **2004**, *43*, 2852–2858.

[39] Ueno, T.; Koshiyama, T.; Ohashi, M.; Kondo, K.; Kono, M.; Suzuki, A.; Yamane, T.; Watanabe, Y. *J. Am. Chem. Soc.* **2005**, *127*, 6556–6562.

[40] Carey, J. R.; Ma, S. K.; Pfister, T. D.; Garner, D. K.; Kim, H. K.; Abramite, J. A.; Wang, Z.; Guo, Z.; Lu, Y. *J. Am. Chem. Soc.* **2004**, *126*, 10812–10813.

[41] Zhang, J.-L.; Garner, D. K.; Liang, L.; Barrios, D. A.; Lu, Y. *Chem. Eur. J.* **2009**, *15*, 7481–7489.

[42] Willner, I.; Zahavy, E.; Heleg-Shabtai, V. *J. Am. Chem. Soc.* **1995**, *117*, 542–543.

[43] Zahavy, E.; Willner, I. *J. Am. Chem. Soc.* **1995**, *117*, 10581–10582.

[44] Sano, Y.; Onoda, A.; Hayashi, T. *Chem. Commun.* **2011**, *47*, 8229–8231.

[45] Mauk, A.; Raven, E. L. In *Advances in Inorganic Chemistry*; Sykes, A. G., Ed.; Academic Press: Hoboken, 2001; Vol. 51, pp 1–49.

5

RATIONAL DESIGN OF PROTEIN CAGES FOR ALTERNATIVE ENZYMATIC FUNCTIONS

Nicholas M. Marshall, Kyle D. Miner, Tiffany D. Wilson, and Yi Lu

5.1 INTRODUCTION

One of the most striking properties of enzymes that has puzzled chemists and biochemists for a long time is their ability to carry out complex reactions while utilizing relatively few functional or prosthetic groups that are also typically poor catalysts when isolated outside the protein environment [1–13]. For example, the heme prosthetic groups that are ubiquitously found throughout nature are utilized for a myriad of catalytic functions, from electron transfer in cytochrome proteins, to the reduction of molecular oxygen in oxidases that allows for energy production in all aerobic life, to highly asymmetric transformation of organic molecules in P450 enzymes [14]. Despite this diverse functionality, the heme itself is effectively identical in all heme proteins, with only slight modifications to the heme edge being the biggest difference between the heme prosthetic groups [7, 11, 15–17]. Furthermore, heme molecules alone are typically unable to perform similar chemistries when removed from the protein scaffold. The fact that the catalytic moiety in these very different systems is nearly identical and only functions within the protein environment exemplifies the critical nature of the "protein cage" in facilitating and determining the function of a given protein.

Coordination Chemistry in Protein Cages: Principles, Design, and Applications, First Edition.
Edited by Takafumi Ueno and Yoshihito Watanabe.
© 2013 John Wiley & Sons, Inc. Published 2013 by John Wiley & Sons, Inc.

Although the importance of the extended protein network in facilitating the function of an enzyme has long been known, many have considered the main function of the protein as a cage to isolate the catalytic portions of the enzyme from other reactive molecules, like water or oxygen. Only in relatively recent years have the effects of the extended protein network around the catalytic site been investigated. This extended protein network, or secondary coordination sphere, has since been shown to be as important as residues directly interacting with the catalytic site in determining reactivity, substrate access and specificity, and electronic properties of the metal site, through noncovalent interactions like hydrogen bonding, hydrophobicity, ionic interactions, and controlled solvent access [6]. Proteins are now considered to be highly tuned ligands, having extended networks that serve many functions in regards to determining the function of an enzyme.

Now that the importance of the protein cage is better understood, attempts to predictably modify the cage in order to alter or impart new functionality to an enzyme are under way. The complexity of such studies is a significant challenge, as mutating even a single amino acid in a protein may have a multitude of unforeseen effects to the structure or function of the protein. To further complicate the matter, multiple mutations may be required to see any of the desired functionality [13]. As such, high throughput, computational and structural determination methods are indispensable in modern protein engineering [6, 10, 13]. In this chapter, we will discuss several studies from various classes of metalloenzymes, in which a functional property of the protein, such as electron transfer or catalysis, was either predictably improved or completely altered, or cases where new functionality was engineered into a protein scaffold by modifying the protein cage, as opposed to the catalytic site itself. As this book focuses on inorganic chemistry, special emphasis will be given to metalloenzymes and, where available, the methods used in each study in order to serve as a guide for future studies in metalloenzyme design and engineering.

5.2   MONONUCLEAR ELECTRON TRANSFER CUPREDOXIN PROTEINS

One striking example of how the protein environment or cage can have a dramatic effect on the functional properties of a metal site in a protein is copper proteins, in particular, type 1 (T1), blue copper proteins, or cupredoxins. Natively, T1 cupredoxins serve as electron transfer (ET) agents, passing a single electron from one protein to another in a diverse range of biological pathways, from photosynthesis to nitrogen or iron respiration [18–20]. These proteins are also exceptionally efficient at ET, which has made them desirable targets for engineering functional properties, such as the predictable modification of redox potential to develop controllable redox reagents for applications in biological fuels or solar cells. Facilitating ET in such a diverse range of biological pathways requires T1 copper sites to natively achieve a wide range of redox potentials. In fact, T1 copper sites are known to exhibit redox potentials from around 0.1 V, in the case of green copper proteins like stellacyanin [21] and mavicyanin [22], up to greater than 1 V in the multicopper oxidase, ceruloplasmin

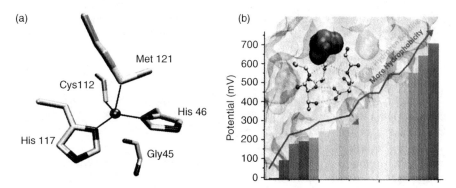

FIGURE 5.1 (a) Native type I copper site in azurin from *Pseudomonas aeruginosa.* (b) The wide range of redox potentials of type I copper sites in designed azurin mutants, highlighting the secondary coordination sphere interactions, such as hydrophobicity and hydrogen bonding, important for such control.

[23], which is responsible for cycling between the oxidation states of iron in many organisms. This observed range of redox potentials covers more than half of the attainable redox potentials in biological systems.

Despite the wide range of redox potentials seen in T1 copper proteins, the ET site is nearly identical in all of the proteins across this broad family, with a single copper ion bound in a trigonal plane by one thiolate from a Cys residue and two imidazolyl nitrogens from two His residues (Fig. 5.1a) [18, 24–26]. Altering any of the three primary ligands drastically alters the site and greatly lowers the ET efficiency. The high degree of similarity in the copper-binding site for this family of proteins also confers similar spectroscopic features to these T1 copper proteins [27–29]. Due to the apparent uniformity of the copper-binding sites, the intra-protein interactions that were responsible for fine-tuning the redox potential of the T1 cupredoxins over such a broad range were elusive, but it has been shown by multiple studies that the surrounding protein environment, or secondary coordination sphere around the copper, is the determining factor in fine-tuning the redox potential of cupredoxin proteins. Properties like the hydrophobicity of amino acids near the copper [30, 31], site distortion from strongly binding axial ligands [32, 33], π–π stacking between the His ligands and other aromatic residues [34], and hydrogen-bonding interactions to and around the primary copper ligands [35–37] have all been shown to have a dramatic effect on the redox potential of the T1 copper site.

In the attempts to answer how such homologous copper sites could exhibit such varied potentials, many techniques were used. The primary means of studying the effects of individual amino acids in these proteins, site-directed mutagenesis, historically ran into many problems, because making one mutation would have a myriad of effects to the surrounding protein, making deconvoluting individual interactions impossible. For example, none of the other 19 canonical amino acids has the same connectivity as the Met residue. Mutating the axial Met ligand in several cupredoxins not only altered the copper ligand at this position but also altered steric contacts with surrounding residues, and/or possible hydrogen-bonding interactions. As such, the use of unnatural amino acids has played a critical part in elucidating the role of certain amino acids in T1 copper proteins. In two studies, a series of unnatural amino acids isostructural to Met were incorporated into the cupredoxin azurin (Az) in place of the axial Met ligand by a technique called expressed protein ligation (EPL) [30, 31]. Because the connectivity of the unnatural amino acids was the same as the native Met residue, secondary effects like distortion of the copper site though steric hindrance were avoided, and a linear correlation between the redox potential of the copper site and the hydrophobicity of the side chain was revealed. As the hydrophobicity of the side chain increased, the redox potential increased and vice versa [30]. EPL with unnatural amino acids was also used to replace the primary Cys ligand in Az with a selenocysteine (SeCys). This substitution was nearly indistinguishable from WT Az in terms of redox potential, but the spectroscopic features of the SeCys-containing variant were significantly altered [38]. Such studies, using incorporation of unnatural amino acids with EPL, have proved invaluable to understanding the properties of such copper proteins and will undoubtedly continue to advance our understanding of the effects of the protein cage in T1 copper and other proteins. More recently-developed techniques for incorporation of unnatural amino acids through *in vivo* and *in vitro* methods [39] will also allow for more thorough characterization of these proteins and for engineering of the functional properties of ET proteins.

In order to study the effect of entire regions of T1 copper proteins on the functional properties of the protein, the copper-binding loops from various cupredoxin proteins constituting upward of 20 amino acids have also been interchanged through loop-directed mutagenesis. In this technique, the copper-binding site itself is left mainly unaltered, but the nearby residues and the solvent accessibility of the copper site are largely modified. These studies have shown that the identity of the copper-binding loop has more bearing on the redox properties of the copper-binding site as compared to the other 80–90% of the protein that was left unaltered [40–42].

In a combination of the aforementioned studies, other previous mutagenesis, bioinformatics analysis and molecular dynamics (MD) simulations, the redox potential of Az was fine-tuned across a 600 mV range by altering the protein cage around the copper site. The observed range of potentials in the variants of this one protein covers the entire range of potentials seen in the family of mononuclear T1 cupredoxins and nearly half of the total range of potentials attainable in biological systems (Fig. 5.1b). Altering the hydrogen-bonding networks near the primary ligands to the copper and the hydrophobicity around the copper site resulted in the observed dramatic changes to the redox potentials. The amino acid changes to be made to the protein were first identified through comparison to other copper sites found in nature and the relationships surrounding their hydrogen-bonding networks and then evaluated in the metal-free protein *in silico* through MD simulation. The promising candidates were then incorporated into the Az gene through site-directed mutagenesis. Although the redox potentials were different from the WT protein, altering the secondary coordination sphere left the spectroscopic features of the copper site mainly unchanged, confirming that the copper-binding site itself is not significantly perturbed. This observation was also confirmed by crystallization of several of the variants. Furthermore, the mutations were found to be additive and could be systematically combined to produce a series of proteins with redox potentials from below 100 mV to nearly 700 mV, with only 50 mV or less between individual variants (Fig. 5.1b) [35], which now allows for the development of designer proteins with specific potentials within this broad range, for biophysical studies or incorporation into bio-inspired devices. Nearly identical mutations to the secondary coordination sphere of the T1 site in the multicopper oxidase laccase, which catalyzes oxygen reduction and has been used to design fuel cell cathodes, have since been shown to tune the redox potential of that T1 site as well [43].

One major challenge that remains in the field of re-engineering copper ET proteins is to tune the redox potentials to the extremes like 1.0 V and −0.5 V. These potentials are required to perform many of the most promising reactions for alternative energy applications, such as water oxidation and oxygen reduction, water reduction to H_2, or CO_2 reduction. Since T1 copper proteins are natively higher redox potential than other ET centers, like Fe–S clusters, the focus with copper proteins is to attain 1.0 V or higher. Copper sites with such redox potentials do exist, but are invariably found in large proteins with multiple copper sites, where the T1 copper sites are typically much more isolated within the hydrophobic core of the protein than small cupredoxins, like Az. Such oxidizing redox potentials are also capable of oxidizing certain amino acids within the protein, like Tyr, Trp, or sulfur-containing residues, so that the protein damaging itself is of great concern when designing proteins with these potentials.

Given that one of the primary ligands to the copper in T1 cupredoxins is a Cys, there is the possibility that it would be oxidized if a redox potential of 1.0 V was obtained. As such, the Cys ligand in Az has been changed to an Asp residue [44, 45], which is less susceptible to oxidation. Typically, copper proteins with an inner coordination sphere consisting of an Asp and two His residues are type 2 (T2) copper sites, which are not efficient long-range ET catalysts and exhibit rates of around $10–100 \ s^{-1}$ for outer sphere ET as compared to $2000 \ s^{-1}$ for T1 copper sites [46]. This was seen to be true for the Cys112Asp variant of Az as well, but when combined with the Met121Leu mutation, the rate of outer sphere ET in this double mutant protein was observed to be on the same order of magnitude as WT Az [46, 47]. This newly engineered cupredoxin, called a type 0 (T0) cupredoxin, illustrates the power of the protein cage in dictating the functional properties of ET copper proteins, as it is believed that the hydrogen-bonding network around the Asp residue in Cys112Asp/Met121Leu Az forces the Asp residue to adopt a coordinative interaction with copper that is similar to the copper–Cys interaction in the native protein (Fig. 5.2) [46]. As such, the protein environment is actually able to compensate for a change as drastic as mutating one of the direct copper ligands and rescue the functional properties of the protein. However, the desired 1.0 V redox potential has, as of yet, not been reported in T1 or T0 mononuclear copper proteins.

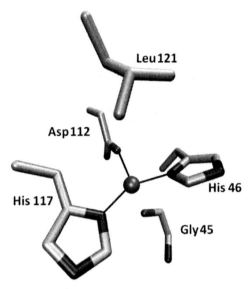

FIGURE 5.2 The copper-binding site of type 0 copper azurin with the Cys112Asp/Met121Leu mutations.

5.3   Cu$_A$ PROTEINS

Modification of protein loops has also been employed with great success for building alternative metal sites within a protein, by taking advantage of the extensive fold degeneracy found among native proteins. Many proteins share similar global structures, with the main variations surrounding the immediate area of the metal site. An archetypal example of using loop modification to achieve altered functionality was that of introducing Cu$_A$ sites into proteins having similar folds but containing different or no copper sites (Fig. 5.3) [48–51]. These efforts were motivated by the difficulty inherent to studying Cu$_A$ sites in their native contexts, where other metallochromophores mask their spectroscopic features and, in most cases, they are anchored to a membrane as part of a much larger complex. It was recognized that Cu$_A$ centers occurred in domains having the Greek key β-barrel fold, which is shared by many other proteins, including small, soluble blue copper proteins. Beyond adopting similar folds, sequence alignment between the Cu$_A$-containing domains and selected blue copper proteins revealed that the majority of the copper-binding residues exist in a single loop between the seventh and eighth strand of the β-barrel. This strategy was first applied to the Greek key β-barrel domain of a bacterial quinol oxidase, CyoA, which natively lacks any copper-binding sites [48]. In this case, the loop between the seventh and eighth strand of the β-barrel was already the same length as the loop in native Cu$_A$ domains, so the researchers made selected mutations to this loop, based on the conserved residues in Cu$_A$ proteins, to create the engineered Cu$_A$ domain.

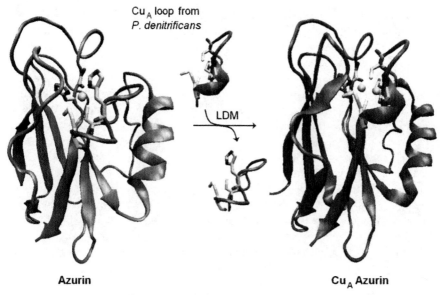

Cu$_A$ loop from
P. denitrificans

LDM

Azurin   **Cu$_A$ Azurin**

LDM = Loop -directed mutagenesis

FIGURE 5.3 Engineering a Cu$_A$ site into the T1 blue copper protein, azurin (Az), using loop-directed mutagenesis. The loop in Az between the seventh and eighth strands of the Greek key β-barrel was replaced with the corresponding loop from the Cu$_A$ domain of CcO in *P. denitrificans*, and the resulting construct, Cu$_A$–A$_z$, formed a Cu$_A$ site.

The copper-incorporated Cu$_A$–CyoA displayed many similarities to native Cu$_A$ centers, as judged by spectroscopic characterizations. This result provided some of the most solid evidence for the ligand set and dinuclear structure of Cu$_A$ centers at the time [48].

Subsequent attempts to engineer a Cu$_A$ center into a different protein focused on the blue copper proteins, since they already contained the majority of the residues required to form the Cu$_A$ site and stably internalized a copper ion. Two reports of loop replacements to yield Cu$_A$ in two different blue copper proteins were published [49, 50]. In these situations, single amino acid mutagenesis was not sufficient to introduce the Cu$_A$ sites, as the copper-binding loops of these blue copper proteins were shorter than those of Cu$_A$ proteins. In the first report, the copper-binding loop from the Cu$_A$–CyoA construct was inserted into the copper-binding loop of the blue copper protein, amicyanin [49]. Again, testifying to the utility of this approach, the copper-incorporated Cu$_A$–amicyanin construct resembled native Cu$_A$ centers. For the second report, the copper-binding loop from a native Cu$_A$, that of cytochrome *c* oxidase from *Paracoccus denitrificans*, was substituted for the copper-binding loop in the well-characterized blue copper protein, Az [50]. Spectroscopic and crystallographic characterizations of Cu$_A$-Az showed that it was an excellent model of native Cu$_A$ sites [52]. Access to a Cu$_A$ site in a water-soluble protein without other metal cofactors yielded many invaluable insights into these unique copper sites. As both Cu_A and blue copper sites natively perform ET functions, the ability to study both of these sites in the same protein framework, that is the Az scaffold, proved an additional boon of the Cu_A–Az model. Studies of the ET rates in both wild type and Cu_A–Az allowed the reorganization energies of the blue copper and Cu_A sites to be compared directly [53]. This comparison unveiled the likely reason for the apparent redundancy in function between these two types of copper sites: the reorganization energy of Cu_A was discovered to be lower than that of blue copper sites, which may be required for ET to occur under the low driving forces in Cu_A's position at the end of the aerobic and anaerobic respiration electron transport chains.

More recently, introduction of Cu_A into the blue copper protein, amicyanin was revisited, this time using the copper-binding loop sequence for the Cu_A center of cytochrome c oxidase from *P. denitrificans* [51]. Introduction of this sequence into the copper-binding loop of amicyanin resulted in a Cu_A site that resembled native Cu_A sites spectroscopically, and was also shown to accept electrons from cytochromes c-551i and c-550, the physiological electron donors to amicyanin and cytochrome c oxidase, respectively. This study demonstrated the power of the loop replacement approach, when the appropriate loop is selected for introduction into a scaffold protein, as it yielded a homogeneous, functional Cu_A site in amicyanin whereas the previously selected loop for Cu_A–amicyanin resulted in a type 2 copper site as well.

5.4 CATALYTIC COPPER PROTEINS

In addition to the ET T1 copper and Cu_A proteins, there are many other catalytic copper proteins that fulfill important roles in nature, such as catalyzing steps of anaerobic respiration [54] and breaking down lignocelluloses [55], a function that has important implications for alternative energy. Given the crucial functions that these catalytic copper proteins perform, it is not surprising that they have been the subject of many protein engineering studies, wherein the native proteins are altered in attempts to improve their functionality, or the copper sites are introduced into new protein cages, having more desirable properties. Many approaches have been employed to achieve results in these studies, where the approach is tailored to the particular problem. For instance, the complexity of the multicopper oxidases and the scarcity of structural information on these enzymes have rendered entirely rational design approaches prohibitively difficult, and in these cases, approaches like directed evolution have prevailed. The studies of these copper enzymes are detailed below, organized by the type of copper protein/site under study.

5.4.1 Type 2 Red Copper Sites

Type 2 (T2) copper sites are distinguished from T1 copper sites by both structure and function. Typically coordinated to the protein through a mixture of His imidizolyls and oxygen donors (such as Asp, Glu, Tyr), T2 copper sites adopt tetragonal geometries with open coordination sites, where substrate binding occurs [56]. Recently, a

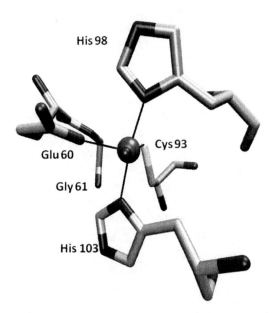

FIGURE 5.4   The type 2 red copper site of nitrosocyanin. His 98, His 103, Cys 93, and a coordinated water make up the base of the square pyramid whereas Glu 60 occupies the apex. Cysteine coordination in this site makes it unique among type 2 copper sites.

T2 copper site possessing Cys thiolate coordination was discovered in the organism *Nitrosomonas europaea*, in a protein called nitrosocyanin, which exhibited a similar fold to the T1 copper cupredoxins and Cu_A domains [57–60]. This site has been dubbed a red copper site, reflecting the intense color imparted by the Cys thiolate coordination. Nitrosocyanin is proposed to have a catalytic function related to the anaerobic metabolism of *N. europaea*. The T2 copper site of nitrosocyanin features a distorted square pyramidal geometry, where the plane is made up of two *trans* His imidizolyls, a water, a Cys thiolate ligand, and a Glu carboxylate that occupies the apex (Fig. 5.4). The similarity of this protein's fold to T1 copper proteins was recognized as an opportunity to introduce models of this thiolate-coordinated T2 site into T1 copper proteins. In one such study, the T1 copper protein Az was transformed into a T2 copper site by introducing either canonical Cys or noncanonical homocysteine (Hcy) into the axial methionine position of the T1 site of Az [61]. The Hcy residue was introduced by EPL, and this variant showed the greatest similarity to the T2 copper site of nitrosocyanin, as judged by spectroscopic characterizations. Presumably, however, the T2 copper site of the Met121Hcy Az construct was coordinatively saturated, given that the remaining ligand set of the original T1 copper site was unperturbed, although the possibility of an open coordination site was not explicitly ruled out by EXAFS structural data. Because the Met121Cys Az mutant displayed intermediate properties between the original T1 copper site and the nitrosocyanin-like T2 copper site in Met121Hcy, this study provided strong experimental evidence for the coupled distortion model within a single protein's active site [29, 61, 62].

Another group employed a different approach to introducing a nitrosocyanin (NC)-like, thiolate-coordinated T2 copper site in the same T1 copper protein, Az. In this case, as with the previous studies incorporating Cu_A into T1 copper proteins (see above), the loop containing the majority of the copper-binding residues in Az was replaced by the equivalent loop from nitrosocyanin [40]. The resulting construct, NC–Az, possessed similar UV-Vis and EPR spectroscopic features to nitrosocyanin. To better model the native nitrosocyanin T2 copper site, the authors additionally attempted to introduce a carboxylate ligand into the primary coordination sphere of the NC–Az T2 copper site, by replacing one of the original His ligands with either Glu or Asp. However, these mutations significantly destabilized the T2 copper site. Thus, while thiolate-coordinated T2 copper sites spectroscopically similar to that in nitrosocyanin have been introduced to T1 copper protein cages, a completely analogous protein model, incorporating all of the features of the native system is yet to be created. Should such a model be made, studies of its reactivity may illuminate the role of this intriguing T2 copper site.

5.4.2   Other T2 Copper Sites

In an effort to realize the potential of T2 copper sites for catalytic transformations in organic synthesis, Reetz and coworkers introduced three amino acids into a cofactorless TIM barrel protein to create a T2 copper Diels–Alder catalyst [63]. The protein cage was selected based upon several criteria, including stability, ease of purification, and ability to contain a copper site with substrate access. To design the T2 copper site, native T2 copper sites were surveyed, and the simplest site was selected, that is, that of ascorbate oxidase, having two His imidizolyls and one or more oxygen ligands. Reetz and coworkers chose to build the site around an existing Asp residue, and used computer modeling to find optimal positions for the two His residues, to build a His_2Asp ligand set. After several rounds of optimization, including removal of His residues outside the designed binding site, the engineered T2 copper site proved to catalyze the Diels–Alder reaction with $\sim$70% conversion and 46%ee. Thus, remarkably, a T2 copper catalyst was introduced through only a handful of mutations.

Similarly, Berry and coworkers introduced a T2 copper site, again into the T1 copper Az protein cage, $\sim$11–13 Å from the T1 copper site, by individual mutation of three amino acids, endeavoring to create a model of the uncoupled dinuclear copper sites found in the enzymes peptidyl α-hydroxylating monooxygenase (PHM), dopamine β-monooxygenase (DβM), and nitrite reductase (NiR) [64]. Two such models were made, one for the PHM/DβM enzymes and the other for the NiR enzymes (Fig. 5.5). To make these models, Berry and coworkers recognized the fold similarity between PHM/DβM domains, NiR, and Az, and searched for an equivalent positioning of the T2 copper site in Az to those in PHM, DβM, and NiR, finding that the T2 sites were formed by antiparallel β-strands in the native proteins. Having found similar positions for the T2 sites in Az, Berry and coworkers made three mutations to introduce two different T2 sites. The resulting constructs bound copper in both the original T1 copper sites and the introduced T2 copper sites, as determined by EPR spectroscopy. Although these models are not functional, this study represents

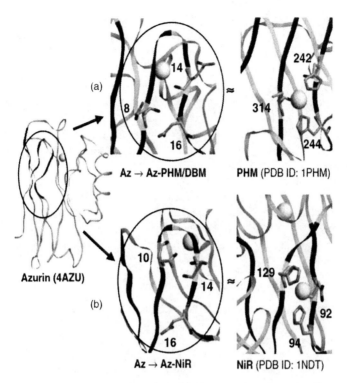

FIGURE 5.5 Introducing a PHM/DβM- and NiR-like uncoupled dinuclear site into the T1 copper protein, azurin (Az). (a) The mutations Gln8Met, Gln14His, and Asn16His were introduced in Az to create a T2 copper site similar to that in PHM. (b) The mutations Asn10His, Gln14Asp, and Asn16His were made in Az to create a T2 copper site analogous to that in NiR. The T1 copper site can be seen behind the target residues. Reproduced with permission from Reference 64. Copyright 2008 Springer.

the first step toward a functional model of the uncoupled dinuclear copper sites in PHM, DβM, and NiR in a different protein cage.

5.4.3 Cu, Zn Superoxide Dismutase

Cu, Zn superoxide dismutases (Cu, ZnSODs) catalyze the dismutation of deleterious superoxide radicals to molecular oxygen and hydrogen peroxide. These enzymes are extremely efficient, catalyzing this reaction at about 10% of what is considered to be the rate of diffusion, as calculated by the collision theory. The efficiency of this catalysis is thought to stem from a potential field distribution around the catalytic copper center that is positive and constant. In one study, Desideri and coworkers recognized a single glutamate close to the active site of *Photobacterium leiognathi* Cu, Zn SOD, and reasoning that this residue interrupted the potential field distribution near the copper site, mutated this glutamate to neutral glutamine [65]. The resulting mutant Glu59Gln Cu, Zn SOD displayed an $\sim$10-fold increase in the rate of catalyzed dismutation of superoxide, a rate comparable to the value calculated for the maximum diffusion-controlled limit, making this mutant Cu, Zn SOD the most efficient enzyme found at the time of this study.

In another study, a prokaryotic SOD homologue that did not natively contain copper was transformed into an active Cu, Zn SOD by mutations to introduce the two missing copper ligands [66]. Bertini and coworkers realized that in *Bacillus subtilis*, an SOD homologue existed (BsSOD), which bound a zinc ion in the conserved zinc site, but lacked the copper site, with a Tyr and a Pro replacing two of the His copper ligands. After mutating each of these two ligands to His, the resulting Tyr88His/Pro104His BsSOD bound copper in a similar T2 copper site to native Cu, Zn SODs, as judged by spectroscopic and structural characterization. Functional characterization revealed that the introduced Cu, Zn active site catalyzed the dismutation of superoxide at $\sim$10% the rate of eukaryotic Cu, Zn SODs. This accomplishment could be considered as a first step in the artificial evolution of a nonfunctional SOD to a fully active one.

5.4.4   Multicopper Oxygenases and Oxidases

The multicopper oxygenases and oxidases contain, as their name suggests, multiple copper sites [67]. Multicopper oxygenases consist of tyrosinases and the copper-containing monooxygenases. Tyrosinase contains a type 3 (T3) copper center, a dinuclear copper site where an oxygen bridges the two copper ions, which are anti-ferromagnetically coupled. These enzymes catalyze the hydroxylation of tyrosine to 3, 4-dihydroxyphenylalanine (L-DOPA) and the following oxidation to dopaquinone. The production of catechols by tyrosinase is of industrial interest. However, due to limited structural information for tyrosinases, completely rational design of modi-fied tyrosinases for industrial applications is challenging. Fishman and coworkers overcame this problem by using directed evolution to select for tyrosinase mutants with an improved monophenolase to diphenolase activity ratio [68]. With native tyrosinases, the monophenolase activity is rate limiting. Thus, to improve the activity of tyrosinase, the monophenolase activity must first be accelerated. Using random mutagenesis and a high-throughput screening method, Fishman and coworkers found a variant of *Bacillus megaterium* tyrosinase, R209H, displaying a 1.7-fold improve-ment in monophenolase activity and a 1.5-fold decrease in diphenolase activity, with an overall 2.6-fold increase in the monophenolase/diphenolase activity ratio. Using computer modeling, Fishman and coworkers hypothesized that the introduced His residue blocks the entrance to the active site, slowing the binding of L-DOPA to the catalytic copper site. Thus, directed evolution promises to be a good approach to tailoring tyrosinase activity for industrial catalysis.

All multicopper oxidases couple the oxidation of substrate to the four-electron reduction of oxygen to water [67]. Among these oxidases are ascorbate oxidases and laccases. For a minimal functional unit, the multicopper oxidases contain one T1 copper center and a trinuclear T2/T3 copper cluster. As with the multicopper oxyge-nases, the complexity of these systems generally excludes completely rational design of their functionality. Given this fact, several directed evolution studies have provided some of the first examples of engineering alternative or improved functionality into multicopper oxidases. In one such study of an ascorbate oxidase from *Acremonium* sp., HI-25, random mutagenesis was performed near the T1 copper site and mutants selected for their pH optima [69]. In general, the pH optima of the selected mutants were shifted upward by 0.5–1 pH units and the mutations were located near the substrate-binding regions. In the end, the authors concluded that the changes were primarily due to the structural perturbations of ion pair networks in the ascorbate oxidase, based upon homology modeling to a structurally characterized ascorbate oxidase from zucchini.

Laccase is responsible for the lignin degradation of lignocelluloses in certain plants and fungi [55]. As the degradation of lignin is a major barrier to harnessing the power of cellulose for alternative energy, much interest has been devoted to improving laccases for industrial use and introducing laccase functionality into related multicopper oxidases. As an added benefit, the major byproduct of laccase catalysis is water, as opposed to polluting organics, making laccase catalysis a very green process. In an early study aimed at tuning laccase reactivity that predated any structural determination of these enzymes, Xu and coworkers introduced specific mutations to laccases from two organisms (*Myceliophthora thermophila* and *Rhizoctonia solani*), by analogy to ascorbate oxidase structures, to residues near or coordinated to the T1 copper center [70]. They found several changes in a variant with three mutations to a segment that they proposed to be between the axial Met ligand and one of the coordinating His residues of the T1 copper site, including a shift of the optimum pH for activity to one unit lower in one organism and one unit higher in the other. Additionally, the triple mutant demonstrated altered K_m, k_{cat}, and fluoride inhibition. Due to the lack of structural information for laccases at that time, the exact cause for these changes was not known, although perturbations to the molecular recognition between the reducing substrate and laccase, as well as to the electron transfer from the substrate to the T1 copper center were suggested.

In another site-directed mutagenesis study of a multicopper oxidase, Wang and coworkers performed a helix deletion on *Klebsiella* sp. 601 multicopper oxidase by analogy to the bacterial laccase CueO, with which it shares high sequence homology, as well as fungal laccases [71]. Several mutants were generated, removing an α-helix that covered the substrate-binding pocket, which potentially interrupted the electron transfer from the substrate to the T1 copper site. One of the variants with the α-helix removed showed improved reactivity with the three substrates tested, but to different extents. The authors concluded that the α-helix may not only block access to the T1 copper site but also determine the substrate selectivity to some extent.

Directed evolution has also been successfully applied to laccases. The first successful instance of directed evolution used to modify a laccase was in a study led by Arnold and coworkers [72]. After 10 rounds of selection, they obtained a laccase with improved activity by 170-fold and better functionality at higher temperatures. In another directed evolution study of a laccase, Alcalde and coworkers selected a laccase for tolerance of organic cosolvent, to improve the utility of laccases in organic synthesis [73, 74]. Five rounds of selection yielded a variant with the capability to withstand various cosolvents at concentrations up to 50% (v/v). Most of the accumulated mutations in the evolved laccase were on the surface of the enzyme, which Alcalde and coworkers suggested stabilizing the protein by adding more electrostatic and hydrogen-bonding contacts.

5.5 HEME-BASED ENZYMES

Myoglobin (Mb) is a small well-characterized heme protein whose natural function is oxygen storage. There are also an extensive number of reported Mb mutants [75], which provides an excellent knowledge base from which to draw when considering mutations to perform. In addition, Mb is relatively easy to crystallize and is commonly used as a reference when employing spectroscopic techniques to characterize other heme proteins [14]. As such, Mb has been widely used as a protein cage for engineering new enzymatic functions.

5.5.1 Mb-Based Peroxidase and P450 Mimics

Work by Watanabe and coworkers focused on the enhancement of Mb-based peroxidase activity [76]. It was demonstrated that the naturally occurring peroxidase activity of sperm whale Mb (swMb) can be enhanced by site-directed mutagenesis. The mutations were chosen based on comparisons between the oxygen-bound structures of swMb and cytochrome c peroxidase (CcP). The distal His residue of Mb was mutated to Leu and a His residue was introduced at either position 29 or 43 (Fig. 5.6). His29 was found to be too far away to enhance activity, but His43 was found to enhance activity and had similar distance to the bound dioxygen as an analogous His in CcP. Repositioning of the His in this fashion altered the hydrogen-bonding distance between the distal His and the ferric-bound hydroperoxo intermediate. Phe43His/His64Leu Mb displays increased peroxidase activity and oxidizes guaiacol and ABTS [77], and the His64Leu mutation has been shown to increase the lifetime of the reactive intermediate, compound I (Cpd I) [78].

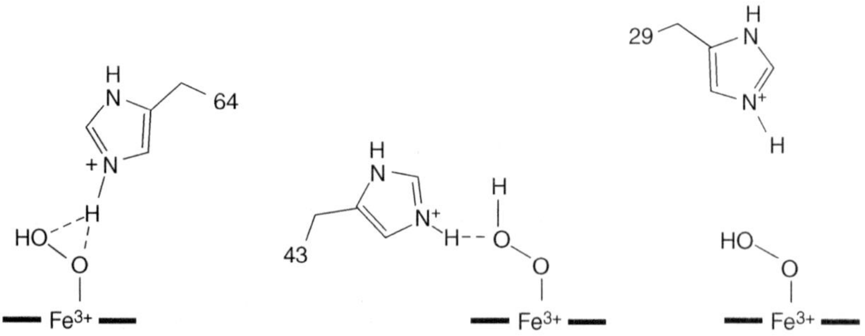

FIGURE 5.6 Proposed hydrogen-binding interactions between the heme-bound dioxygen in myoglobin and native distal histidine, His 64, or introduced distal histidines, His29 or His 43. Reproduced with permission from Reference 77. Copyright 1999 American Society for Biochemistry and Molecular Biology.

Additional studies by Watanabe and coworkers in Mb include the engineering of P450-like, or peroxygenase chemistry, in which O_2 is reduced and then cleaved to form the reactive Cpd I intermediate. Designing P450-like chemistry into Mb was accomplished by rational selection of residues to alter the hydrogen-bonding network around the heme. An Asp residue, which is known to be important to the function of chloroperoxidase, was first introduced at the position of the distal His residue in Mb, resulting in 78- and 580-fold increases in peroxidase and peroxygenase activities, respectively [79]. Thus, with only one carefully selected point mutation, the function of Mb was substantially changed. Based on observations that position 68 can affect the binding properties of ligands in Mb [80, 81], the reactivity of the His64Asp Mb mutant was fine-tuned by changing a Val residue at position 68 to an Ile [82, 83]. Mutation of Val 68 to Ala, Ser, Leu, and Ile was performed and the outcome showed that the identity of the residue at position 68 could adjust the reactivity of His64Asp Mb toward either peroxidase or peroxygenase activity. For instance, the His46Asp/Val68Leu variant had a rate of peroxidase activity twice that of His64Asp alone but was the only mutant studied that showed this effect. This study highlights how even subtle differences in amino acids can affect reactivity. Further mutation of Mb, specifically the incorporation of the Phe43Trp mutation, resulted in increased peroxidase and peroxygenase activity in Mb of 3- and 20-fold, respectively, over both wild type and His64Leu Mb [84].

5.5.2   Mimicking Oxidases in Mb

The relatively large distal pocket above the heme of Mb also allows for the incorporation of larger functional elements, such as secondary metal-binding sites, which are found in many complex proteins, such as heme copper oxidases (HCOs). Such bimetallic sites catalyze critical biological reactions, like reduction of water or NO. Similarly to the Cu_A sites mentioned above, the heme and non-heme metal sites described here are ubiquitously found in very large membrane-bound proteins that are more difficult to purify and study in comparison to myoglobin and other soluble monomeric proteins. For this reason, numerous routes have been conceived to model the bimetallic, catalytic site of HCO in smaller, water-soluble proteins.

In one such example, the mutations Leu29His and Phe43His were introduced into Mb [85], which when combined with the native distal His64 in the Mb, form a secondary metal-binding site in the heme pocket of Mb, similar to the secondary metal-binding site in HCOs (Fig. 5.7). The mutations were chosen based on overlays of the crystal structures of WTswMb and bovine HCO. The copper in the so called Cu_B site of HCOs has been proposed to provide a second electron to the bound oxygen in the reduction of oxygen to water. Unlike native HCOs, this model, called Cu_BMb, is not purified with a metal in the Cu_B site. This feature allows for the placement of redox-inactive mimics of copper in the Cu_B site, namely Ag(I) and Zn(II), which is yet to be done in native HCOs to test the role of the metal ion. Positioning of copper in Cu_BMb was confirmed using EPR spectroscopy, by demonstrating antiferromagnetic coupling between cyanide-bound heme iron and added Cu [85].

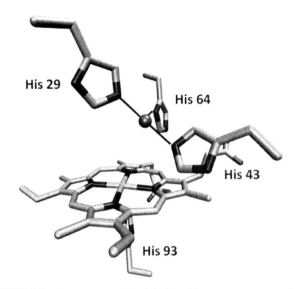

FIGURE 5.7 Computer model of designed heme-copper center in Mb.

More important than simply binding copper, Cu_BMb also displays copper-dependent oxygen chemistry that WTswMb does not [86]. When Mb is reduced to the Fe(II) state, it natively binds oxygen and forms oxyMb. The introduction of the two additional distal His residues in Cu_BMb did not prevent oxygen binding; however, the oxygen affinity of Cu_BMb is lower [86]. The addition of Ag(I) was seen to restore native-like oxygen binding. In the presence of Cu(II) and reductant, heme oxygenase (HO) chemistry, where the heme cofactor is degraded to verdoheme, was observed [86]. This copper-dependent chemistry is interesting, not only because WT Mb does not display this activity, but also because HO chemistry shares a ferric hydroperoxo intermediate with HCO enzymes [86–88]. The factor believed to determine whether HCO chemistry or HO chemistry proceeds is protonation of this intermediate.[87, 88]. If the ferric hydroperoxo intermediate becomes protonated, it can form a ferryl species and proceed through the remaining HCO chemistry, but if it remains unprotonated, HO chemistry dominates.

A clue to solving this proton delivery issue is that different HCOs vary in the type of heme that is located in the Heme-Cu_B site. The a/o-type hemes typically found in oxidases contain a hydroxyfarnesyl group that is both within hydrogen-bonding distance of the novel, cross-linked His-Tyr, and positioned at the end of the K-channel, which delivers protons to the active site [89–91]. Mechanistic studies of native HCOs suggest that a hydrogen-bonding network forms with the hydroxyl group of hydroxyfarnesyl and water molecules bridging the bound oxygen and the tyrosine hydroxyl. Conveniently, the heme cofactor of myoglobin can be replaced by other hemes, in order to study the effects that different substituent groups on the porphyrin ring have on their oxygen chemistry. This feature of myoglobin was utilized in one study, wherein the b-type heme of myoglobin was replaced with a mimic of o-type heme (Fe(III)-2,4 (4, 2-hydroxyethyl vinyl deuteroporphyrin IX)), in which a vinyl group of heme b is substituted with a hydroxyethyl group, by analogy to the hydroxyl of o-type heme [92]. Cu_BMb-containing heme o mimic, denoted as Cu_BMb(o), slowed the copper-dependent HO chemistry by $\sim$19-fold, supporting the conclusion that the heme type plays a role in determining the chemistry of heme proteins.

5.5.3 Mimicking NOR Enzymes in Mb

A high level of sequence similarity exists between HCOs and bacterial nitric oxide reductases (NORs). However, based on modeling and sequence homology, additional Glu residues are positioned near the heme cofactor, and rather than a secondary copper-binding site, NORs contain a non-heme iron site, called the Fe_B site [93–95]. Therefore, by introducing a Glu residue to Cu_BMb at position 68, the Fe_B site was also modeled into Mb, based on the similarities between NORs and HCOs [96]. As there was no crystal structure of bacterial NORs at the time for overlay and comparison, an energy-minimized computer model of the construct, called Fe_BMb, was first made to aid in designing the metal site, and then a crystal structure of Fe-bound Fe_BMb was later obtained that confirmed the computer modeling (Fig. 5.8). Fe_BMb displays NOR activity, that is, the conversion of two NO molecules to N_2O and H_2O, making Fe_BMb the first structural and functional model of NORs. The high degree of structural similarity between Fe_BMb and c-type NORs (cNOR) was later confirmed when the crystal structure of the NOR from *Pseudomonas aeruginosa* was crystallized [97]. As a further extension of cNOR modeling, an additional Glu residue was added at position 107 in the heme pocket to investigate how NOR activity was affected [98], as previous studies of NORs suggested that multiple Glu residues may be present in the heme-Fe_B site of NOR [93–95]. The main

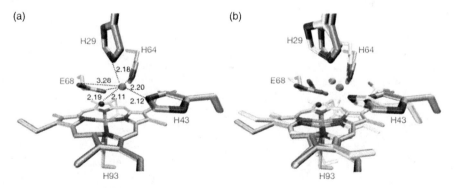

FIGURE 5.8 Crystal structures of designed heme–non-heme iron centers in Mb that mimic nitric oxide reductase. (a) Crystal structure of Fe_BMb with iron in the non-heme site (gray sphere) and (b) overlay of (a) with an energy-minimized computer model of Fe_BMb. Reproduced with permission from Reference 96. Copyright 2009 Nature Publishing Group.

difference between the Glu residue at position 68 and the additional Glu at position 107 was that the new Glu residue was not a ligand to the metal site. The effect of the mutation, however, was substantial, in that it doubled the rate of NOR activity compared to the original Fe_BMb. As the I107E mutant does not significantly alter the redox potential of the heme, it is likely involved in proton transfer via a hydrogen-bonding network.

5.5.4   Engineering Peroxidase Proteins

Engineering of heme proteins will also allow for application of proteins to practical functions, such as the degradation of difficult to oxidize compounds, like lignin, and various environmental pollutants, as well as the production of alternative carbon fuel sources and the asymmetric synthesis of pharmaceuticals. Considering the increasing interest in alternative energy sources to petroleum products and the need to clean up pollutants related to petroleum, the potential for such chemistry from proteins able to be expressed recombinantly and in high yields is an attractive prospect. The development of an efficient method for converting lignin to more desirable molecules, able to be utilized by other microorganisms, is a key challenge in the development of biofuels, such as ethanol, from plant materials [99–101]. Removal of lignin using current biofuel-producing methods involves treatment with physical (milling, grinding), physiochemical (steam or ammonia explosion), or chemical (addition of organic solvents or oxidizing agents, and/or acids/bases) methods [102], which are less environmentally friendly than protein catalysts that degrade lignin. In nature, the degradation of lignin is performed by enzymes that are produced by white-rot fungi, namely lignin peroxidase (LiP), manganese peroxidase (MnP) [103], and laccase [55].

MnP is an attractive target to mimic via protein engineering, as the mechanism of MnP involves the oxidation Mn(II) to Mn(III) and release of the Mn(III) to oxidize substrate, which eliminates the need for a substrate-binding pocket for lignin. Numerous attempts to recombinantly express native MnP proteins in high yield for the purpose of degrading lignin have been unsuccessful. As such, multiple groups have worked toward the goal of engineering MnP activity into CcP, a heme protein with a similar structure to MnP that can be expressed in high yield. In one such study, a Mn-binding site was designed into yeast CcP for the purpose of using highly oxidized intermediates of heme to oxidize Mn(II) to Mn(III), so that Mn(III) could be utilized to degrade either lignin or pollutants. In the engineered yeast CcP, denoted MnCcP, containing a Mn-binding site [104, 105], oxidation of Mn(II) to Mn(III) was observed. To design MnCcP, introduction of three carboxylate-containing residues were chosen based on sequence comparison between MnP and yeast CcP. Removal of Trp residues natively occurring in CcP and replacement with Phe (called MnCcP2) led to an approximately fivefold faster rate than MnCcP by removing alternate routes through which the oxidizing equivalents decomposed. Later, the Mn(II)-binding site in MnCcP was repositioned based on the structural studies performed by Poulos and coworkers comparing CcP and MnP, and resulted in a variant with a 2.5-fold increase in rate relative to the original

MnCcP construct without the need for further removal of radical-transferring residues [106, 107].

The work of Goodin and coworkers has also focused on mimicking MnP activity; their study involved the incorporation of an Mn(II)-binding site into CcP (called Mn6.8) as well [108]. This MnP model also oxidized Mn(II) to Mn(III) at rates comparable to those of MnCcP.1, without removal of the Trp residues mentioned above. Both designed enzymes, MnCcP.1 and Mn6.8, have been crystallized. Structures of MnCcP.1 were obtained both without metal in the designed site and with Co as an Mn analog in the designed Mn site. Mn6.8 was crystallized with Mn in the designed Mn site. The obtained crystal structures of both Mn6.8 and MnCcP.1 show similarities to the Mn site of MnP. For practical applications, these engineered proteins have some significant advantages over MnP. These CcP mutants can be recombinantly expressed in *E. coli* in high yields whereas MnP has proven difficult to express recombinantly, and the white-rot fungi that produce it are not well suited to laboratory growth conditions. This relative ease of obtaining functional protein and activity improvements through manipulation of the protein cage should allow such systems to be utilized for biofuel production.

5.5.5 Engineering Cytochrome P450s

Another major class of heme proteins, the cytochromes P450, is well known for rapid hydroxylation of many pharmaceutical compounds and other xenobiotic compounds, as well as the interconversion of sterols in the body [109]. Their general utility in activating C–H bonds has made P450 enzymes, one of the types of proteins that have been extensively studied for a variety of synthetic purposes. However, many of the compounds natively oxidized are large, so P450 enzymes have difficulties hydroxylating smaller organic ligands. P450 enzymes also require specific redox partners and expensive reductants, such as NADPH [110] to provide reducing equivalents for reduction of O_2; these requirements are a major hurdle in biotechnological applications of P450 enzymes. P450 BM3 (also known as CYP102A1) is a well-studied P450 enzyme, naturally fused to a flavin reductase domain, and is one of the most active P450 enzymes known [111]. Fusion proteins of P450 enzymes with reductase domains have the promise of increasing activity and avoiding the need to express both proteins separately, and such fusions to non-P450 reductase domains do allow for the reduction of P450 enzymes, but there is a high level, greater than 80%, of uncoupling, translating to the reducing equivalents being wasted. Arnold and coworkers overcame the NADPH limitation by using directed evolution (Fig. 5.9) to create a P450 enzyme that uses hydrogen peroxide in the place of oxygen, NADPH, and the reductase domain, that is, P450 peroxygenases [112, 113]. The ability to use peroxide allowed for much easier assaying of the enzyme for the purpose of further reengineering the substrate specificity.

As a result of P450 enzymes being fairly promiscuous with respect to substrates, it is reasonable to expect that useful products could be obtained by altering them to accept different substrates [114]. In general, the coupling efficiency of many P450s with novel substrates is often lower 10%, meaning that at least 90% of the

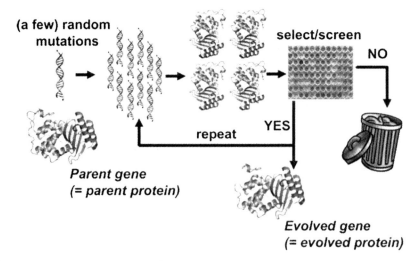

FIGURE 5.9   General overview of the directed evolution process. Reproduced with permission from Reference 113. Copyright 2009 Proceedings of the National Academy of Sciences (PNAS).

reducing equivalents are wasted and reactive oxygen species are produced. Arnold and coworkers have focused on the hydroxylation of propane and ethane to produce propanol and ethanol [115]. Using directed evolution, P450 BM3 has been engineered to convert ethane to ethanol and propane to propanol, with total turnover numbers of 250 and 6000, respectively. Evaluating activity by focusing on total turnover number lead to an improvement in substrate specificity and coupling efficiency of over 95%, as these factors are interrelated. Research performed in parallel by Wong and coworkers also demonstrated ethanol production using P450cam from *Pseudomonas putida* [116], as an extension of previous work on engineering mutants for higher rates of propanol production. In addition, Arnold and coworkers have been able to produce biological compounds, derived from propranolol, that are useful for testing drug interactions, using later generations of the above-mentioned P450 peroxygenases that have been further engineered to use propranolol as a substrate [117].

Independent studies by Watanabe [118] and Reetz [119] have succeeded in using organic molecules as co-substrates, in order to trick P450 enzymes into hydroxylating methane to form methanol. This is highly promising, as methane oxidation is one of the most difficult transformations to perform. Astonishingly, the work presented by both labs shows that methanol-production activity requires no point mutations to the native enzyme, but merely a careful choice of a "surrogate ligand" to allow conformational changes to occur that normally take place with the native fatty acid substrates of the enzyme (Fig. 5.10). Reetz and coworkers utilized chemically inert perfluorocarboxylic acids as activating guests, having lengths between 2 and 14 carbons whereas Watanabe and coworkers used perfluorocarboxylic acids with lengths between 8 and 14 carbons.

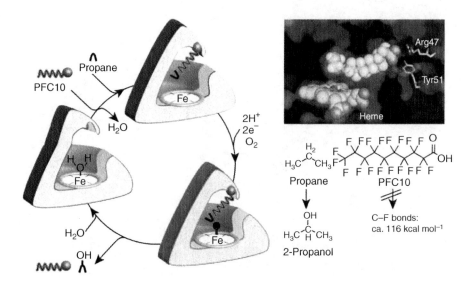

FIGURE 5.10   The proposed reaction of P450 BM3 with propane in the presence of a 10-carbon perfluorocarboxylic acid (PFC10) as a surrogate ligand. Inset shows manually docked propane and PFC10 in the active site of P450 BM3. Reproduced with permission from Reference 118. Copyright 2011 John Wiley and Sons.

5.6   NON-HEME ET PROTEINS

Non-heme iron ET proteins constitute a wide variety of proteins including mononuclear ferredoxins, Fe–S clusters and Reiske-type Fe–S clusters [120, 121]. Similarly to cupredoxin ET proteins, predictably altering the redox potentials of these proteins has applications toward developing redox reagents for biophysical studies or ET agents in bio-inspired devices. Due to the huge diversity of this class of proteins and the large amount of work that has gone into redox tuning of such proteins, two studies will be the focus of this section: one on the high-potential iron–sulfur protein (HiPIP) and one with a Reiske-type Fe–S cluster protein.

HiPIP is a 4Fe–4S cluster protein, which, as the name implies, has a substantially higher redox potential as compared to other 4Fe–4S ferredoxins. The Fe–S cluster in HiPIP is effectively indistinguishable from those found in other 4Fe–4S ferredoxins, despite the difference in redox potential. Through a series of in-depth spectroscopic and computational studies, solvent accessibility to the Fe–S cluster was shown to affect the redox potential of such proteins by as much as 300 mV, by altering hydrogen-bonding networks near the solvent-exposed ET site. This redox tuning by controlled solvent access through the protein cage was also proposed to be important in protein–protein interactions and regulation of ET events. As ferredoxins with nearly solvent-exposed redox sites interact with their substrates, solvent is excluded from the Fe–S site, which will cause the redox potential to change. Such dynamic processes may prohibit Fe–S proteins from transferring electrons to the wrong partners, but effects

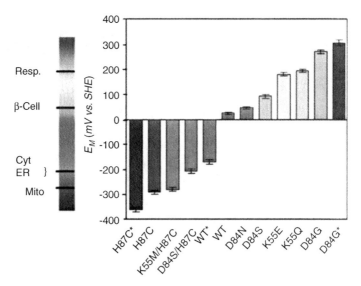

FIGURE 5.11 Redox potentials of the 2Fe–2S cluster in engineered variants of mitoNEET. Reproduced with permission from Reference 123. Copyright 2010 American Chemical Society.

of protein–protein interactions and solvent exposure complicate the engineering of new protein-based redox partners, and must be taken into account when designing proteins as redox reagents [122].

In another study, the redox potential of the Reiske-type 2Fe–2S cluster of the protein mitoNEET was fine-tuned across about 600 mV (Fig. 5.11) by altering hydrogen-bonding interactions between the iron ligands and two residues in the secondary coordination sphere around the ET site [123]. The results show that factors like hydrogen bonding to ligands can be ubiquitously used across multiple types of metal sites to alter functional properties, like redox potential. Such long-range interactions also tend to have much less of an effect on the other characteristics of the metal sites, such as the activation barrier for ET, the overall structure of the metal site, and the degree of reconstitution that can be achieved relative to WT.

5.7 Fe AND Mn SUPEROXIDE DISMUTASE

Superoxide dismutase (SOD) proteins natively degrade the reactive oxygen species, superoxide, to oxygen and hydrogen peroxide, which can further be degraded by other enzymes, like catalase. There are many varieties of SOD that use various metals, such as copper, zinc, nickel, manganese, and iron. The iron and manganese varieties of these proteins are homologous, and the metal-binding site consists of three His residues, a carboxylate from an Asp and an open binding site occupied by water in the resting state [124]. This metal-bound solvent molecule is surrounded by an intricate hydrogen-bonding network with the rest of the protein. The extensive structural information available for Fe-SOD was used to alter the hydrogen-bonding network around the metal-bound water, which caused a 1 V change in the redox potential of the metal site [125]. The redox potential of an SOD enzyme is arguably its most important functional property, and this study exemplifies the dramatic effects that the surrounding protein cage can have on a metal-binding site.

5.8   NON-HEME Fe CATALYSTS

Another striking example of the role that the protein environment plays in dictating the functionality of enzymes is the non-heme iron catalytic proteins. This broad family of proteins comprises mononuclear and multinuclear iron proteins and catalyzes a myriad of functions, from methane oxidation, to hydrogen production and oxidation, to synthesis of important biomolecules. Within the mononuclear iron catalysts, non-heme iron proteins are responsible for a huge variety of reactions, such as ring opening, hydroxylation, dihydroxylation, cyclization, dehydration, epoxidation, and halogenation [126–128]. The diversity in the natural reactivity of these proteins makes them another attractive target for reengineering to produce new pharmaceuticals, degrade pollutant molecules in the environment, or convert abundant carbon sources into viable fuels. The ability to reengineer such proteins is, however, contingent upon understanding the mechanisms of catalysis and the factors that govern the reactivity of different non-heme iron sites.

The iron-binding catalytic site of most non-heme iron enzymes are often strikingly similar, with the 2-His, 1-carboxylate facial triad proteins representing a huge portion, with high diversity to their reactivity [129, 130]. Enzymes that catalyze ring openings of aromatic substrates may proceed via two distinct mechanisms: intradiol or extradiol. Those that proceed by the extradiol mechanism bind iron in this 2-His 1-carboxylate motif, but those that natively cleave aromatic rings by an intradiol mechanism bind iron with four ligands, two Tyr residues and two His ligands. Despite the differences in the primary coordination spheres of the native enzymes, it has been shown that extradiol dioxygenases can proceed by an intradiol mechanism. The key feature that allows for the interconversion of these mechanisms is a His residue in the secondary coordination sphere of the iron, which is believed to both destabilize the O–O bond of iron-bound oxygen and facilitate general acid–base chemistry, to shuttle protons to and from substrate. This His residue was originally identified through structural analysis [131]. The role of secondary coordination sphere residues in determining the mechanism through which reactions proceed again shows the power of the protein cage, and how an understanding of such interactions in proteins will allow for better control of their reactivity.

Another example of the protein environment at work in non-heme iron proteins is the case of non-heme iron halogenases. These proteins are again very similar to other non-heme iron proteins containing the 2-His 1-carboxylate facial triad, but are missing the carboxylate ligand, which is replaced with a halogen (Fig. 5.12). Halogenases also use molecular oxygen as an oxidant and proceed through many of the same intermediates as the non-heme iron hydroxylating enzymes. Rather than

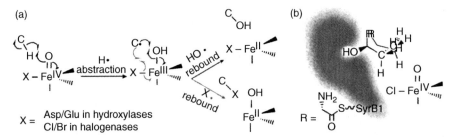

FIGURE 5.12 (a) Proposed mechanism for hydroxylation and halogenation in non-heme Fe oxidases and (b) cartoon representing the critical substrate localization that determines hydroxylation versus halogenation in SyrB2. Reproduced with permission from Reference 133. Copyright 2009 Proceedings of the National Academy of Sciences (PNAS).

adding a hydroxyl to an organic substrate, however, halogenases add a halogen. It was previously proposed that the absence of the carboxylate ligand was the determining factor between hydroxylation and halogenation, but adding a carboxylate ligand to halogenase proteins did not result in an effective hydroxylase [132].

At the critical branching point, where the halogen is inserted, there is also the possibility to add a hydroxyl group, as in the native hydroxylating enzymes. The puzzling feature of halogenases is that the hydroxyl radical is present at the same time as the halogen radical, such that it could be added to the substrate, and is actually much more reactive than the halogen radical. From a kinetic standpoint, halogenases should hydroxylate substrates, rather than halogenate them. In a recent study, it was shown that the key to driving the reaction toward the halogenated product is actually the extended protein environment and its effect on substrate orientation relative to the iron site. The presence of a hydroxyl group on the native substrate of the halogenase SyrB2 was observed to have a substantial effect on whether the hydroxylated or the halogenated product formed. When this hydroxyl was removed, the major product could be changed to the hydroxylated product. As such, it was proposed that the hydroxyl group on the substrate interacts with a residue near the iron site, likely through hydrogen bonding, and positions the substrate in proximity to the halogen ligand and away from the hydroxyl on the iron [133]. However, no structural information is currently available for where substrate binds in SryB2 and which residues may be interacting with it. An understanding of the determining factors that govern chemoselectivity in such proteins, combined with detailed structural analyses, will undoubtedly allow for reengineering of such proteins in the future.

5.9 ZINC PROTEINS

The redesign of activities in metalloproteins is not limited to the redox-active metals. Zinc-binding proteins, for example, can also be engineered to change activity. Conversion of glyoxidase II (gly II)—which hydrolyzes the thioester bond of

S-D-lactoylglutathione during a step in the detoxification of 2-oxoaldhyde—into a β-metallolactamase (BML) was reported by Park et al [134]. The authors of this study used simultaneous incorporation and adjustment of functional elements (SIAFE) in combination with directed evolution to achieve the altered functionality. This method involves deletion, insertion, and substitution of a portion of the protein sequence with point mutations introduced via directed evolution. The sequence portions that sterically limited alternative substrates in Gly II were deleted. Point mutations for zinc binding were introduced based on comparison between the loops of Gly II and the BML family. A library of $\sim 2.0 \times 10^7$ mutants was produced, using error prone PCR and multiple sequences for loops 1, 4, and 6. The initial library showed 13 positive colonies, and subsequent rounds of directed evolution resulted in a protein, evMBL8, which increases *E. coli* resistance to the selected antibiotic, cefotaxime, by 100-fold over cells not expressing evMBL8.

The DNA-binding specificities of engineered zinc fingers, selected by phage display, have also been extensively studied to the point that zinc fingers for almost any sequence of interest can be produced [135]. The resulting zinc finger proteins have great promise for use as either repressors and/or activators of genes *in vivo* [136]. The combination of zinc fingers with naturally occurring nucleases [137–139], recombinase domains [140], and integrases [141] has been performed. In addition, evolution of these non-zinc finger domains can improve the efficiency and specificity of these chimeric proteins [138, 142]. These new enzymes are well suited for practical use in the biosciences and in gene therapies in humans, as they display up to 98% specificity. The problems associated with these constructs include controlling low levels of unintended cleavage of DNA, especially for use in humans. Methods for controlling activity, including some utilizing unnatural amino acids, have recently shown promise for inhibiting activity until it is desired in mammalian cell lines [143].

5.10   OTHER METALLOPROTEINS

Although the most common transition metals occurring in metalloenzymes are copper, iron, and zinc, other metals such as manganese, molybdenum, and vanadium, catalyze important and novel reactions, such as water oxidation during photosynthesis, nitrogenation, and halogenation. The use of rational design and directed evolution approaches to improve or alter the native enzymatic function of these metalloenzymes, containing metals other than copper, iron, and zinc, are discussed below.

5.10.1   Cobalt Proteins

Sometimes, albeit in rare cases, improving the functionality of a metalloenzyme is as simple as substituting the active site metal with a different one. In one very successful example of this, Georgiou and coworkers rationalized that substitution of Co^{2+} for the naturally occurring Mn^{2+} in human arginase I (hArgI) would enhance the activity of this enzyme at physiological pH [144]. Improving the activity of hArgI at physiological pH had important therapeutic implications, since some cancers, including renal, prostate, and hepatocellular, are auxotrophic with respect to L-Arg. In the mechanism of hArgI at physiological pH, a metal-bound hydroxide from water attacks the guanidinium carbon of L-Arg. Given this mechanism, the authors hypothesized that Co^{2+}, having a lower pK_a for the hexaquo cation than Mn^{2+} by ~ 1 unit, may be more active at a lower pH than the Mn^{2+}-bound hArgI, which has an optimum pH of ~ 9.5. Additionally, previous substitutions of Co^{2+} for Mn^{2+} in other enzymes had resulted in lower pH optima, and notably, an arginase from *Helicobacter pylori* had been demonstrated to bind Co^{2+} instead of Mn^{2+} and possessed a more acidic pH optimum. As a final inspiration and clue that Co^{2+} substitution for Mn^{2+} may improve the activity of hArgI at lower pH, He and Lippard had made synthetic models displaying arginase activity, and after testing the activity with Mn^{2+}, Zn^{2+}, Ni^{2+}, and Co^{2+}, found that only Co^{2+} catalyzed the hydrolysis of aminoguanidinium. Thus, Georgiou and coworkers substituted Co^{2+} for Mn^{2+} in hArgI and found that the activity of the enzyme at pH 7.4, near physiological pH, was improved 10-fold. Moreover, the Co^{2+}-substituted hArgI demonstrated improved serum stability, an added benefit for therapeutic applications. For *in vivo* cytotoxicity experiments with hepatocellular carcinoma and melanoma cell lines, Co^{2+}-substituted hArgI yielded 12- and 15-fold lower IC_{50} values, respectively. The authors accounted for the increased activity with Co^{2+} based on L-Arg coordinating directly to one of the Co^{2+} ions during the reaction.

5.10.2 Manganese Proteins

Engineering proteins with increased optimal activity at a different pH can be important for industrial processes as well as therapeutic applications. Lambeir and coworkers made a single amino acid substitution near the active site of glucose isomerase, to lower the pH optimum of its activity, for industrial application in the conversion of glucose to fructose [145]. They found that mutation of a Glu residue to Gln significantly improved the catalytic activity of the enzyme at acidic pH, shifting the pH optimum by ~ 2 units, and the variant was now also specific to Mn^{2+}, whereas before it was active with Co^{2+}, Mg^{2+}, and Mn^{2+}. The mutated Glu residue was chosen because it was conserved amongst glucose isomerases, but was not directly involved in metal or substrate binding, and maintained a negative dipole near the active site, which might be expected to affect metal binding affinity and pH dependence. The authors also created a variant in which the same Glu residue was substituted with Asp, leaving a negative charge near the same position. This variant showed intermediate behavior between the wild type and Glu186Gln glucose isomerases, supporting the author's explanation.

A holy grail of the protein engineering approach is to recreate the activity of the water oxidation complex in photosystem II (PSII), which, so far, has been unmatched by other water oxidation methods. With this long-term goal in mind, Allan and coworkers designed a manganese center into the bacterial reaction center of *Rhodobacter sphaeroides* by site-directed mutagenesis [146]. By analogy to PSII and through the use of computer modeling, four variants were generated, each having two to three residues mutated to Glu or Asp around an existing Glu residue. Upon characterization of the Mn^{2+}-binding properties of these variants, the authors found that one variant bound Mn^{2+} with a relatively high affinity, two with similar, lower affinities, and one did not bind Mn^{2+} with any appreciable affinity. The variants that bound Mn^{2+} reduced the oxidized bacteriochlorophyll dimer, demonstrating the effectiveness of the designed site. While this system is far from being able to affect water oxidation, it gives a starting point for future studies of light-driven metal catalysis.

5.10.3   Molybdenum Proteins

In a study of dimethyl sulfoxide reductase (DMSO reductase) from *R. sphaeroides*, Rajagopalan and coworkers made a single mutation of the serine anchoring the molybdenum cofactor in the protein to cysteine, and found that the mutation completely changed the substrate specificity of the enzyme [147]. With a cysteine attaching the molybdenum cofactor to the protein, the modified DMSO reductase became more active with adenosine N^1-oxide than the wild-type enzyme had been with any substrate. Although the authors observed spectroscopic changes with this mutant, indicating a change in the geometry of the molybdenum, they could not pinpoint the cause for the change in selectivity.

Another major interest for alternative energy applications is generation of hydrogen gas as a fuel from the nitrogen-fixing activity of nitrogenase. One group sought to improve the hydrogen production activity of nitrogenase in the cyanobacterium, *Anabaena* sp. strain 7120, by performing mutagenesis of six residues to a subset of the amino acids within 5 Å of the FeMo-co active site, and selecting for the mutations that yielded higher hydrogen production activity for organisms grown on N_2 [148]. Several of the variants did increase hydrogen production in their strains. The authors noted that all of the highly active variants for hydrogen production introduced a His residue in proximity to another His, which was proposed to be the obligate proton donor to nitrogenous substrates. The introduced His residues may participate in this proton donation.

5.10.4   Nickel Proteins

The ability to turn on or off the function of an enzyme by simple addition or chelation of a metal ion is an attractive prospect, with the possibility to be realized through the allosteric regulation of an enzyme by a designed metal-binding site. Toward this goal, Fletterick and coworkers introduced a metal-binding site at the dimer interface of glycogen phosphorylase [149]. To do so, the authors used the crystal structure of phosphorylated phosphorylase to search several positions at the dimer interface where two amino acids could be replaced with histidine, such that the imidazole side chains could orient to coordinate a transition metal. This initial screen narrowed the possible amino acid pairs from 200 to 70, and of these remaining 70, only 29 pairs would result in structural movement upon metal binding. After examining these candidates by molecular modeling, the authors chose three pairs of histidines to study in detail. They measured the activation of the mutated phosphorylases upon the addition of Ni^{2+}, Zn^{2+}, and Co^{2+}. For one of the mutants, addition of metal resulted in activation of the phosphorylase. The authors were able to pinpoint the source of allosteric activation in this mutant as the movement of just a handful of residues upon binding of the metal.

Another group sought to introduce metal-based allosteric regulation of β-lactamase [150]. To accomplish this goal, Fastrez and coworkers randomized three neighboring loops and used phage display to select first for metal binding, and then for increased β-lactamase activity. Through this process, the authors found a few mutants having metal-specific activity that was enhanced by factors up to three or inhibited by factors greater than 10, including one unusual mutant that was differentially regulated, contingent upon the metal added. This mutant was activated by Ni^{2+} binding, inhibited by Cu^{2+} binding, and essentially unaffected by Zn^{2+} binding. This study illustrated the possibility of regulating an enzyme activity by introducing a remote metal-binding site.

Later, this same group targeted the hinge region of a β-lactamase to introduce metal-based allosteric regulation [151]. They took a bifurcated approach, both rationally designing a metal-binding site and selecting mutants through directed evolution. The most successful mutant was selected by directed evolution, and was allosterically downregulated by the addition of Zn^{2+}, Ni^{2+}, and Co^{2+}.

5.10.5   Uranyl Proteins

Some of the more novel enzymes can be effective bioremediation agents, and protein engineering approaches can improve their applications in bioremediation. Matin and coworkers applied directed evolution, using error-prone PCR, to a chromate reductase, ChrR, to select for variants with improved chromate and uranyl reductase activities [152]. Their selections resulted in a variant that contained a single mutation, Tyr128Arg, which yielded enhanced kinetics for both chromate and uranyl reductase activity. However, structural information was not available, so the enhancement in activity could not be rationalized.

As a first step toward engineering a system for uranyl bioremediation, He and coworkers rationally designed a uranyl-binding site into NikR [153]. They did so by comparison to other known uranyl-binding biomolecules, which they observed to mostly bind uranyl through carboxylate groups. The existing square planar ligand set of the Ni^{2+} ion in NikR was used as the basis for the uranyl site, where the original His and Cys ligands were mutated to aspartate. A nearby valine was also mutated to serine, to accommodate the oxo groups of the uranyl cation and introduce a group with the possibility to hydrogen bond to one of the oxo groups. The resulting construct bound uranyl stoichiometrically with nanomolar affinity, and also bound DNA only in the presence of the uranyl ion, mimicking the function of native NikR in the presence of Ni(II).

5.10.6   Vanadium Proteins

Of great interest to the biosynthesis of halogenated natural products are the haloperoxidases, found in marine organisms. Littlechild and coworkers observed that the catalytic core for the vanadate active sites of chloro- and bromoperoxidases was very similar, and reasoned that small differences in the surrounding residues must dictate the halide specificity [154]. They noticed that an amino acid close to the vanadate cofactor was a tryptophan in two fungal chloroperoxidases, whereas it was an arginine in the bromoperoxidases from two algal organisms. To probe whether this difference dictated halide specificity, Littlechild and coworkers mutated the arginine in one bromoperoxidase from *Corallina pilulifera* to the other 19 amino acids, and looked for changes in the halide specificity. Two variants, Arg397Phe and Arg397Trp, were found to have enhanced rates of chloroperoxidation. The authors concluded that the halide selectivity was determined by the affinity of the active site for a particular halide.

In another study of a haloperoxidase, Wever and coworkers used a combination of directed evolution, through error-prone PCR, saturation mutagenesis of a "hot spot" proline residue, and rational recombination of mutants, to select for a chloroperoxidase with 100-fold higher activity at pH 8 [155]. The authors rationalized the effects of the variant as alterations to the electron density of the oxygen on the vanadate cofactor and modifications to the electrostatics close to the halide-binding residue, Phe397, as well as a change in this residue's position.

5.11   SUMMARY AND PERSPECTIVES

The recent progress in rational design of protein cages for alternative enzymatic functions, and in particular, studies that have strived to predictably alter or impart new functionality to metalloenzymes, have led to important advances in our knowledge of how metalloenzymes work and have set the groundwork for producing designer proteins with a selected function and high rates of catalysis. A key to the success of all these projects is the recognition of the surroundings and spatial position of the desired metal site in its native fold, and of the other proteins in the same fold class, including the similarity of the protein cage itself, the three-dimensional position of the site in the protein cage, and the positions of the ligands. Of particular importance is the presence of noncovalent interactions, such as hydrophobicity and hydrogen bonding, in conferring and fine-tuning protein function, as, for example, in the case of tuning the reduction potentials of type 1 blue copper azurins [35], and iron–sulfur clusters [123]. In many of these examples, computer modeling aided with the design of the site, providing rough measures of how destabilizing the mutations would be and whether they were positioned correctly. Three-dimensional structural studies, such as X-ray crystallography and NMR, are playing an increasing role in the design and proof of how these structural features affect the protein function, as few other techniques are able to provide the level of information about such subtle differences.

When a large portion of the ligands to a metal site originates in a loop, the strategy of loop-directed mutagenesis is a useful tool to keep in the protein design box, as has been demonstrated for introducing both Cu_A and nitrosocyanin-like sites into alternative protein cages. In other cases, single amino acid mutagenesis is enough to dramatically improve the activity of a catalytic metal site, as was true for improving the activity of a Cu, Zn SOD to the diffusion-controlled limit. In this instance, the authors recognized the features of this enzyme that allowed it to be so efficient and searched for amino acids that went against these features, here a negatively charged amino acid in the middle of a positive channel that facilitated catalysis.

For complicated, large enzymes and enzyme complexes, such as the nitrogenases, multicopper oxidases and oxygenases, and chromate reducing proteins, structural information can be a limiting barrier to rational design of improved or altered functionalities. In these cases, sequence alignments and threading of amino acid sequences onto structures of similar proteins can guide design. Alternatively, the use of directed evolution to select for certain characteristics can inform future design attempts. Therefore, many design strategies are available that can facilitate successful improvement or alteration of enzymatic activities, and the approach chosen primarily should reflect the goal of the work, the particular requirements of the metal site, and the available knowledge of that metal site and enzyme class.

Many challenges remain, however, and relatively few examples of studies where a metalloenzyme was predictably altered to achieve a desired function can be found in the literature. Furthermore, those studies that resulted in predictably altered function most often result in enzymes with very low rates of catalysis as compared to the native function of the enzyme. The major difficulty in reengineering metalloenzymes is their inherent complexity. Even the smallest metalloenzymes have at least 100 individual residues, all of which will have multiple noncovalent interactions that affect the functionality of a protein. As we have shown here, this extended protein network is critical to tuning and facilitating protein function, but accurately predicting the outcome of even a single mutation or how an altered substrate will interact with the catalytic site and surrounding residues is a large challenge. Metals with partially occupied d orbitals also have preferred geometries, as well as ligand and metal orbital overlaps to consider. Furthermore, the canonical 20 amino acids limit the choices for tuning the properties of a metal site by altering the electronic properties of the ligands.

To meet the challenges of modern metalloenzyme engineering, the first requirement is improved methods of crystallization and structural determination for metalloenzymes. Such knowledge is indispensable, as it guides decisions as to what residues to mutate and what residues can be substituted to have the desired effect to, for example, the hydrogen-bonding network around a metal site. In beam spectrometers that are currently being integrated into high-intensity radiation sources will also become essential for identifying reactive intermediates. However, for many proteins, obtaining a three-dimensional structure is a daunting task by itself, which can be overcome through high level homology modeling of structures based upon sequence similarity to other proteins and then molecular dynamics to predict the structure. However, there is no substitute for an actual crystal or NMR structure. It is also highly advantageous to solve structures for any variants made, in order to verify the effect of mutations.

Predicting mutations that will impart a desired function to a metalloenzyme simply through structural analysis and chemical intuition is prohibitively time-consuming, particularly for large metalloenzymes. Furthermore, predicting the effects of a single mutation on protein stability is also difficult, which means that a rationally designed variant may have the desired function, but not fold properly. As such, metalloenzyme design and engineering require computational programs for predicting the effects of mutations on the metal-binding site, substrate binding and protein stability. Several programs, such as Rosetta or NAMD, have had tremendous success at predicting non-metalloenzyme structures and stability, and have even been able to design enzymes with novel functions. Such programs, as of yet, have been applied to metalloenzymes only in limited cases. The major reason for this is that the behavior of a metal is difficult to mimic, *in silico*, with limited computational power. Other methods, such as density functional theory (DFT), have been successfully applied to predicting the electronic structures of small metal complexes, but are computationally taxing to be applied to large metalloproteins. Programs that combine the ability of software like Rosetta and NAMD to predict protein structures with DFT- like computations on metal sites are needed in the field of metalloenzyme engineering, and much attention has been devoted to developing QM/MM programs to accomplish this goal. One promising *ab initio* program, called GAMESS, has already seen success in predicting redox potentials in T1 copper sites where the secondary coordination sphere is altered. Another *ab initio* program, called ORCA, is specially targeted to reproducing the spectroscopic features of transition metal centers in proteins [156, 157]. On the basis of these well-reproduced spectroscopic features, the structures computed by ORCA are assigned a high level of relevance to the actual structure of the metal center, and as such, can facilitate future designs of that site.

Another area that is greatly advancing the field of metalloenzyme engineering is high-throughput screening. Such screening has already been put to use in engineering certain metalloenzyme catalysts, such as P450 enzymes mentioned in this chapter. Using high-throughput screening is highly effective for catalyst design where, for example, a nonfluorescent substrate can be converted into a fluorescent one, thereby providing a spectroscopic handle for screening activity. Screening millions of protein variants for a certain redox potential, however, is much more difficult as a tag for screening may not exist. Therefore, developing new and innovative ways to screen for a wider variety of properties will be essential in the future.

Finally, the ability to efficiently produce proteins incorporating unnatural amino acids will greatly expand the toolbox for metalloenzyme engineering. As has already been discussed, the limited number of natural ligands among the 20 canonical amino acids may be overcome in nature by fine-tuning electronic properties of the ligands through altering the surrounding protein network. Accurately altering the surrounding environment to tune the electronic properties of a metal site is very difficult. With the ability to incorporate unnatural amino acids, a His residue could be substituted with a residue having a substituted imidazole, a pyrazole, or even a pyridine side chain, thereby combining the advantages of changing the ligand type, as can be done with synthetic metal complexes, with the advantages of having the extended protein network as well. Such technology would also allow for greater flexibility in designing steric or hydrogen-bonding interactions to and around the catalytic site, to influence the metal site or control substrate access.

REFERENCES

[1]   Abe, S.; Ueno, T.; Watanabe, Y. *Top. Organomet. Chem.* **2009**, *25*, 25–43.

[2]   Borovik, A. S. *Acc. Chem. Res.* **2005**, *38*, 54–61.

[3]   Cherry, J. R.; Fidantsef, A. L. *Curr. Opin. Biotechnol.* **2003**, *14*, 438–443.

[4]   Kannt, A.; Lancaster, C. R. D.; Michel, H. *J. Bioenerg. Biomembr.* **1998**, *30*, 81–87.

[5]   Thomas, C. M.; Ward, T. R. *Appl. Organomet. Chem.* **2005**, *19*, 35–39.

[6]   Lu, Y.; Yeung, N.; Sieracki, N.; Marshall, N. M. *Nature* **2009**, *460*, 855–862.

[7]   Matsuo, T.; Hayashi, A.; Abe, M.; Matsuda, T.; Hisaeda, Y.; Hayashi, T. *J. Am. Chem. Soc.* **2009**, *131*, 15124–15125.

[8]   Maglio, O.; Nastri, F.; de Rosales, R. T. M.; Faiella, M.; Pavone, V.; DeGrado, W. F.; Lombardi, A. *Cr. Chim.* **2007**, *10*, 703–720.

[9]   DeGrado, W. F.; Woolfson, D. N. *Curr. Opin. Struct. Biol.* **2006**, *16*, 505–507.

[10] Baker, D. *Protein Sci.* **2010**, *19*, 1817–1819.

[11] Lombardi, A.; Nastri, F.; Pavone, V. *Chem. Rev.* **2001**, *101*, 3165–3189.

[12] Dutton, P. L.; Moser, C. C. *Faraday Discuss.* **2011**, *148*, 443–448.

[13] Arnold, F. H. *Acc. Chem. Res.* **1998**, *31*, 125–131.

[14] Messerschmidt, A.; Huber, R.; Wieghardt, K.; Poulos, T. Eds. *Handbook of Metalloproteins.* Wiley: Chichester, **2001**.

[15] Mogi, T.; Saiki, K.; Anraku, Y. *Mol. Microbiol.* **1994**, *14*, 391–398.

[16] Bowman, S.E.; Bren, K. L. *Nat. Prod. Rep.* **2008**, *25*, 1118–1130.

[17] Loewen, P. *Gene* **1996**, *179*, 39–44.

[18] Lu, Y. Electron transfer: Cupredoxins. In *Biocoordination Chemistry*; Que, J. L.; Tolman, W. B. Eds; Elsevier: Oxford, UK, 2004; pp 91–122.

[19] Vila, A. J.; Fernández, C. O. Copper in Electron Transfer Proteins. In *Handbook on Metalloproteins;* Bertini, I.; Sigel, A.; Sigel, H. Eds; Marcel Dekker: New York, NY, 2001; pp 813–856.

[20] Gray, H. B.; Malmström, B. G.; Williams, R. J. P. *J. Biol. Inorg. Chem.* **2000**, *5*, 551–559.

[21] Malmström, B. G.; Reinhammar, B.; Vänngård, T. *Biochim. Biophys. Acta.* **1970**, *205*, 48–57.

[22] Schinina, M. E.; Maritano, S.; Barra, D.; Mondovi, B.; Marchesini, A. *Biochim. Biophys. Acta.* **1996**, *1297*, 28–32.

[23] Lindley, P. F. Ceruloplasmin. In *Handbook of Metalloproteins;* Messerschmid, A.; Huber, R.; Poulos, T.; Wieghard, K., Eds; Wiley: Chichester, 2001; pp 1369–1380.

[24] Dennison, C. *Coord. Chem. Rev.* **2005**, *249*, 3025–3054.

[25] Freeman, H. C. *ACS Symp. Ser.* **1998**, *692*, 62–95.

[26] Adman, E. T. *Adv.Protein Chem.* **1991**, *42*, 145–197.

[27] Solomon, E. I.; Baldwin, M. J.; Lowery, M. D. *Chem. Rev.* **1992**, *92*, 521–542.

[28] Solomon, E. I.; Szilagyi, R. K.; DeBeer George, S.; Basumallick, L. *Chem. Rev.* **2004**, *104*, 419–458.

[29] Solomon, E. I. *Inorg. Chem.* **2006**, *45*, 8012–8025.

[30] Garner, D. K.; Vaughan, M. D.; Hwang, H. J.; Savelieff, M. G.; Berry, S. M.; Honek, J. F.; Lu, Y. *J. Am. Chem. Soc.* **2006**, *128*, 15608–15617.

[31] Berry, S. M.; Ralle, M.; Low, D. W.; Blackburn, N. J.; Lu, Y. *J. Am. Chem. Soc.* **2003**, *125*, 8760–8768.

[32] Hart, P. J.; Nersissian, A. M.; Herrmann, R. G.; Nalbandyan, R. M.; Valentine, J. S.; Eisenberg, D. *Protein Sci.* **1996**, *5*, 2175–2183.

[33] Romero, A.; Hoitink, C. W. G.; Nar, H.; Huber, R.; Messerschmidt, A.; Canters, G. W. *J. Mol. Biol.* **1993**, *229*, 1007–1021.

[34] Yanagisawa, S.; Sato, K.; Kikuchi, M.; Kohzuma, T.; Dennison, C. *Biochemistry* **2003**, *42*, 6853–6862.

[35] Marshall, N. M.; Garner, D. K.; Wilson, T. D.; Gao, Y.-G.; Robinson, H.; Nilges, M. J.; Lu, Y. *Nature* **2009**, *462*, 113–116.

[36] Yanagisawa, S.; Banfield, M. J.; Dennison, C. *Biochemistry* **2006**, *45*, 8812–8822.

[37] Pascher, T.; Karlsson, B. G.; Nordling, M.; Malmström, B.G.; Vänngård, T. *Eur. J. Biochem.* **1993**, *212*, 289–296.

[38] Berry, S. M.; Gieselman, M. D.; Nilges, M. J.; Van der Donk, W. A.; Lu, Y. *J. Am. Chem. Soc.* **2002**, *124*, 2084–2085.

[39] Young, T. S.; Schultz, P. G. *J. Biol. Chem.* **2010**, *285*, 11039–11044.

[40] Berry, S. M.; Bladholm, E. L.; Mostad, E. J.; Schenewerk, A. R. *J. Biol. Inorg. Chem.* **2011**, *16*, 473–480.

[41] Yanagisawa, S.; Dennison, C. *J. Am. Chem. Soc.* **2004**, *126*, 15711–15719.

[42] Canters, G. W.; Buning, C.; Comba, P.; Dennison, C.; Jeuken, L.; Melter, M.; Sanders-Loehr, J. *J. Am. Chem. Soc.* **2000**, *122*, 204–211.

[43] Kataoka, K.; Hirota, S.; Maeda, Y.; Kogi, H.; Shinohara, N.; Sekimoto, M.; Sakurai, T. *Biochemistry* **2011**, *50*, 558–565.

[44] Lancaster, K. M.; Yokoyama, K.; Richards, J. H.; Winkler, J. R.; Gray, H. B. *Inorg. Chem.* **2009**, *48*, 1278–1280.

[45] Lancaster, K. M.; DeBeer George, S.; Yokoyama, K.; Richards, J. H.; Gray, H. B. *Nat. Chem.* **2009**, *1*, 711–715.

[46] Lancaster, K. M.; Farver, O.; Wherland, S.; Crane, E. J.; 3rd, Richards, J. H.; Pecht, I.; Gray, H. B. *J. Am. Chem. Soc.* **2011**, *133*, 4865–4873.

[47] Lancaster, K. M.; Sproules, S.; Palmer, J. H.; Richards, J. H.; Gray, H. B. *J. Am. Chem. Soc.* **2010**, *132*, 14590–14595.

[48] van der Oost, J.; Lappalainen, P.; Musacchio, A.; Warne, A.; Lemieux, L.; Rumbley, J.; Gennis, R. B.; Aasa, R.; Pascher, T.; Malmström, B. G.; Saraste, M. *EMBO J.* **1992**, *11*, 3209–3217.

[49] Dennison, C.; Vijgenboom, E.; de Vries, S.; van der Oost, J.; Canters, G. W. *FEBS Lett.* **1995**, *365*, 92–94.

[50] Hay, M.; Richards, J. H.; Lu, Y. *Proc. Natl. Acad. Sci. USA* **1996**, *93*, 461–464.

[51] Jones, L. H.; Liu, A.; Davidson, V. L. *J. Biol. Chem.* **2003**, *278*, 47269–47274.

[52] Robinson, H.; Ang, M. C.; Gao, Y.-G.; Hay, M. T.; Lu, Y.; Wang, A. H. J. *Biochemistry* **1999**, *38*, 5677–5683.

[53] Farver, O.; Lu, Y.; Ang, M. C.; Pecht, I. *Proc. Natl. Acad. Sci. USA* **1999**, *96*, 899–902.

[54] Suzuki, S.; Kataoka, K.; Yamaguchi, K. *Acc. Chem. Res.* **2000**, *33*, 728–735.

[55] Dwivedi, U. N.; Singh, P.; Pandey, V. P.; Kumar, A. *J. Mol. Catal. B: Enzym.* **2011**, *68*, 117–128.

[56] MacPherson, I. S.; Murphy, M. E. *Cell Mol. Life Sci.* **2007**, *64*, 2887–2899.

[57] Whittaker, M.; Bergmann, D.; Arciero, D.; Hooper, A. B. *Biochim. Biophys. Acta.* **2000**, *1459*, 346–355.

[58] Lieberman, R. L.; Arciero, D. M.; Hooper, A. B.; Rosenzweig, A. C. *Biochemistry* **2001**, *40*, 5674–5681.

[59] Arciero, D. M.; Pierce, B. S.; Hendrich, M. P.; Hooper, A. B. *Biochemistry* **2002**, *41*, 1703–1709.

[60] Basumallick, L.; Sarangi, R.; DeBeer George, S.; Elmore, B.; Hooper, A. B.; Hedman, B.; Hodgson, K. O.; Solomon, E. I. *J. Am. Chem. Soc.* **2005**, *127*, 3531–3544.

[61] Clark, K. M.; Yu, Y.; Marshall, N. M.; Sieracki, N. A.; Nilges, M. J.; Blackburn, N. J.; van der Donk, W. A.; Lu, Y. *J. Am. Chem. Soc.* **2010**, *132*, 10093–10101.

[62] Solomon, E. I.; Hare, J. W.; Dooley, D. M.; Dawson, J. H.; Stephens, P. J.; Gray, H. B. *J. Am. Chem. Soc.* **1980**, *102*, 168–178.

[63] Podtetenieff, J.; Taglieber, A.; Bill, E.; Reijerse, E. J.; Reetz, M. T. *Angew. Chem. Int. Ed.* **2010**, *49*, 5151–5155.

[64] Berry, S.; Mayers, J.; Zehm, N. *J. Biol. Inorg. Chem.* **2009**, *14*, 143–149.

[65] Folcarelli, S.; Venerini, F.; Battistoni, A.; O'Neill, P.; Rotilio, G.; Desideri, A. *Biochem. Biophys. Res. Commun.* **1999**, *256*, 425–428.

[66] Banci, L.; Bertini, I.; Ciofi-Baffoni, S.; Katsari, E.; Katsaros, N.; Kubicek, K.; Mangani, S. *Proc. Natl. Acad. Sci. USA* **2005**, *102*, 3994–3999.

[67] Solomon, E. I.; Sundaram, U. M.; Machonkin, T. E. *Chem. Rev.* **1996**, *96*, 2563–2605.

[68] Shuster Ben-Yosef, V.; Sendovski, M.; Fishman, A. *Enzyme Microb. Technol.* **2010**, *47*, 372–376.

[69] Sugino, M.; Kajita, S.; Banno, K.; Shirai, T.; Yamane, T.; Kato, M.; Kobayashi, T.; Tsukagoshi, N. *Biochim. Biophys. Acta.* **2002**, *1596*, 36–46.

[70] Xu, F.; Berka, R. M.; Wahleithner, J. A.; Nelson, B. A.; Shuster, J. R.; Brown, S. H.; Palmer, A. E.; Solomon, E. I. *Biochem. J.* **1998**, *334*, 63–70.

[71] Li, Y.; Gong, Z.; Li, X.; Wang, X. G. *BMC Biochem.* **2011**, *12*, 30.

[72] Bulter, T.; Alcalde, M.; Sieber, V.; Meinhold, P.; Schlachtbauer, C.; Arnold, F. H. *Appl. Environ. Microbiol.* **2003**, *69*, 987–995.

[73] Zumarraga, M.; Bulter, T.; Shleev, S.; Polaina, J.; Martinez-Arias, A.; Plou, F. J.; Ballesteros, A.; Alcalde, M. *Chem. Biol.* **2007**, *14*, 1052–1064.

[74] Zumarraga, M.; Camarero, S.; Shleev, S.; Martinez-Arias, A.; Ballesteros, A.; Plou, F. J.; Alcalde, M. *Proteins* **2008**, *71*, 250–260.

[75] Scott, E. E.; Gibson, Q. H.; Olson, J. S. *J. Biol. Chem.* **2001**, *276*, 5177–5188.

[76] Matsui, T.; Ozaki, S.; Watanabe, Y. *J. Biol. Chem.* **1997**, *272*, 32735–32738.

[77] Matsui, T.; Ozaki, S.-I.; Liong, E.; Phillips, G. N. Jr.; Watanabe, Y. *J. Biol. Chem.* **1999**, *274*, 2838–2844.

[78] Ozaki, S.-I.; Matsui, T.; Watanabe, Y. *J. Am. Chem. Soc.* **1997**, *119*, 6666–6667.

[79] Matsui, T.; Ozaki, S.-I.; Watanabe, Y. *J. Am. Chem. Soc.* **1999**, *121*, 9952–9957.

[80] Quillin, M. L.; Arduini, R. M.; Olson, J. S.; Phillips, G. N., Jr. *J. Mol. Biol.* **1993**, *234*, 140–155.

[81]   Quillin, M. L.; Li, T.; Olson, J. S.; Phillips, G. N., Jr., Dou, Y.; Ikeda-Saito, M.; Regan, R.; Carlson, M.; Gibson, Q. H.; Li, H. *J. Mol. Biol.* **1995**, *245*, 416–436.

[82]   Yang, H. J.; Matsui, T.; Ozaki, S.; Kato, S.; Ueno, T.; Phillips, G. N.; Fukuzumi, S.; Watanabe, Y. *Biochemistry* **2003**, *42*, 10174–10181.

[83]   Kato, S.; Yang, H. J.; Ueno, T.; Ozaki, S.; Phillips, G. N.; Fukuzumi, S.; Watanabe, Y. *J. Am. Chem. Soc.* **2002**, *124*, 8506–8507.

[84]   Ozaki, S.; Hara, I.; Matsui, T.; Watanabe, Y. *Biochemistry* **2001**, *40*, 1044–1052.

[85]   Sigman, J. A.; Kwok, B. C.; Lu, Y. *J. Am. Chem. Soc.* **2000**, *122*, 8192–8196.

[86]   Sigman, J. A.; Kim, H. K.; Zhao, X.; Carey, J. R.; Lu, Y. *Proc. Natl. Acad. Sci. USA* **2003**, *100*, 3629–3634.

[87]   Ortiz de Montellano, P. R. *Acc. Chem. Res.* **1998**, *31*, 543–549.

[88]   Ortizde Montellano, P. R.; Wilks, A. *Adv. Inorg. Chem.* **2001**, *51*, 359–407.

[89]   Das, T. K.; Pecoraro, C.; Tomson, F. L.; Gennis, R. B.; Rousseau, D. L. *Biochemistry* **1998**, *37*, 14471–14476.

[90]   Blomberg, M. R. A.; Siegbahn, P. E. M.; Wikstroem, M. *Inorg. Chem.* **2003**, *42*, 5231–5243.

[91]   Cukier, R. I. *Biochim. Biophys. Acta.* **2005**, *1706*, 134–146.

[92]   Wang, N.; Zhao, X.; Lu, Y. *J. Am. Chem. Soc.* **2005**, *127*, 16541–16547.

[93]   Butland, G.; Spiro, S.; Watmough, N. J.; Richardson, D. J. *J. Bacteriol.* **2001**, *183*, 189–199.

[94]   Flock, U.; Thorndycroft, F. H.; Matorin, A. D.; Richardson, D. J.; Watmough, N. J.; Ardelroth, P. *J. Biol. Chem.* **2008**, *283*, 3839–3845.

[95]   Flock, U.; Lachmann, P.; Reimann, J.; Watmough, N. J.; Aedelroth, P. *J. Inorg. Biochem.* **2009**, *103*, 845–850.

[96]   Yeung, N.; Lin, Y.-W.; Gao, Y.-G.; Zhao, X.; Russell, B. S.; Lei, L.; Miner, K. D.; Robinson, H.; Lu, Y. *Nature* **2009**, *462*, 1079–1082.

[97]   Hino, T.; Matsumoto, Y.; Nagano, S.; Sugimoto, H.; Fukumori, Y.; Murata, T.; Iwata, S.; Shiro, Y. *Science* **2010**, *330*, 1666–1670.

[98]   Lin, Y.-W.; Yeung, N.; Gao, Y.-G.; Miner, K. D.; Tian, S.; Robinson, H.; Lu, Y. *Proc. Natl. Acad. Sci. USA* **2010**, *107*, 8581–8586.

[99]   Gold, M. H.; Wariishi, H.; Valli, K. *ACS Symp. Ser.* **1989**, *389*, 127–140.

[100]   Tien, M.; Cai, D. *Biol. Oxid. Syst.* **1990**, *1*, 433–451.

[101]   Gold, M. H.; Youngs, H. L.; Gelpke, M. D. S. *Met. Ions Biol. Syst.* **2000**, *37*, 559–586.

[102]   Kumar, P.; Barrett, D. M.; Delwiche, M. J.; Stroeve, P. *Ind. Eng. Chem. Res.* **2009**, *48*, 3713–3729.

[103]   Hofrichter, M. *Enzyme Microb. Technol.* **2002**, *30*, 454–466.

[104]   Yeung, B. K. S.; Wang, X.; Sigman, J. A.; Petillo, P. A.; Lu, Y. *Chem. Biol.* **1997**, *4*, 215–221.

[105]   Gengenbach, A.; Syn, S.; Wang, X.; Lu, Y. *Biochemistry* **1999**, *38*, 11425–11432.

[106]   Sundaramoorthy, M.; Kishi, K.; Gold, M. H.; Poulos, T. L. *J. Biol. Chem.* **1994**, *269*, 32759–32767.

[107]   Sundaramoorthy, M.; Youngs, H. L.; Gold, M. H.; Poulos, T. L. *Biochemistry* **2005**, *44*, 6463–6470.

[108] Wilcox, S. K.; Putnam, C. D.; Sastry, M.; Blankenship, J.; Chazin, W. J.; McRee, D. E.; Goodin, D. B. *Biochemistry* **1998**, *37*, 16853–16862.

[109] Ortiz de Montellano, P. R. *Cytochrome P-450: Structure, Mechanism, and Biochemistry*, Plenum Press: New York, NY, 1986; p 556.

[110] Hlavica, P. *Biotechnol. Adv.* **2009**, *27*, 103–121.

[111] Miles, C. S.; Ost, T. W.; Noble, M. A.; Munro, A. W.; Chapman, S. K. *Biochim. Biophys. Acta.* **2000**, *1543*, 383–407.

[112] Cirino, P. C.; Arnold, F. H. *Angew. Chem. Int. Ed.* **2003**, *42*, 3299–3301.

[113] Bloom, J. D.; Arnold, F. H. *Proc. Natl. Acad. Sci. USA* **2009**, *106* Suppl 1, 9995–10000.

[114] Jung, S. T.; Lauchli, R.; Arnold, F. H. *Curr. Opin. Biotechnol.* **2011**, *22*, 809–817.

[115] Meinhold, P.; Peters, M. W.; Chen, M. M.; Takahashi, K.; Arnold, F. H. *Chembiochem.* **2005**, *6*, 1765–1768.

[116] Xu, F.; Bell, S. G.; Lednik, J.; Insley, A.; Rao, Z.; Wong, L. L. *Angew. Chem. Int. Ed.* **2005**, *44*, 4029–4032.

[117] Otey, C. R.; Bandara, G.; Lalonde, J.; Takahashi, K.; Arnold, F. H. *Biotechnol. Bioeng.* **2006**, *93*, 494–499.

[118] Kawakami, N.; Shoji, O.; Watanabe, Y. *Angew. Chem. Int. Ed.* **2011**, *50*(23), 5315–5318.

[119] Zilly, F. E.; Acevedo, J. P.; Augustyniak, W.; Deege, A.; Haeusig, U. W.; Reetz, M. T. *Angew. Chem., Int. Ed.* **2011**, *50*, 2720–2724.

[120] Beinert, H. *J. Biol. Inorg. Chem.* **2000**, *5*, 2–15.

[121] Beinert, H.; Holm, R. H.; Munck, E. *Science* **1997**, *277*, 653–659.

[122] Dey, A.; Jenney, F. E., Jr.; Adams, M. W. W.; Babini, E.; Takahashi, Y.; Fukuyama, K.; Hodgson, K. O.; Hedman, B.; Solomon, E. I. *Science* **2007**, *318*, 1464–1468.

[123] Zuris, J. A.; Halim, D. A.; Conlan, A. R.; Abresch, E. C.; Nechushtai, R.; Paddock, M. L.; Jennings, P. A. *J. Am. Chem. Soc.* **2010**, *132*, 13120–13122.

[124] Miller, A.-F. *Acc. Chem. Res.* **2008**, *41*, 501–510.

[125] Yikilmaz, E.; Porta, J.; Grove, L.; Vahedi-Faridi, A.; Bronshteyn, Y.; Brunold, T. C.; Borgstahl, G. E. O.; Miller, A.-F. *J. Am. Chem. Soc.* **2007**, *129*, 9927–9940.

[126] Solomon, E. I.; Brunold, T. C.; Davis, M. I.; Kemsley, J. N.; Lee, S.-K.; Lehnert, N.; Neese, F.; Skulan, A. J.; Yang, Y.-S.; Zhou, J. *Chem. Rev.* **2000**, *100*, 235–349.

[127] Costas, M.; Mehn, M. P.; Jensen, M. P.; Que, L., Jr. *Chem. Rev.* **2004**, *104*, 939–986.

[128] Krebs, C.; Fujimori, D. G.; Walsh, C. T.; Bollinger, J. M. *Acc. Chem. Res.* **2007**, *40*, 484–492.

[129] Koehntop, K. D.; Emerson, J. P.; Que, L. Jr. *J. Biol. Inorg. Chem.* **2005**, *10*, 87–93.

[130] Que, L. Jr. *Nat. Struct. Biol.* **2000**, *7*, 182–184.

[131] Bugg, T. D. H.; Lin, G. *Chem. Commun.* **2001**, 941–952.

[132] Blasiak, L. C.; Vaillancourt, F. H.; Walsh, C. T.; Drennan, C. L. *Nature* **2006**, *440*, 368–371.

[133] Matthews, M. L.; Neumann, C. S.; Miles, L. A.; Grove, T. L.; Booker, S. J.; Krebs, C.; Walsh, C. T.; Bollinger, J. M. Jr. *Proc. Natl. Acad. Sci. USA* **2009**, *106*, 17723–17728.

[134] Park, M. J.; Lee, J.; Suh, Y.; Kim, J.; Nam, W. *J. Am. Chem. Soc.* **2006**, *128*, 2630–2634.

[135] Dreier, B.; Segal, D. J.; Barbas, C. F., 3rd. *J. Mol. Biol.* **2000**, *303*, 489–502.

[136] Beerli, R. R.; Barbas, C. F. *Nat. Biotechnol.* **2002**, *20*, 135–141.

[137] Carroll, D.; Bibikova, M.; Beumer, K.; Trautman, J. K. *Science* **2003**, *300*, 764–764.

[138] Guo, J.; Gaj, T.; Barbas, C. F., 3rd. *J. Mol. Biol.* **2010**, *400*, 96–107.

[139] Kim, Y. G.; Cha, J.; Chandrasegaran, S. *Proc. Natl. Acad. Sci. USA* **1996**, *93*, 1156–1160.

[140] Akopian, A.; He, J.; Boocock, M. R.; Stark, W. M. *Proc. Natl. Acad. Sci. USA* **2003**, *100*, 8688–8691.

[141] Gordley, R. M.; Gersbach, C. A.; Barbas, C. F. 3rd. *Proc. Natl. Acad. Sci. USA* **2009**, *106*, 5053–5058.

[142] Gaj, T.; Mercer, A. C.; Gersbach, C. A.; Gordley, R. M.; Barbas, C. F., 3rd. *Proc. Natl. Acad. Sci. USA* **2011**, *108*, 498–503.

[143] Chou, C.; Deiters, A. *Angew. Chem. Int. Ed.* **2011**, *50*, 6839–6842.

[144] Stone, E. M.; Glazer, E. S.; Chantranupong, L.; Cherukuri, P.; Breece, R. M.; Tierney, D. L.; Curley, S. A.; Iverson, B. L.; Georgiou, G. *ACS Chem. Biol.* **2010**, *5*, 333–342.

[145] Van Tilbeurgh, H.; Jenkins, J.; Chiadmi, M.; Janin, J.; Wodak, S. J.; Mrabet, N. T.; Lambeir, A. M. *Biochemistry* **1992**, *31*, 5467–5471.

[146] Thielges, M.; Uyeda, G.; Camara-Artigas, A.; Kalman, L.; Williams, J. C.; Allen, J. P. *Biochemistry* **2005**, *44*, 7389–7394.

[147] Hilton, J. C.; Temple, C. A.; Rajagopalan, K. V. *J. Biol. Chem.* **1999**, *274*, 8428–8436.

[148] Masukawa, H.; Inoue, K.; Sakurai, H.; Wolk, C. P.; Hausinger, R. P. *Appl. Environ. Microbiol.* **2010**, *76*, 6741–6750.

[149] Browner, M. F.; Hockos, D.; Fletterick, R. *Nat. Struct. Biol.* **1994**, *1*, 327–333.

[150] Mathonet, P.; Barrios, H.; Soumillion, P.; Fastrez, J. *Protein Sci.* **2006**, *15*, 2335–2343.

[151] Mathieu, V.; Fastrez, J.; Soumillion, P. *Protein Eng. Des. Sel.* **2010**, *23*, 699–709.

[152] Barak, Y.; Ackerley, D. F.; Dodge, C. J.; Banwari, L.; Alex, C.; Francis, A. J.; Matin, A. *Appl. Environ. Microbiol.* **2006**, *72*, 7074–7082.

[153] Wegner, S. V.; Boyaci, H.; Chen, H.; Jensen, M. P.; He, C. *Angew. Chem. Int. Ed.* **2009**, *48*, 2339–2341.

[154] Ohshiro, T.; Littlechild, J.; Garcia-Rodriguez, E.; Isupov, M. N.; Iida, Y.; Kobayashi, T.; Izumi, Y. *Protein Sci.* **2004**, *13*, 1566–1571.

[155] Hasan, Z.; Renirie, R.; Kerkman, R.; Ruijssenaars, H. J.; Hartog, A. F.; Wever, R. *J. Biol. Chem.* **2006**, *281*, 9738–9744.

[156] Neese, F.; Petrenko, T.; Ganyushin, D.; Olbrich, G. *Coord. Chem. Rev.* **2007**, *251*, 288–327.

[157] Neese, F. *Coord. Chem. Rev.* **2009**, *253*, 526–563.

PART III

COORDINATION CHEMISTRY OF PROTEIN ASSEMBLY CAGES

6

METAL-DIRECTED AND TEMPLATED ASSEMBLY OF PROTEIN SUPERSTRUCTURES AND CAGES

F. Akif Tezcan

6.1 INTRODUCTION

In a landmark study in 2010 [1], Adams and colleagues reported that at least 60% of the *Pyrococcus furiousus* protein content was associated with metal ions, with the same likely holding true for other organisms. Clearly, metal ions play irreplaceable roles in biological systems, and in hindsight, the findings by Adams and colleagues are hardly surprising: metal ions lend not only their capability of performing vital tasks such as redox and Lewis-acid catalysis that would be unimaginable with organic building blocks, but also their unparalleled bonding properties to stabilize biological architectures and many other types of interfaces formed during cellular processes. In return, the biological scaffolds provide layers of control over the reactivities of the metal ions they contain, such that they can performs tasks as challenging as water oxidation and nitrogen fixation, and as diverse as gene regulation and neurotransmission. The study of this synergy between metal ions and biological scaffolds—particularly proteins—has been the driving force for the field of Bioinorganic Chemistry over the last half century. As a part of a natural progression, the questions about the amazingly complex metalloproteins and metalloenzymes like "what do they look like" and "how do they work" evolved into "can we mimic them synthetically", "can we surpass their capabilities" or "can we use them for different purposes than what they were originally intended for?" Efforts along these lines have produced protein and

Coordination Chemistry in Protein Cages: Principles, Design, and Applications, First Edition.
Edited by Takafumi Ueno and Yoshihito Watanabe.
© 2013 John Wiley & Sons, Inc. Published 2013 by John Wiley & Sons, Inc.

peptide scaffolds with new inorganic functionalities [2–6], and artificially "evolved" enzymes with new or improved activities [7–11], thanks in great part to advances in synthetic chemistry, molecular biology, computational design, and physical and analytical techniques.

In our work, we started with another interesting, but not as much studied, question: "How did metalloproteins and metalloenzymes evolve?" Whereas in theory evolution cannot be directly proven, we imagined that this question would lead us to alternative ways of creating novel bioinorganic scaffolds and functionalities. Some metallo-proteins likely evolved from pre-existing protein folds that acquired the ability to bind metals through random genetic events and subsequently attained their current structures and functions in the course of natural selection [12]. Support for such an evolutionary pathway comes from phylogenetic and structural/functional analyses [13] as well as many examples of stable protein scaffolds that have been reengineered to house new or alternative metal coordination sites [6]. In an alternative pathway, metal ions could first have directed the formation of a protein or peptide assembly, followed by the evolution of the protein structure around the metal ion [14]. Although this pathway would not have the benefit of re-using existing genetic material and a preformed scaffold, it may lead to greater flexibility toward generating bioinorganic diversity. With this grand scheme in mind, two different, but related approaches were developed [15, 16]: Metal-Directed Protein Self-Assembly (MDPSA), which exploits design metal coordination on protein surfaces to direct the formation of pro-tein assemblies and protein interfaces (i.e., the formation of a "protein cage" around the metal), and Metal-Templated Interface Redesign (MeTIR), which involves the reconstruction of the interfacial protein environment around the metal ions to tune their properties. This Chapter provides background and some highlights of these two approaches.

6.2 METAL-DIRECTED PROTEIN SELF-ASSEMBLY

6.2.1 Background

Aside from the goal of forming a cage around metal ions, there is another significant motivation in being able to direct the self-assembly of proteins through coordina-tion chemistry. Namely, protein–protein interactions (PPIs) are the chief contributors to cellular complexity. Transient PPIs are responsible for cellular dynamics, com-munication, and transfer of chemical currency (e.g., electrons, phosphate groups), whereas permanent PPIs are central to the construction of cellular machines and scaffolds. These key roles of PPIs have motivated intense efforts to engineer them *de novo* [17]; however, this task is complicated by the fact that PPIs are composed of weak, noncovalent bonds distributed over large protein surfaces that often exceed 1000 Å^2 [18].

MDPSA, inspired by supramolecular coordination chemistry [19–21], is based on the premise that metal bonding interactions capture all salient features of protein–protein interactions on a much smaller surface: First, metal–ligand bonds are significantly stronger than the noncovalent bonds that make up protein interfaces, circumventing the need to design large molecular surfaces for stability. Second, metal ions can show distinct preferences for the types of ligands they coordinate (e.g., based on hardness and softness), providing a handle for introducing specificity into PPIs. Third, metal-ligand bonds are highly directional, and the stereochemical preference and symmetry of metal coordination may be imposed onto PPIs. Fourth, metal–ligand bonds can be kinetically labile, allowing protein self-assembly to proceed under thermodynamic control. Fifth, metal coordination can be formed or broken through pH changes or external ligands, rendering protein self-assembly responsive to external stimuli. Sixth, and importantly for bioinorganic applications, metal ions bring along intrinsic reactivity (Lewis acidity, redox reactivity), which may be incorporated into protein interfaces.

In theory, the replacement of extensive noncovalent interactions with a few metal coordination bonds should be straightforward to accomplish. Yet, this strategy is still plagued by the chemical heterogeneity of protein surfaces, which are replete with carboxylates, amines, imidazoles, and thiol groups, rendering the site-selective localization of metals very challenging. A second hurdle stems from the large sizes of proteins, especially with respect to the metal ions that would link them together. Upon metal crosslinking, protein surfaces are bound to come into extensive contact through their amino acid side chains, making the outcome of MDPSA challenging to predict.

6.2.2   Design Considerations for Metal-Directed Protein Self-Assembly

In order to help circumvent the aforementioned challenges, a small, compact protein, cyt cb_{562}, was used as a model building block. Cyt cb_{562} is a variant of the four-helix bundle hemeprotein cyt b_{562} that contains genetically engineered c-type linkages between the heme and the protein backbone [22]. These bonds render cyt cb_{562} significantly more stable toward unfolding and thereby more resistant to structural perturbation. Cyt cb_{562} has a rigid, cylindrical shape (Fig. 6.1a), which is ideal as a building block for large assemblies, and makes such assemblies more amenable to crystallization and structural analysis. Importantly, cyt cb_{562} is monomeric even at millimolar concentrations, meaning that its surface does not carry any bias toward oligomerization.

Even when proteins possess a simple shape and a uniform topology like cyt cb_{562}, it is challenging to visualize protein–protein docking geometries that would be amenable to metal crosslinking. In this regard, crystal packing interactions (CPIs) provide a source of feasible protein–protein docking geometries. CPIs are typically not extensive ($<1000 \, \text{Å}^2$) and only stable in combination with other CPIs under crystallization conditions. At the same time, each CPI provides a metastable arrangement of proteins in which there are no steric clashes between them [23]. Importantly, CPIs may contain twofold or higher symmetry, which minimizes the number of metal coordination motifs that need to be engineered onto protein surfaces, providing a starting point to design high-order protein oligomers and superstructures.

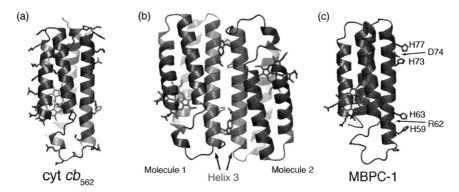

FIGURE 6.1   (a) Crystal structure of the cytochrome cb_{562} monomer, highlighting polar surface residues capable of metal coordination. (b) Antiparallel arrangement of two cyt cb_{562} molecules in the crystal lattice along their Helices 3. (b) Model of MBPC1, where key residues involved in metal binding and secondary interactions are depicted as sticks. Adapted with permission from Reference 15. Copyright 2010 American Chemical Society.

An examination of CPIs in cyt cb_{562} crystals reveals that each monomer is paired in an antiparallel fashion with another monomer along their third helices (Fig. 6.1b) [22]. This relatively large (775 $\overset{\circ}{A}{}^{2}$), C_2 symmetrical interface is clearly nonphysiological since cyt cb_{562} is monomeric in solution at concentrations used for crystallization (2–5 mM). At the same time, this close-packed arrangement of monomers shows a route for the self-assembly of cyt cb_{562}, whereby two metal-binding motifs can be incorporated near each end of Helix3 for metal-mediated crosslinking.

To overcome the first challenge of specific metal localization, chelating motifs were built on the cyt cb_{562} surface, in analogy to the small-molecule ligand platforms in synthetic inorganic chemistry. The all-α-helical constitution of cyt cb_{562} is particularly well suited for the purpose of installing metal-chelating motifs, owing to the regular spacing of the amino acids on helices. In nature, metal-binding amino acid residues (His, Glu, Asp) placed in $i/i+3$ and $i/i+4$ patterns on α-helices (thus pointing in the same direction) are quite regularly used to construct bidentate metal binding sites, such as those in Zn-finger domains, and di-iron and di-copper centers [24], among others. Inspired by these natural examples, chemists have employed $i/i+3$ and $i/i+4$ bidentate motifs consisting of natural or unnatural metal ligands to stabilize α-helical protein folds [25, 26], to build *de novo* metalloproteins with stable mono- and binuclear metal centers [23, 26], to build substrate-selective metallopeptide catalysts [27], and to facilitate purification by immobilized metal affinity chromatography [28].

The imidazole side chain of His is an ideal component for a surface chelating motif, as its borderline soft nitrogen atoms possess a high affinity for most transition metals, and in an $i/i+4$ pattern, the bis-His motif provides metal dissociation constants that are in the low μM regime for late first-row transition metals [26, 29]. In initial studies, two bis-His motifs (His59/His63 and His73/His77) were incorporated near the ends

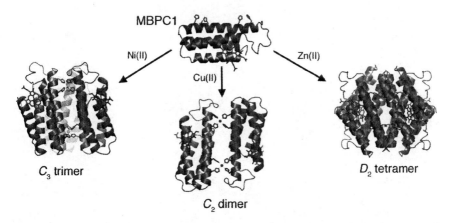

FIGURE 6.2 MBPC1 can self-assemble into discrete structures in a metal-dependent fashion based on the stereochemical preference of the added metal ion (a C_3 trimer with Ni^{2+}, a C_2 dimer with Cu^{2+} or Pd^{2+}, a D_2 tetramer with Zn^{2+}). Adapted with permission from Reference 31. Copyright 2009 American Chemical Society.

of Helix3 of cyt cb_{562} to make the construct MBPC1 (Fig. 6.1c), with the idea that metal coordination would lead to the oligomerization of this monomeric protein [30]. Indeed, the addition of equimolar Ni^{2+}, Cu^{2+}, and Zn^{2+} to MBPC1 results in the formation of discrete assemblies whose supramolecular arrangements are determined by the stereochemical preferences of these metal ions (Fig. 6.2). Octahedral Ni^{2+} coordination yields a C_3-symmetric trimer (Ni_2:MBPC1$_3$) with the Ni's coordinated to three bis-His motifs donated by all three protomers; tetragonal Cu^{2+} produces a C_2-symmetric dimer (Cu_2:MBPC1$_2$) with two bis-His motifs forming the equatorial coordination plane; tetrahedral Zn^{2+} coordination yields a D_2-symmetric tetramer (Zn_4:MBPC1$_4$) where the Zn ligand set consists of a bis-His motif (H73/H77) from one protomer, a single His (H63) from a second, and an Asp (D74) from a third [30, 31]. These findings firmly indicate that the supramolecular arrangement of MPBC-1 can be controlled by metal coordination geometry.

6.2.3   Interfacing Non-Natural Chelates with MDPSA

Although even simple combinations of natural metal-coordinating residues can lead to considerable diversity, the structural and functional scope of MDPSA can be significantly broadened if synthetic metal ligands are interfaced with proteins. From a structural standpoint, a multidentate ligand incorporated into a protein's surface via a single point attachment can offer more structural flexibility than, for instance, a bis-His motif. In the meantime, the large number of ligands available in the synthetic inorganic toolbox would offer great diversity in terms of tuning metal reactivity. In the past, non-natural chelating groups have been extensively used in metalloprotein engineering [8, 10, 32]. Chelates like 1,10-phenanthroline [8, 33], 2,2′-bipyridine [34, 35], Schiff base complexes [36], and diphosphines [37] have been covalently incorporated

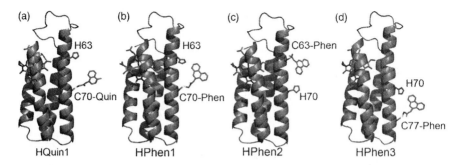

FIGURE 6.3 Cartoon representation of cyt cb_{562} variants functionalized with HCMs. Residues comprising the HCMs are shown as sticks. Adapted with permission from Reference 39. Copyright 2010 American Chemical Society.

into proteins using chemical modification, solid-phase synthesis, supramolecular anchoring or through *in vivo/in vitro* translation of non-natural amino acids. These chelates have been particularly useful for generating coordinatively unsaturated metal centers that can perform selective chemical transformations within the complex environment of the surrounding protein scaffolds.

It was envisioned that bidentate chelates such as 1,10-phenanthroline (Phen) and 8-hydroxyquinoline (Quin) covalently linked to surface Cys's could be combined in an $i/i+7$ pattern—corresponding to a two-helix-turn separation—with a His on an α-helical surface to yield tridentate hybrid coordination motifs (HCMs). Such tridentate HCMs should provide improved metal affinities and selectivities compared, for example, to bis-His chelates, and enable better control over protein oligomerization. Using Helix3 of cyt cb_{562} as an anchoring platform, several Phen- and Quin-bearing HCM variants were constructed (Fig. 6.3) [38, 39]. The Phen and Quin ligands were prepared in high yields as thiol-reactive iodoacetamide derivatives (IA-Phen and IA-Quin). Although IA-Phen and IA-Quin have low solubility in water, they are easily introduced into cyt cb_{562} solutions after being solubilized in dimethylformamide (DMF) or dimethyl sulfoxide (DMSO). Overall yields between 70% and 95% were routinely obtained for both the Phen- and Quin-labeled cyt cb_{562} variants.

The tridentate nature of the Phen and Quin HCMs yield metal dissociation constants ranging from nM to fM, which are at least three orders of magnitude lower than those of the $i/i+4$ bis-His motifs for each metal [38, 39]. They are also considerably lower (by ~2 orders of magnitude) than those for free Phen and Quin ligands, which strongly suggests the participation of the His component of the HCMs in metal binding. Moreover, the affinities of $i/i+7$ HCMs placed in different locations on the cyt cb_{562} surface for divalent metals vary little, indicating that metal-binding ability is not very sensitive to helix location.

In the process of binding metals with high affinity, $i/i+7$ HCMs also crosslink a two-turn helical segment, which imparts stability on the helix and thereby the entire protein. Metal crosslinking of both natural and non-natural residues at $i/i+4$ positions has extensively been shown to induce α-helicity in peptides and significantly

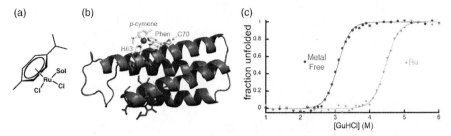

FIGURE 6.4   (a) Solvated monomer of the dichloro(p-cymene)Ru(II) dimer. (b) A model for the proposed coordination for the Ru(p-cymene):HPhen1 adduct. (c) Chemical unfolding HPhen1 in the presence (black diamonds) and the absence (gray circles) of the Ru(p-cymene) group. Adapted with permission from Reference 39. Copyright 2010 American Chemical Society.

stabilize helical protein structures [26, 29]. Likewise, covalent cross-linking of side chain functionalities in $i/i+4$, $i/i+7$, or $i/i+11$ positions can lock small peptides in α-helical conformations [40–42], which have proven to be promising pharmaceutical agents that disrupt protein–protein interactions and display increased resistance to proteases *in vivo* [41]. Along these lines, a Ru(p-cymene) adduct of a cyt cb_{562} variant with a surface His-Phen HCM was prepared (Figs. 6.4a and 6.4b). Ru(p-cymene) is a piano-stool-type complex, which presents its exchangeable coordination sites in the right geometry to accommodate a tridentate $i/i+7$ HCM. This Ru-construct was found to be significantly more stabilized compared to the unmodified protein, with a corresponding free energy of stabilization ($\Delta\Delta G_{folding}$) of 4.1 kcal/mol (Fig. 6.4c) [39]. This result highlights the potential that HCMs may provide in selectively localizing metal-based probes.

Upon establishing that $i/i+7$ HCMs coordinate metals in the tridentate fashion as planned, their utility in controlling protein–protein interactions (PPIs) was examined. In particular, the case of homodimerization was considered, which is the simplest and most prevalent form of protein self-assembly, extensively utilized throughout cellular signaling pathways [43]. It was imagined that the tridentate HCMs we have available would promote efficient protein dimerization in response to an octahedral metal coordination geometry and impose a strict supramolecular geometry owing to their two-point attachment to the protein surface. Sedimentation velocity (SV) experiments show that cyt cb_{562} variants featuring His-Phen and His-Quin HCMs readily dimerize in the presence of half an equivalent of Ni^{2+} [38, 39]. The dissociation constants for the Ni:HPhen1$_2$ and Ni:HQuin1$_2$ dimers were determined by sedimentation equilibrium (SE) experiments to be ~9 μM and 42 μM, respectively [38, 39]. These dimerization constants closely approximate that of bZIP family transcription factors, which use peripheral leucine zippers domains for dimerization, with K_d's in the low micromolar range [44].

The crystal structure of the Ni:HQuin1$_2$ complex was determined at 2.3 Å resolution (Fig. 6.5a) [38], which reveals a V-shaped dimer with a parallel arrangement of two HQuin1 protomers. The acute angle (~50°) between the protomers results

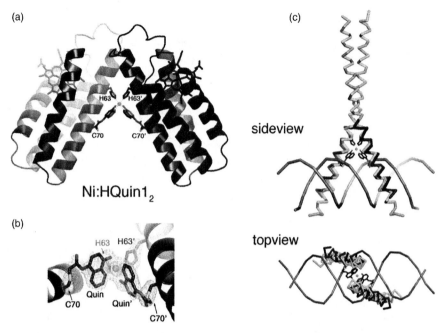

FIGURE 6.5   (a) Crystal structure of the Ni:HQuin1 dimer. (b) A close-up view of the Ni^{2+} coordination environment in the Ni:HQuin1 dimer. (c) A superimposition of Helix3 from the Ni:HQuin1 dimer (black) and the basic domain of the transcription factor GCN4 (light gray). Adapted with permission from Reference 38. Copyright 2010 American Chemical Society.

in minimal contact (~300 Å^2 buried surface) between their surfaces and is entirely enforced by Ni coordination to His-Quin HCMs in a distorted octahedral geometry (Fig. 6.5b). The observed bond metrics closely approximate those of Ni^{2+} complexes with free quinolate and amine-type ligands [45], suggesting that Ni coordination in the Ni:HQuin1$_2$ complex is free from steric constraints that may be imposed by the covalent attachment of the His-Quin HCM to the protein surface.

A goal of using metal coordination to direct protein self-assembly is to generate biologically functional structures in an easily predictable manner. This goal would permit the construction of biologically active structures with novel or expanded functionalities. The V-shaped architecture of Ni:HQuin1$_2$ closely approximates the DNA-binding domains of the bZIP-family transcription factors. The bZIP proteins contain a flexible basic domain that interacts with the DNA major groove and a helical leucine-zipper domain whose dimerization induces the preorganization of the basic domain for stable DNA binding [46]. DNA recognition by bZIP proteins has been shown to be sensitive to the dimeric orientation of the basic domains [47, 48]. A structural superposition of the Helix3 regions of the Ni:HQuin1$_2$ dimer with the DNA-bound basic domain of a representative bZIP protein reveals a very close match, with a root-mean-square deviation of 1.6 Å (Fig. 6.5c). This example demonstrates the potential utility of surface HCMs in directing the formation of protein structures that are poised to recognize biological targets without the need to engineer extensive protein surfaces.

6.2.4   Crystallographic Applications of Metal-Directed Protein Self-Assembly

Aside from providing access to discrete supramolecular architectures in solution, another advantage of MDPSA is that it facilitates protein crystallization. The primary roadblock in protein X-ray crystallography is the ability to grow diffraction-quality crystals. Since lattices are generally built from a combination of symmetry elements, the pre-organization or biasing of protein–protein interfaces to conform to a certain symmetry would narrow down the "search space" for molecular organization and thus increase the probability and speed of lattice formation. This has indeed been recently demonstrated by Yeates and colleagues, who adopted MDPSA to aid in protein crystallization [49]. Toward this end, they designed various T4 lysozyme (T4L) and maltose-binding protein (MBP) mutants to contain $i/i+3$ and $i/i+4$ bis-His and bis-Cys motifs on the surface. These T4L and MBP variants were co-crystallized with Cu^{2+}, Ni^{2+}, or Zn^{2+}, which resulted in 8 new crystal structures for T4L and 8 for MBP, representing in total 13 new and distinct crystal forms for these proteins. In line with these findings, over 50 different metal-mediated crystal forms and lattice-packing arrangements were observed with cyt cb_{562} variants in our lab. One of these variants, the so-called CFMC1, led to the formation of Zn-mediated tetrahedral cages in the crystal lattice, which were subsequently utilized to encapsulate and structurally characterize a heme-microperoxidase at atomic resolution for the first time as detailed below [50].

Structural characterization of reactive or conformationally fluxional species is hampered by their resistance to crystallization or their short lifetimes, which prevents their examination by diffraction methods or NMR spectroscopy. Time-resolved X-ray crystallography has provided detailed structural snapshots of short-lived reaction intermediates [51]; however, it is technically challenging and restricted to the study of photo-triggerable chemical events. Alternatively, unstable species have been isolated and stabilized through encapsulation in molecular hosts, which has subsequently allowed the determination of their structures by NMR methods or crystallography [52, 53]. These molecular hosts are assemblies of small organic building blocks stitched together by noncovalent or metal-mediated interactions, and have sizes that arc in the order of a few nanometers. Owing to their small sizes, these synthetic hosts allow extensive interactions with the encapsulated molecular guest(s) and the subsequent trapping of reactive species for prolonged periods, which otherwise would not persist in bulk solution [54–61]. By the same token, it should be possible to use the large and chemically rich interiors of protein cages to trap and characterized larger biological or nonbiological molecules. Such a cage, termed $Zn_{30}:CFMC1_{12}$, was constructed within a crystal lattice.

The structural evolution of a series of metal-directed protein assemblies that led to the discovery of $Zn_{30}:CFMC1_{12}$ is illustrated in Figure 6.6. The parent species, $Zn_4:MBPC1_4$ (Fig. 6.6a), is, as previously described, a D_2-symmetrical tetramer of cytochrome cb_{562} building blocks held together by four interfacial Zn^{2+} ions

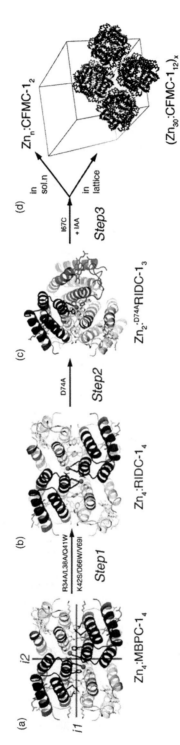

FIGURE 6.6 Progression from the parent Zn_4:$MBPC1_4$ complex (a) to the Zn_{30}:$CFMC1_{12}$ tetrahedral units (d) formed in the rhombohedral crystal lattice. Interfaces 1 (*i1*) and 2 (*i2*) in the Zn_4:$MBPC1_4$ tetramer are indicated on the former structure. Only four Zn_{30}:$CFMC1_{12}$ units are shown in (d) for clarity. IAA is used for the covalent functionalization of C67 to yield CFMC1. Reprinted with permission from Reference 50. Copyright 2010 Wiley-VCH Verlag GmbH & Co. KGaA.

coordinated to three histidines (63/73/77) and one aspartate (74) side chain [30]. In Step 1, six hydrophilic-to-hydrophobic mutations (R34A/L38A/Q41W/K42S/D66W/V69I) were introduced into one of the three twofold symmetric interfaces ($i1$) in Zn_4:$MBPC1_4$ to stabilize this interface [62] and produce the stabilized tetrameric complex Zn_4:$RIDC1_4$ (Fig. 6.6b). In Step 2, one of the Zn ligands (D74) in Zn_4:$RIDC1_4$ was eliminated through the D74A mutation to open up the Zn coordination sphere and split the tetrameric architecture into two, whereby each half contains the stabilized interface $i1$. This modification led to the formation of a trimeric architecture, Zn_2:$^{D74A}RIDC1_3$ (Fig. 6.6c), with two coordinatively saturated Zn ions. In Step 3, unfavorable, sterically demanding interactions were introduced into the protein surfaces in Zn_2:$^{D74A}RIDC1_3$ to prevent the third protein monomer from saturating the Zn coordination sphere, which was accomplished through the incorporation of a surface cysteine residue through the I67C mutation and its covalent modification with an iodoacetic acid (IAA) to yield CFMC1.

SV measurements indicated that CFMC1 exclusively produced a dimeric species in the presence of Zn in solution. Yet, the crystal structure of the Zn-CFMC1 adduct at 2.5-Å resolution revealed a cage-like arrangement of CFMC1 molecules in the lattice (Fig. 6.6d), which was exploited as a crystallographic host. The rhombohedral (*H*32) lattice contains four protein molecules per asymmetric unit, where chains A, B, and C are related by threefold noncrystallographic symmetry (NCS), and chains B and D are related by a twofold NCS. The crystallographic threefold symmetry operation produces a dodecamer (Zn_{30}:$CFMC1_{12}$) that has the shape of a truncated tetrahedron measuring approximately 80 Å at the edges, with an interior cavity width of 35 Å (Figs. 6.7a and 6.7b). Each dodecameric unit in the lattice is assembled through an extensive network of interfacial Zn ions, which are sufficiently strong to expose the engineered hydrophobic residues as well as the uncoordinated H73 side chain toward the inside of the cage cavity. It was envisioned that the resulting hydrophobic patches and H73 near the large threefold NCS pore could together provide a

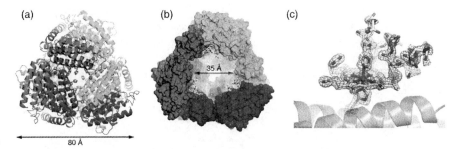

FIGURE 6.7 (a) Tetrahedral architecture of each Zn_{30}:$CFMC1_{12}$ unit as viewed down the crystallographic threefold symmetry axis. (b) Surface representation for the inner cavity of Zn_{30}:$CFMC1_{12}$ showing the three immobilized $MP9_{cb562}$ molecules (sticks). (c) Closeup view of the immobilized $MP9_{cb562}$ molecule with the corresponding $2F_o-F_c$ electron density map (1.1 σ). Adapted with permission from Reference 50. Copyright 2010 Wiley-VCH Verlag GmbH & Co. KGaA.

small microenvironment to immobilize a microperoxidase (MP) and determine its crystal structure. MPs are small proteolytic digest products of c-type cytochromes that contain the heme group attached to a conserved Cys-X-X-Cys-His motif, whereby the two Cys residues form covalent thioether bonds (c-type linkages) to the heme vinyl groups and the His side chain acts as an axial ligand to Fe. Owing to their small sizes and high solubilities, MPs have been extensively utilized as structural and chemical models for the inner-sphere coordination environments of heme proteins [63]. Yet, no full atomic-resolution structure for any MP is available due to the conformational flexibility of their peptide portion. In order to obtain an MP which could be used as a guest inside Zn_{30}:$CFMC1_{12}$, cytochrome cb_{562} was subjected to tryptic digestion, which yielded a nine-amino-acid-long peptide fragment ($MP9_{cb562}$; $K_{95}TTCNACHQ_{103}$) attached to the heme cofactor.

CFMC1 was co-crystallized with equimolar Zn and substoichiometric amounts of $MP9_{cb562}$ to ensure that the latter does not capture the histidines (other than H73) that are involved in interprotein Zn coordination. The resulting crystals maintained the original rhombohedral space group with similar unit cell dimensions indicating the formation of the Zn_{30}:$CFMC1_{12}$ cage. The 1.9-Å resolution structure clearly showed the presence of three symmetry-related $MP9_{cb562}$ molecules (one per asymmetric unit) within the cage (Fig. 6.7c). The quality of the data has the unambiguous placement of the heme group and all nine amino acids in the electron density. The structure of $MP9_{cb562}$ clearly indicated that the planarity of the heme group was not significantly perturbed by c-type linkages. It was previously suggested that the H-bonds between backbone carbonyl of the first Cys in the CXXCH motif and the amide nitrogens of the second Cys and the His residue may be important contributors to heme ruffling [64], which in turn may influence heme redox potentials [65]. Comparison of H-bonding interactions in $MP9_{cb562}$ with those from the corresponding fragments in the other c-type cytochromes indicated that there was little variance between the distances of the aforementioned H-bonds despite significant differences in the extent of heme ruffling.

Aside from the structural details of $MP9_{cb562}$, this study highlighted the potential of cage-like protein architectures in hosting large and flexible targets for structural interrogation. A particular advantage of the Zn_{30}:$CFMC1_{12}$ cage in this regard is that it is formed within an ordered lattice, which would enable the structural characterization of any guest molecule provided that they are properly immobilized within the cage cavity through careful surface design. It can also be envisioned how this strategy can be extended to natural protein cages like ferritins and virus capsids, which are readily crystallized [66, 67] and whose cavity surfaces can withstand significant modifications [68, 69] toward designing specific interactions with guest molecules.

6.3   METAL-TEMPLATED INTERFACE REDESIGN

6.3.1   Background

Despite having access to only a handful of metal-coordinating groups, the ability of proteins to control and harness the reactivity of metal centers is unmatched in

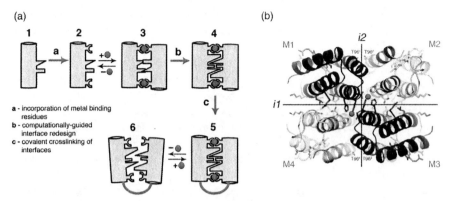

FIGURE 6.8 (a) A cartoon scheme for MeTIR using a monomeric protein as a starting building block. (b) Interprotomeric interfaces in the D_2-symmetric Zn_4:MBPC1$_4$ complex. Residues subjected to redesign as well as those that coordinate Zn^{2+} ions are shown as sticks. Reprinted with permission from Reference 72. Copyright 2010 American Chemical Society.

synthetic metal complexes. This is primarily due to the highly evolved noncovalent bonding networks of proteins, within which the coordination and solvation environments of metals can be exquisitely tuned. Because the surface of a monomeric protein such as cyt cb_{562} is not evolved to function as part of an assembly, the interfaces in the described protein assemblies above are not well packed and do not contain any chemical information aside from the metal-coordinating residues. Nevertheless, all of these interfaces bury large molecular surfaces with numerous amino acid side chain functionalities, which can be redesigned to introduce such chemical information relevant to the interfacial metal ions. This approach, based on the initial formation of metal-directed protein complexes, followed by the redesign of the interfacial interactions was termed Metal-Templated Interface Redesign (MeTIR). MeTIR is inspired by the early work of Busch and Sargeson [70, 71], in which metal-templating was exploited to construct macrocylic ligands with enforced coordination geometries that provided stable and specific metal binding. This strategy is illustrated in Figure 6.8a.

6.3.2 Construction of a Zn-Selective Tetrameric Protein Complex Through MeTIR

Because Zn_4:MBPC1$_4$ (Fig. 6.8b) possesses the largest of all protein interfaces ($\sim$5000 Å^2) described earlier in the Chapter, it was chosen as the initial system to demonstrate the feasibility of imparting metal specificity on a protein platform using MeTIR. Owing to its twofold dihedral symmetry, Zn_4:MBPC1$_4$ presents three pairs of C_2-symmetric interfaces ($i1$, $i2$, $i3$) between its four protomeric constituents. Of these interfaces, only $i1$ presents an extensive surface (>1000 Å^2) with close protein–protein contacts. Therefore, the computationally guided redesign of $i1$ was undertaken to examine if a favorable set of interactions can be built into $i1$ to stabilize the entire Zn-driven assembly (Step b in Fig. 6.8a). A construct, RIDC1, which features six mostly hydrophobic mutations (R34A/L38A/Q41W/K42S/D66W/V69I) in *i1*, was found to form a considerably stabilized tetrameric assembly (Zn$_4$:RIDC1$_4$, **4**) with an identical supramolecular geometry to the parent tetramer [62]. Despite this stabilization, Zn$_4$:RIDC1$_4$ (like Zn$_4$:MBPC1$_4$) remained a dynamically exchanging assembly that does not stay intact as a tetramer in solution. Consequently, it was not possible to uncouple protein oligomerization from Zn binding and directly assess whether interface redesign has led to improvements in Zn affinity and selectivity.

In order to obtain a stable tetrameric complex that would also form in the absence of Zn, RIDC1 was further engineered to include a disulfide bond across *i2* through the T96C mutation, which produced the construct ^{C96}RIDC1. Under oxidizing conditions ^{C96}RIDC1 readily forms an SS-crosslinked dimer. SV experiments indicate that the predominant oligomeric form of ^{C96}RIDC1 in solution is a tetramer (^{C96}RIDC1$_4$, Fig. 6.9a), even in the absence of metals. The tetramer–dimer dissociation constant ($K_{d(4mer–2mer)}$) for ^{C96}RIDC1$_4$ has been determined by SE measurements to be <100 nM, which makes it one of the most stable engineered protein complexes.

Upon addition of Zn to ^{C96}RIDC1$_4$, a shift in the sedimentation coefficient from 4.25 to 4.5 S was observed, which suggested that the complex remained tetrameric upon Zn binding, but underwent a large rearrangement. To elucidate this conformational change, crystal structures of ^{C96}RIDC1$_4$ and its Zn adduct were determined at 2.1 and 2.4-Å resolution [72]. These structures revealed a remarkable double-clothespin motion of the four protomers upon Zn coordination, measuring ∼16 Å at the N-terminus of Helix3 (Fig. 6.9). The simultaneous stability and conformational plasticity—hallmarks of many natural metalloproteins—displayed by ^{C96}RIDC1$_4$ is afforded by a combination of covalent and noncovalent interactions. The hydrophobic interactions built into *i1* are stable, yet flexible enough to adopt multiple conformations. At the same time, the covalent disulfide bonds incorporated into *i2* close up the protein macrocycle, while still allowing the protein monomers to undergo a significant translational and rotational motion relative to one another. The resulting architecture of Zn$_4$:^{C96}RIDC1$_4$ (Fig. 6.9a) is superposable onto the parent species, Zn$_4$:MBPC1$_4$, with a root-mean-square deviation of 0.59 Å over 424 C$_a$'s. Significantly, the coordination environment of the four Zn ions (formed by H63/H73/H77/D74) is maintained after the interfacial modifications, as intended by the "template-and-stabilize" approach.

With protein oligomerization now uncoupled from metal binding, it was examined if the templating strategy indeed provided increased Zn affinity and selectivity. Competitive metal-binding titrations demonstrated the expected binding of four Zn equivalents to ^{C96}RIDC1$_4$ with an average dissociation constant in the low nM regime [72]. Given that the parent *i*, *i+4* bis-His motifs on α-helices display Zn dissociation constants in the low mM range [29], the Zn-binding titrations suggest that the pre-formation of a tetrameric, templated acceptor complex results in a ≥1000-fold increase in Zn-binding affinity relative to the monomeric parent species, MBPC1.

To elucidate if ^{C96}RIDC1$_4$ also displays increased selectivity for Zn binding, its interactions with several other divalent metal ions (M^{2+}) were examined [72], including the neighboring Co^{2+}, Ni^{2+}, and Cu^{2+}, which typically are effective competitors

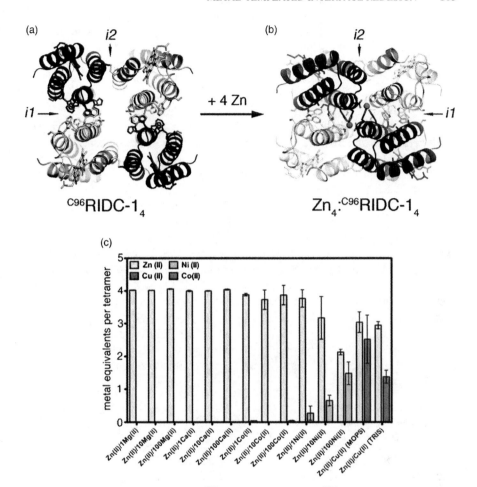

FIGURE 6.9 Crystal structures of ^{C96}RIDC1$_4$ (a) and Zn$_4$:^{C96}RIDC1$_4$ (b). Redesigned residues in $i1$ and $i2$ are shown as sticks. (c) Extent of divalent metal ion binding to ^{C96}RIDC1$_4$ in competition experiments. Adapted with permission from Reference 72. Copyright 2010 American Chemical Society.

for Zn-binding sites. These experiments reveal that ^{C96}RIDC1$_4$ displays significant Zn selectivity over all ions except Cu^{2+}, especially relative to its parent structure, MBPC1 (Fig. 6.9c). Previous studies have shown the affinity of the i, $i+4$ bis-His motif for Zn^{2+} to be comparable to that for Ni^{2+} and 5- to 10-fold higher than that for Co^{2+} [26, 29], in accordance with the Irving–Williams(IW) series [73]. In contrast, Zn^{2+} completely outcompetes Co^{2+} for ^{C96}RIDC1$_4$ binding at all ratios measured (up to 100 Co:1 Zn), and has an effective affinity roughly 100-fold higher than Ni^{2+}.

Due to a combination of its d^9 configuration and high Lewis acidity, Cu^{2+} is situated at the top of the Irving–Williams series, leading to its higher affinity for most ligand platforms designed for specific Zn binding [74, 75] and even natural Zn enzymes [76]. Initial results indicated that neither Cu^{2+} nor Zn^{2+} outcompeted each other for $^{C96}RIDC1_4$ binding. Rather, each $^{C96}RIDC1_4$ unit appeared to bind ~ 3 equivalents of each ion in the noncoordinating 3-(N-morpholino)propanesulfonic (MOPS) buffer solution. To examine whether this apparent oversaturation of $^{C96}RIDC1_4$ was due to the binding of Cu^{2+} or Zn^{2+} ions to the $^{C96}RIDC1_4$ surface, crystals of $^{C96}RIDC1_4$ were grown in the presence of equimolar amounts of both ions. The resulting 2.1-Å resolution diffraction data revealed a structure identical to that of $Zn_4{:}^{C96}RIDC1_4$. Anomalous difference maps calculated from data sets collected at the Zn and Cu K-absorption edges unambiguously indicated the presence of Zn and not Cu at core metal-binding sites [72].

To describe the Zn selectivity of $^{C96}RIDC1_4$ quantitatively, its affinity for Co^{2+}, Ni^{2+}, and Cu^{2+} was examined [72]. Titrations with these ions indicated that $^{C96}RIDC1_4$ possessed one weak Co^{2+} binding site, and that it could accommodate two equivalents of Ni^{2+} or Cu^{2+} with affinities that were either similar to those for Zn^{2+} (in the case of Ni^{2+}) or two to three orders of magnitude higher (in the case of Cu^{2+}) on a per site basis. However, the higher multiplicity for Zn^{2+} binding (4 equivalents) resulted in a considerably more favorable overall free energy (~ 190 kJ/mol for Zn^{2+} compared to 140 kJ/mol for Cu^{2+} and 90 kJ/mol for Ni^{2+}), and ultimately in the Zn selectivity of $^{C96}RIDC1_4$ over these ions. Thus, templated interface redesign led to increased bias not only toward Zn coordination geometry but also toward Zn binding multiplicity, which, to our knowledge, is a rare, and perhaps unique, case in designed/synthetic systems.

6.3.3   Construction of a Zn-Selective Protein Dimerization Motif Through MeTIR

Although $^{C96}RIDC1_4$ undergoes significant conformational changes upon metal binding, it still displays some structural order due in large part to the short disulfide bonds installed in $i2$. It was envisioned that if the interfacial covalent linkages are lengthened and rendered more flexible, this may in turn increase the flexibility of the assembly and result in the formation of alternative oligomeric forms while still retaining metal selectivity, which is the most basic but essential metal-based function (Fig. 6.10). An appropriate location to insert long covalent linkages into the tetrameric Zn-RIDC1 complex is across the third interface, $i3$; $i3$ is a minimal interface (490 Å^2 buried surface) formed entirely by the inner Zn coordination sphere and is not amenable to redesign or crosslinking through a disulfide bond. Instead, a long, flexible chemical linker may be incorporated across $i3$ into RIDC1, which could result in the formation of either a folded-over dimer (stabilized by the hydrophobic mutations in $i1$) (Fig. 6.10) or a D_2-symmetrical tetramer in the absence of Zn, in effect mimicking what nature accomplishes through genetically encoded peptide linkers for fusing proteins in order to realize the "chelate effect" [13, 77]. Toward this end, the symmetrically related pairs of Gly82 residues near the vertex of $i3$ were replaced with cysteines for chemical coupling through bis-maleimide functionalized crosslinkers [78].

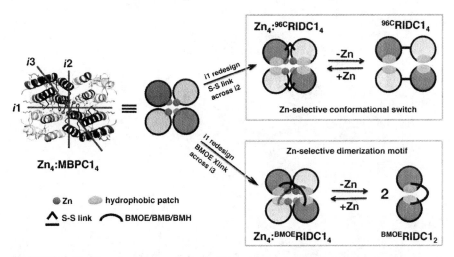

FIGURE 6.10 Templated construction of Zn-selective protein complexes (top pathway). The short and relatively rigid disulfide bonds across the $i2$ interface in combination with fluid; hydrophobic interactions in $i1$ yield a tetramer (^{96C}MBPC1$_4$) that undergoes a Zn-selective conformational change 1$_4$ (bottom pathway). In contrast, the combination of fluid $i1$ interactions with a flexible linker across $i3$ yields a Zn-selective dimerization motif. Reprinted with permission from Reference 78. Copyright 2011 American Chemical Society.

The crosslinkers used, bis-maleimidoethane (BMOE), bis-maleimidobutane (BMB), and bis-maleimidohexane (BMH), have spacer arm lengths of 8.0 Å, 10.9 Å, and 13.0 Å, respectively, and readily cover the anticipated ^{C82}S–$^{C82'}$S distance of 10–11 Å. It was first probed whether the crosslinked dimers (^{BMOE}RIDC1$_2$, ^{BMB}RIDC1$_2$, and ^{BMH}RIDC1$_2$) assemble into the same Zn-mediated D_2-symmetric tetrameric architecture that formed the basis of templated redesign. SV experiments showed that even at a concentration of 2.5 μM (lowest dimeric concentration measurable), ^{BMOE}RIDC1$_2$, ^{BMB}RIDC1$_2$, and ^{BMH}RIDC1$_2$ all were entirely tetrameric in the presence of Zn^{2+}, with a peak sedimentation coefficient of ~4.5 S, matching that of the parent complex, Zn$_4$:RIDC1$_4$ [62]. The crystal structures of the Zn adducts of ^{BMOE}RIDC1$_2$, ^{BMB}RIDC1$_2$, and ^{BMH}RIDC1$_2$ were determined at 2.3 Å, 1.8 Å, and 2.6 Å resolution, respectively. The supramolecular architectures of the resulting complexes were identical to one another (Fig. 6.11a), and to Zn$_4$:MBPC1$_4$ [30] and Zn$_4$:RIDC1$_4$ [62]. Importantly, the Zn-coordination environments, which lie in close proximity to the crosslinking sites, were unchanged as intended by the use of flexible linkers (Fig. 6.11b). The electron density for the entire length of the BMOE linker was evident, whereas for the longer BMB and BMH linkers only the Cys82-thioether-succinamide moiety could be discerned (Fig. 6.11c) [78].

The structures of ^{BMOE}RIDC1$_2$, ^{BMB}RIDC1$_2$, and ^{BMH}RIDC1$_2$ were then investigated in the absence of Zn. SV measurements indicated that dimers were the predominant oligomeric form under the conditions tested for all crosslinkers. At low proteins concentrations (2.5 μM), a small population of possibly tetrameric species

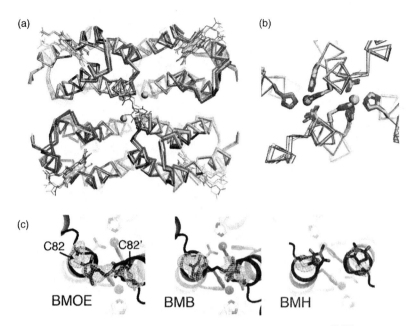

FIGURE 6.11 (a) Structural superposition of Zn_4:RIDC1$_4$, Zn_4:^{BMOE}RIDC1$_4$, Zn_4:^{BMB}RIDC1$_4$ and Zn_4:^{BMH}RIDC1$_4$. (b) Superposition of the Zn coordination environments. (c) F_o-F_c omit electron density maps (2.5 σ) of the BMOE, BMB, and BMH crosslinkers. Reprinted with permission from Reference 78. Copyright 2011 American Chemical Society.

(4.5–4.8 S) was observed (Fig. 6.12a), whereas at higher concentrations (> 25 μM), the SV distributions for such high-order aggregates became considerably broader and tailed into larger sedimentation coefficients that were suggestive of linear polymeric chains (Fig. 6.12b). These results suggested that the crosslinked dimers prefer to adopt the folded-over conformation rather than a pre-organized tetramer reminiscent of the Zn-induced structure.

Given that the crosslinked dimers in their folded form should have many metal-coordinating residues (His, Glu, and Asp) in close proximity and be relatively flexible, they could be expected to coordinate different transition metal ions. Indeed, Co^{2+}, Ni^{2+}, or Cu^{2+} binding to ^{BMOE}RIDC1$_2$ all induced the formation of a tetrameric species (Fig. 6.12c), with a maximal sedimentation coefficient of 4.5 identical to that of Zn_4:^{BMOE}RIDC1$_4$. However, since the redesign of $i1$ and the placement of crosslinkers across $i3$ were based on the four tetrahedral Zn templates, it was anticipated that there should be increased affinity for Zn^{2+} binding relative to the other divalent metal ions. This expectation was borne out by metal-binding titrations of ^{BMOE}RIDC1$_2$, which yielded half-saturation concentrations of 9 μM for Co^{2+}, 0.5 μM for Ni^{2+}, and 0.04 μM for Zn^{2+}. In accord with the high affinity of Cu^{2+} for protein-based donor atoms, the titrations showed that ^{BMOE}RIDC1$_4$ had at least 12 binding sites for Cu^{2+}, some likely on the surface and some in the core. The high plurality of Cu^{2+} binding to ^{BMOE}RIDC1$_4$ and the availability of surface-binding sites made

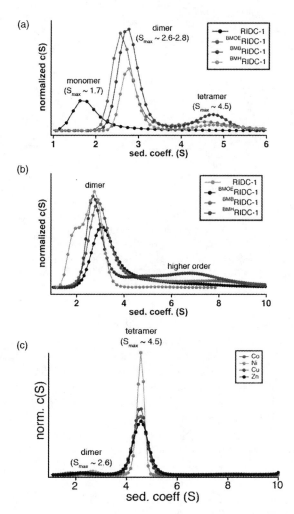

FIGURE 6.12 (a) SV profiles for various crosslinked (2.5 μM dimer) and a noncrosslinked construct (5 μM monomer) in the absence of metals. (b) Same as in (a), but at 25 μM and 50 μM concentrations, respectively, highlighting the presence of higher-ordered species formed even in the absence of any metal ions. (c) SV profiles for 2.5 μM ^{BMOE}RIDC1$_2$ dimer in the presence of 5 μM Co^{2+}, Ni^{2+}, Cu^{2+}, and Zn^{2+}, reflecting the formation of a tetrameric species on binding all four metal ions. Reprinted with permission from Reference 78. Copyright 2011 American Chemical Society.

it challenging to assess the Zn^{2+} vs. Cu^{2+} selectivity of the internal coordination sites through titrations. Similar to the case of ^{C96}RIDC1$_4$, ^{BMOE}RIDC1$_2$ was crystallized in the presence of a mixture of Zn^{2+} and Cu^{2+}. The structure of the resulting complex was determined at 2.6 Å resolution, which revealed that it possesses the same supramolecular architecture as Zn$_4$:^{BMOE}RIDC1$_4$, with four metals associated with the core and seven with the surface of the tetramer (Fig. 6.13). The anomalous

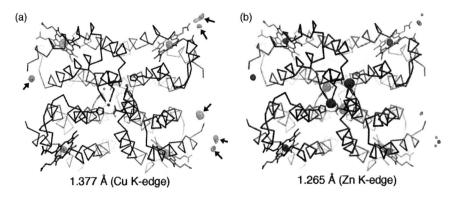

1.377 Å (Cu K-edge)   1.265 Å (Zn K-edge)

FIGURE 6.13 Structure of Zn_4:$^{BMOE}RIDC1_4$ grown in the presence of Cu^{2+}. A ribbon representation of Zn_4:$^{BMOE}RIDC1_4$ crystallized in the presence of $CuSO_4$ is shown along with the anomalous difference maps generated from data collected (a) at the Cu K-edge (5 σ) and (b) at the Zn K-edge (5 σ). The surface-associated Cu ions are shown with arrows in (a). Reprinted with permission from Reference 78. Copyright 2011 American Chemical Society.

difference maps computed using data collected at the Zn and Cu K-edges (1.265 and 1.377 Å) indicated the core ions to be Zn and the surface ions to be Cu. Again, the Zn selectivity was attributed not only to the formation of a covalently and noncovalently stabilized macrocycle built around the tetrahedral Zn coordination sites, but also to the fact that this macrocycle was programmed to house four Zn ions in its interior, all of which contribute to the overall stability of Zn_4:$^{BMOE}RIDC1_4$ formation.

Thus, both MeTIR-generated protein complexes described in this chapter ($^{C96}RIDC1_4$ and $^{BMOE}RIDC1_2$) simultaneously display plasticity and metal selectivity. The combination of these two traits is a hallmark of all cellular signaling systems, and particularly of those that couple selective metal binding to conformational changes or oligomerization to activate downstream processes, such as the prototypical EF-hand signaling proteins (calmodulin [79], S100 [80]) or any DNA-binding metalloregulatory protein [81].

6.4 SUMMARY AND PERSPECTIVES

Nature has been utilizing proteins as coordination platforms for billions of years, and chemists are increasingly exploiting the interiors of stable proteins and peptide assemblies for the same purpose. Yet, proteins had not been considered as building blocks for inorganic chemistry in the traditional sense. The approaches described in this Chapter (MDPSA and MeTIR) show that proteins can indeed be treated as such. With the proper placement of metal-binding side chain functionalities on the surface, proteins form discrete complexes dictated by the coordination preferences of the metal ions. They can be assembled into ordered metal-driven frameworks just like their small organic counterparts. They can be arranged into supramolecular architectures that display exquisite metal selectivity, which is one of the primary metal-based functions of proteins. The caveats in using proteins literally as coordination platforms are that they are large and possess significant chemical and structural heterogeneity. These features render tasks like creating coordinatively unsaturated interfacial metal sites with tailored chemical reactivities extremely challenging. Nevertheless, the potential payoff of creating entirely new metalloproteins and enzymes (perhaps through the pathways that evolution also took), coupled with how much we have learned about metal-directed protein assembly in so little time, is strong motivation to keep pursuing the strategies described in this Chapter.

ACKNOWLEDGMENTS

The author would like to express his gratitude to the many graduate and undergraduate workers who are responsible for the work described in this Chapter. The author would also like to acknowledge the National Science Foundation (CHE-0908115), the Department of Energy, Division of Materials Sciences, Office of Basic Energy Sciences (DE-FG02-10ER46677), the Sloan Foundation, and the Beckman Foundation for financial support.

REFERENCES

[1] Cvetkovic, A.; Menon, A. L.; Thorgersen, M. P.; Scott, J. W.; Poole, F. L.; Jenney, F. E.; Lancaster, W. A.; Praissman, J. L.; Shanmukh, S.; Vaccaro, B. J.; Trauger, S. A.; Kalisiak, E.; Apon, J. V.; Siuzdak, G.; Yannone, S. M.; Tainer, J. A.; Adams, M. W. W. *Nature* **2010**, *466*, 779–782.

[2] Hill, R. B.; Raleigh, D. P.; Lombardi, A.; Degrado, W. F. *Acc. Chem. Res.* **2000**, *33*, 745–754.

[3] Barker, P. D. *Curr. Opin. Struct. Biol.* **2003**, *13*, 490–499.

[4] Case, M. A.; McLendon, G. L. *Acc. Chem. Res.* **2004**, *37*, 754–762.

[5] Ghosh, D.; Pecoraro, V. L. *Curr. Opin. Chem. Biol.* **2005**, *9*, 97–103.

[6] Lu, Y.; Yeung, N.; Sieracki, N.; Marshall, N. M. *Nature* **2009**, *460*, 855–862.

[7] Wilson, M. E.; Whitesides, G. M. *J. Am. Chem. Soc.* **1978**, *100*, 306–307.

[8] Qi, D. F.; Tann, C. M.; Haring, D.; DiStefano, M. D. *Chem. Rev.* **2001**, *101*, 3081–3111.

[9] Yamaguchi, H.; Hirano, T.; Kiminami, H.; Taura, D.; Harada, A. *Org. Biomol. Chem.* **2006**, *4*, 3571–3573.

[10] Ward, T. R. *Acc. Chem. Res.* **2011**, *44*, 47–57.

[11] Reetz, M. T. *Angew. Chem. Int. Edit.* **2011**, *50*, 138–174.

[12] Blundell, T. L. *The Evolution of Metal-Binding Sites in Proteins*; Symposium Press London: University of Sussex, 1977.

[13] Armstrong, R. N. *Biochemistry* **2000**, *39*, 13625–13632.

[14] Liu, C. L.; Xu, H. B. *J. Inorg. Biochem.* **2002**, *88*, 77–86.

[15] Salgado, E. N.; Radford, R. J.; Tezcan, F. A. *Acc. Chem. Res.* **2010**, *43*, 661–672.

[16] Radford, R. J.; Brodin, J. D.; Salgado, E. N.; Tezcan, F. A. *Coord. Chem. Rev.* **2011**, *255*, 790–803.

[17] Kortemme, T.; Baker, D. *Curr. Opin. Chem.Biol.* **2004**, *8*, 91–97.

[18] Fletcher, S.; Hamilton, A. D. *J. R. Soc. Interface* **2006**, *3*, 215–233.

[19] Caulder, D. L.; Raymond, K. N. *Acc. Chem. Res.* **1999**, *32*, 975–982.

[20] Holliday, B. J.; Mirkin, C. A. *Angew. Chem. Int. Ed.* **2001**, *40*, 2022–2043.

[21] Leininger, S.; Olenyuk, B.; Stang, P. J. *Chem. Rev.* **2000**, *100*, 853–907.

[22] Faraone-Mennella, J.; Tezcan, F. A.; Gray, H. B.; Winkler, J. R. *Biochemistry* **2006**, *45*, 10504–10511.

[23] Janin, J.; Rodier, F. *Proteins* **1995**, *23*, 580–587.

[24] Lombardi, A.; Summa, C. M.; Geremia, S.; Randaccio, L.; Pavone, V.; DeGrado, W. F. *Proc. Natl. Acad. Sci. USA* **2000**, *97*, 6298–6305.

[25] Ruan, F. Q.; Chen, Y. Q.; Hopkins, P. B. *J. Am. Chem. Soc.* **1990**, *112*, 9403–9404.

[26] Ghadiri, M. R.; Choi, C. *J. Am. Chem. Soc.* **1990**, *112*, 1630–1632.

[27] Popp, B. V.; Ball, Z. T. *J. Am. Chem. Soc.*, *132*, 6660–6662.

[28] Todd, R. J.; Johnson, R. D.; Arnold, F. H. *J. Chromatogr. A* **1994**, *662*, 13–26.

[29] Krantz, B. A.; Sosnick, T. R. *Nat. Struct. Biol.* **2001**, *8*, 1042–1047.

[30] Salgado, E. N.; Faraone-Mennella, J.; Tezcan, F. A. *J. Am. Chem.Soc.* **2007**, *129*, 13374–13375.

[31] Salgado, E. N.; Lewis, R. A.; Mossin, S.; Rheingold, A. L.; Tezcan, F. A. *Inorg. Chem.* **2009**, *48*, 2726–2728.

[32] Lu, Y. *Curr. Opin. Chem. Biol.* **2005**, *9*, 118–126.

[33] Chen, C. H. B.; Milne, L.; Landgraf, R.; Perrin, D. M.; Sigman, D. S. *ChemBioChem* **2001**, *2*, 735–740.

[34] Ghadiri, M. R.; Soares, C.; Choi, C. *J. Am. Chem. Soc.* **1992**, *114*, 825–831.

[35] Cheng, R. P.; Fisher, S. L.; Imperiali, B. *J. Am. Chem. Soc.* **1996**, *118*, 11349–11356.

[36] Carey, J. R.; Ma, S. K.; Pfister, T. D.; Garner, D. K.; Kim, H. K.; Abramite, J. A.; Wang, Z. L.; Guo, Z. J.; Lu, Y. *J. Am. Chem. Soc.* **2004**, *126*, 10812–10813.

[37] Heinisch, T.; Ward, T. R. *Curr. Opin. Chem. Biol.* **2010**, *14*, 184–199.

[38] Radford, R. J.; Nguyen, P. C.; Ditri, T. B.; Figueroa, J. S.; Tezcan, F. A. *Inorg. Chem.* **2010**, *49*, 4362–4369.

[39] Radford, R. J.; Nguyen, P. C.; Tezcan, F. A. *Inorg. Chem.* **2010**, *49*, 7106–7115.

[40] Schafmeister, C. E.; Po, J.; Verdine, G. L. *J. Am. Chem. Soc.* **2000**, *122*, 5891–5892.

[41] Verdine, G. L.; Walensky, L. D. *Clin. Cancer Res.* **2007**, *13*, 7264–7270.

[42] Zhang, F. Z.; Sadovski, O.; Xin, S. J.; Woolley, G. A. *J. Am. Chem. Soc.* **2007**, *129*, 14154–14155.

[43] Klemm, J. D.; Schreiber, S. L.; Crabtree, G. R. *Ann. Rev. Immun.* **1998**, *16*, 569–592.

[44] Kohler, J. J.; Metallo, S. J.; Schneider, T. L.; Schepartz, A. *Proc. Natl. Acad. Sci. USA* **1999**, *96*, 11735–11739.

[45] Crispini, A.; Puccis, D.; Sessa, S.; Cataldi, A.; Napoli, A.; Valentini, A.; Ghedini, M. *New J. Chem.* **2003**, *27*, 1497–1503.

[46] Ellenberger, T. E.; Brandl, C. J.; Struhl, K.; Harrison, S. C. *Cell* **1992**, *71*, 1223–1237.

[47] Talanian, R. V.; McKnight, C. J.; Kim, P. S. *Science* **1990**, *249*, 769–771.

[48] Cuenoud, B.; Schepartz, A. *Science* **1993**, *259*, 510–513.

[49] Forse, G. J.; Ram, N.; Banatao, D. R.; Cascio, D.; Sawaya, M. R.; Klock, H. E.; Lesley, S. A.; Yeates, T. O. *Prot. Sci.* **2011**, *20*, 168–178.

[50] Ni, T. W.; Tezcan, F. A. *Angew. Chem. Int. Ed. Eng.* **2010**, *49*, 7014–7018.

[51] Moffat, K. *Chem. Rev.* **2001**, *101*, 1569–1582.

[52] (a) Yoshizawa, M.; Klosterman, J. K.; Fujita, M. *Angew. Chem. Int. Ed.* **2009**, *121*, 3470–3490. (b) *Angew. Chem. Int. Ed.* **2009**, *48*, 3418–3438.

[53] (a) Schmuck, C. *Angew. Chem.* **2007**, *119*, 5932–5935. (b) *Angew. Chem. Int. Ed.* **2007**, *46*, 5830–5833.

[54] (a) Cram, D. J.; Tanner, M. E.; Thomas, R. *Angew. Chem.* **1991**, *103*, 1048–1051. (b) *Angew. Chem. Int. Ed.* **1991**, *30*, 1024–1027.

[55] (a) Warmuth, R.; Marvel, M. A. *Angew. Chem.* **2000**, *112*, 1168–1170. (b) *Angew. Chem. Int. Ed.* **2000**, *39*, 1117–1119.

[56] Ajami, D.; Rebek, J. *Nat. Chem.* **2009**, *1*, 87–90.

[57] (a) Fiedler, D.; Bergman, R. B.; Raymond, K. N. *Angew. Chem.* **2006**, *118*, 759–762. (b) *Angew. Chem. Int. Ed.* **2006**, *45*, 745–748.

[58] Kawano, M.; Kobayashi, Y.; Ozeki, T.; Fujita, M. *J. Am. Chem. Soc.* **2006**, *128*, 6558–6559.

[59] (a) Tashiro, S.; Tominaga, M.; Yamaguchi, Y.; Kato, K.; Fujita, M. *Angew. Chem.* **2006**, *118*, 247–250. (b) *Angew. Chem. Int. Ed.* **2006**, *45*, 241–244.

[60] Tashiro, S.; Kobayashi, M.; Fujita, M. *J. Am. Chem. Soc.* **2006**, *128*, 9280–9281.

[61] (a) Hatakeyama, Y.; Sawada, T.; Kawano, M.; Fujita, M. *Angew. Chem.* **2009**, *121*, 8851–8854. (b) *Angew. Chem. Int. Ed. Eng.* **2009**, 48, 8695–8698.

[62] Salgado, E. N.; Ambroggio, X. I.; Brodin, J. D.; Lewis, R. A.; Kuhlman, B.; Tezcan, F. A. *Proc. Natl. Acad. Sci. USA* **2010**, *107*, 1827–1832.

[63] Adams, P. A.; Baldwin, D. A.; Marques, H. M. In The Hemepeptides from Cytochrome c: Preparation, Physical and Chemical Properties, and Their Use as Model Compounds for the Hemoproteins. *Cytochrome c - A Multidisciplinary Approach*; Scott, R. A., Mauk, A. G., Eds.; University Science Books: Sausalito, 1996; pp 635–692.

[64] Ma, J. G.; Laberge, M.; Song, X. Z.; Jentzen, W.; Jia, S. L.; Zhang, J.; Vanderkooi, J. M.; Shelnutt, J. A. *Biochemistry* **1998**, *37*, 5118–5128.

[65] Bowman, S. E. J.; Bren, K. L. *Nat. Prod. Rep.* **2008**, *25*, 1118–1130.

[66] Lawson, D. M.; Artymiuk, P. J.; Yewdall, S. J.; Smith, J. M. A.; Livingstone, J. C.; Treffry, A.; Luzzago, A.; Levi, S.; Arosio, P.; Cesareni, G.; Thomas, C. D.; Shaw, W. V.; Harrison, P. M. *Nature* **1991**, *349*, 541–544.

[67] Golmohammadi, R.; Valegard, K.; Fridborg, K.; Liljas, L. *J. Mol. Biol.* **1993**, *234*, 620–639.

[68] Swift, J.; Wehbi, W. A.; Kelly, B. D.; Stowell, X. F.; Saven, J. G.; Dmochowski, I. J. *J. Am. Chem. Soc.* **2006**, *128*, 6611–6619.

[69] Uchida, M.; Klem, M. T.; Allen, M.; Suci, P.; Flenniken, M.; Gillitzer, E.; Varpness, Z.; Liepold, L. O.; Young, M.; Douglas, T. *Adv. Mater.* **2007**, *19*, 1025–1042.

[70] Thompson, M. C.; Busch, D. H. *J. Am. Chem. Soc.* **1964**, *86*, 3651–3656.

[71] Creaser, I. I.; Geue, R. J.; Harrowfield, J. M.; Herlt, A. J.; Sargeson, A. M.; Snow, M. R.; Springborg, J. *J. Am. Chem. Soc.* **1982**, *104*, 6016–6025.

[72] Brodin, J. D.; Medina-Morales, A.; Ni, T.; Salgado, E. N.; Ambroggio, X. I.; Tezcan, F. A. *J. Am. Chem. Soc.* **2010**, *132*, 8610–8617.

[73] Frausto da Silva, J. J. R.; Williams, R. J. P. *The Biological Chemistry of the Elements*; Oxford University Press: Oxford, 2001.

[74] Nolan, E. M.; Ryu, J. W.; Jaworski, J.; Feazell, R. P.; Sheng, M.; Lippard, S. J. *J. Am. Chem. Soc.* **2006**, *128*, 15517–15528.

[75] Walkup, G. K.; Imperiali, B. *J. Am. Chem. Soc.* **1997**, *119*, 3443–3450.

[76] Hunt, J. A.; Ahmed, M.; Fierke, C. A. *Biochemistry* **1999**, *38*, 9054–9062.

[77] Goodsell, D. S.; Olson, A. J. *Annu. Rev. Biophys. Struct.* **2000**, *29*, 105–153.

[78] Salgado, E. N.; Brodin, J. D.; To, M. M.; Tezcan, F. A. *Inorg. Chem.* **2011**, *50*, 6323–6329.

[79] Nelson, M. R.; Chazin, W. J. *Prot. Sci.* **1998**, *7*, 270–282.

[80] Fritz, G.; Botelho, H. M.; Morozova-Roche, L. A.; Gomes, C. M. *FEBS J.* **2010**, *277*, 4578–4590.

[81] Reyes-Caballero, H.; Campanello, G. C.; Giedroc, D. P. *Biophys. J.* **2011**, *156*, 103–114.

7

CATALYTIC REACTIONS PROMOTED IN PROTEIN ASSEMBLY CAGES

Takafumi Ueno and Satoshi Abe

7.1 INTRODUCTION

The coordination chemistry of proteins continues to influence applications to materials science and nanotechnology, and fundamental research in understanding various metal-promoted biological events. Nowadays, the structures and coordination characteristics of metal centers in native proteins can be directly investigated as a result of the rapid advances in biophysical and spectroscopic techniques [1, 2]. Chemists and materials scientists are able to use a wide variety of proteins as organic ligands or template molecules for controlling the properties of metal-coordination compounds [3–6]. The number of reports on metal complex/protein or nanoparticle (NP)/protein composites has rapidly increased since the preparation of silk fibril protein with Pd-NPs by Akabori [7].

The chemistry of protein cages has recently gained a lot of attention with respect to their use as nanoreactors, regardless of their original functions, because of their unique properties such as self-assembly of large numbers of protein monomers, precise three-dimensional structures in the size range of several to hundreds of nanometers, and highly monodispersed size distributions [8, 9]. The reason for this interest is that it is still difficult to reconstruct such assemblies using synthetic molecules and polymers. From the bioinorganic chemistry point of view, the preparation, yields, and stabilities of these protein cages make them suitable as molecular templates for accumulating metal ions, complexes, and NPs, not only on their outer surfaces but also in their discrete internal spaces. The coordination characters of these cages can be precisely

Coordination Chemistry in Protein Cages: Principles, Design, and Applications, First Edition.
Edited by Takafumi Ueno and Yoshihito Watanabe.
© 2013 John Wiley & Sons, Inc. Published 2013 by John Wiley & Sons, Inc.

changed using the techniques described below. Protein cages can, therefore, provide unique coordination environments unattainable using other biomolecules.

7.1.1 Incorporation of Metal Compounds

As shown in Figure 7.1, several typical protein cages have been used for reactions of inorganic materials and metal complexes [8, 10–19]. Composites of proteins and metals have been prepared using various elegant methods, namely (1) the introduction of peptide fragments [20–25], (2) chemical conjugation of metal complexes [26–28], (3) association–dissociation and swelling reactions of protein assembly cages [29–37], and (4) the use of the original metal-storage properties of native protein cages [8, 17, 38–41], and combinations of these [42–44]. Method 1 is the most common technique because it is useful for retaining protein cages containing metals by preventing denaturation caused by random binding of metal ions to proteins. In general, His-tags, Cys-tags, negatively charged peptide fragments, or selected peptide fragments with specific affinities toward inorganic materials displayed on/in protein

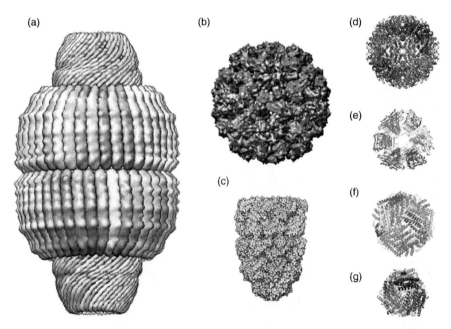

(a)   (b)   (d)   (e)   (c)   (f)   (g)

FIGURE 7.1 Structure images of typical protein cages reconstructed with X-ray structure analyses for vault, 67 nm height and 40 nm diameter (a) [10], cowpea chlorotic mottle virus, 28 nm diameter (b) [11], GroEL, 20 nm height and 14 nm diameter (c) [12], Lumazine synthase, 15 nm diameter (d) [13], a small heat-shock protein, 12 nm diameter (e) [14], ferritin, 12 nm diameter (f) [15], and Dps, 8.5 nm diameter (g) [16]. Reproduced with permission from RCSB PDB (for a and c) and VIPERdb (for b, at http://viperdb.scripps.edu/acknowledge.php, PDB ID:1CWP).

cages are changed or added to the original sequences. Method 2 is used to introduce metal complexes with organic ligands. This enables us to display such ligands in controlled numbers and at specific locations. It can be applied to protein cages that have other functions besides the preparation of inorganic materials. Other metal complexes and metal NPs, which do not have specific coordination sites on protein cages, are incorporated using the third method. Interestingly, an intact metalloprotein, horseradish peroxidase, has been encapsulated in a virus-like particle by this method [37]. The fourth method has been used for a well-known and intensively studied protein cage, ferritin (Fr), and the related protein cage Dps (DNA-binding proteins from starved cells), and depositions of unnatural metal oxides in Fr cages have been reported [38, 41, 45]. Various metal ions are accumulated within the Fr cage and then these metals can form NPs, as Fe ions are converted into Fe oxides within Fr.

7.1.2 Insight into Accumulation Process of Metal Compounds

As described above, various methods for the synthesis of nano-biomaterials using protein cages have been reported. The details of the mechanisms of these reactions are not well understood because they are complicated reactions involving different coordinations and interactions of individual metal ions co-existing in a single protein cage. Single-particle analysis of high angle annular dark field scanning transmission electron microscopy images of Fr-containing Fe ions shows that several Fe-oxide NPs are formed and aggregate to generate a single NP in the protein cage [46]. X-ray absorption near edge structure and small angle x-ray scattering experiments on Fr indicated that the NPs formed in apo-Fr consisted of polyphasic structures of Fe oxide [47]. The intermediate process from Fe ion accumulation to Fe-oxide NP formation was monitored using electrospray ionization time-of-flight mass spectrometry [45].

The initial steps of metal ion accumulation in protein cages remain obscure because of the challenges and difficulties in obtaining high-resolution structural data to explain the coordination process of each metal ion deposited on a single protein cage during the reaction steps [15, 48–61]. Although single-crystal X-ray structural analysis is a very useful tool for elucidating these reactions, the preparation of model composites has not yet been achieved. Recently, metal-substituted Frs have been used for observing reactions with the crystal structure of Fr involving large numbers of metal ions [62–65].

In this chapter, we focus on the recent progress in catalytic reactions of metal complexes deposited in Fr, and insights into the accumulation mechanism, leading to coordination design of the discrete interiors of Frs. We then discuss coordination design in protein cages and future perspectives.

7.2 FERRITIN AS A PLATFORM FOR COORDINATION CHEMISTRY

Fr is an Fe-storage protein and comprises 24 subunits that assemble to form a hollow cage structure with an outer diameter of 12 nm, as shown in Figure 7.2 [53, 66].

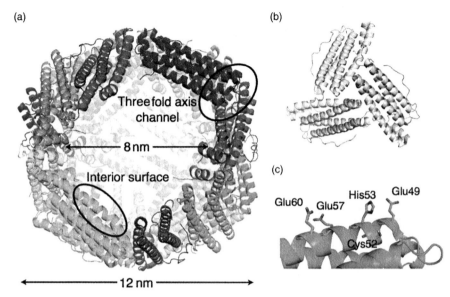

FIGURE 7.2 Whole structure of apo-Fr (a), the threefold axis channel (b), and the interior surface (c). The apo-Fr structure is taken from PDB ID 1AEW [15].

Fe atoms are stored as a cluster of ferric oxyhydroxide within a cavity of inner diameter 8 nm formed by protein subunits, as described in Chapter 1 (Fig. 7.2a). The cage remains intact over a wide range of pH values (2–11) and temperatures (< *ca.* 80°C). A procedure for trapping small molecules inside the interior of apo-Fr has been reported [67]. The procedure consists of the dissociation of apo-Fr into subunits at pH 2, followed by reforming at pH 7. The loaded apo-Fr is then separated from the excess molecules that are not trapped inside the cage by dialysis or size-exclusion column chromatography. This function has been used for trapping metal complexes, such as magnetic resonance imaging reagents or metal–drug systems [32, 33]. Another important property of this protein cage is the perforation of the protein shell by small channels located at the junctions of three subunits (Fig. 7.2b). The channels are required for the transport of metal ions, that is, small metal complexes, and then they finally interact with coordinating amino acid residues exposed on the interior surface (Fig. 7.2c) [68]. These unique properties have been used for the preparation of various metal NPs, such as pure metals, metal oxides, and quantum dots, and manganese and uranyl oxides have been synthesized in Fr cages [8, 17, 41]. The penetration of organic molecules has been confirmed by size-selective insertion into the holo-Fr interior through such channels [69, 70]. These results imply that holo-Fr could accommodate foreign organic compounds in its cavities, even in the presence of metal NPs. We have used these unique properties of Fr to demonstrate catalytic reactions for organic synthesis in the Fr cage based on the co-existence of catalytic NPs and organic substrates, as described below.

7.3   CATALYTIC REACTIONS IN FERRITIN

7.3.1   Olefin Hydrogenation

Hydrogenation is one of the most important reactions and is used in the production of numerous pharmaceuticals and other chemicals [71, 72]. This reaction has been one of the target reactions for developing heterogeneous and homogeneous biocatalysts since Akabori's report in 1956 that metallic Pd drawn on silk fibril protein catalyzes the asymmetric (heterogeneous) hydrogenation of oximes and oxazolones [7]. Whitesides later reported the same asymmetric hydrogenation reactions using avidin/biotin conjugated with an Rh complex, as described in Chapter 8 [73]. Although apo-Fr is one of the most attractive protein cages for nanoreactors, providing unique functions and versatile molecular architectures for the preparation of inorganic NPs, difficulties in using composite materials for catalytic reactions remain. This first motivated us to synthesize a zerovalent Pd-NP *in situ* by chemical reduction of Pd ions accumulated in an apo-Fr cage, and then promoting the hydrogenation reaction using both the NP and the interior space (Fig. 7.3) [40]. A composite of Pd-NP and apo-horse spleen Fr (**Pd•apo-Fr**) was prepared by adding 500 equivalent of K_2PdCl_4 to a buffered solution of apo-Fr, and then adding $NaBH_4$ to the mixture. The size of the Pd-NPs synthesized in apo-Fr was almost monodispersed (2.0 ± 0.3 nm) because the reaction proceeded in the size-restricted apo-Fr interior space, as described in Chapter 7.4. More importantly, the composite can catalyze olefin hydrogenation of acrylamide (**1**) with a high turnover frequency (TOF) of 33,000/**Pd•apo-Fr**/h under 1 atm of H_2 at $7°C$. This rate is almost comparable to that of a dendrimer-encapsulated Pd catalyst in water [74]. As shown in Table 7.1, the TOF for acrylic acid (**2**) catalyzed by **Pd•apo-Fr** was eight times smaller than that of (**3**), although the TOF of (**2**) in the absence of apo-Fr is almost identical to that of (**3**). The low TOF of (**2**) with **Pd•apo-Fr** is expected to be caused by charge repulsion during penetration through threefold axis

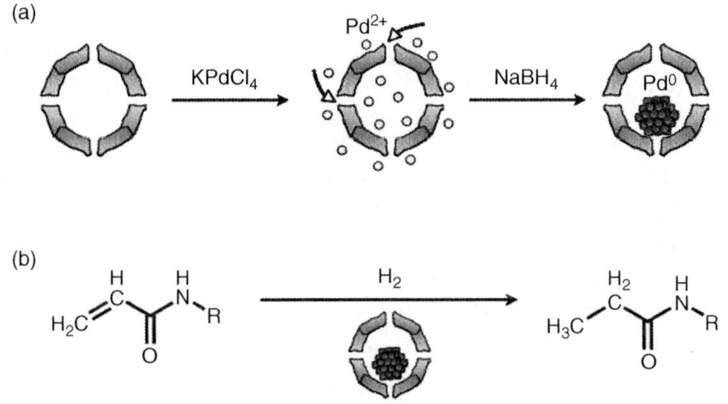

FIGURE 7.3   Schematic drawings of preparation of **Pd•apo-Fr** (a) and olefin hydrogenation by **Pd•apo-Fr** (b).

TABLE 7.1   Hydrogenation Activity of Pd•apo-ferritin

Olefins	TOF for **Pd•apo-Fr**[a,b]	TOF for **Pd-NP**[b]
$CH_2=CHCONH_2$ (**1**)	72 ± 0.7	58 ± 5.9
$CH_2=CHCOOH$ (**2**)	6.3 ± 1.1	12 ± 2.6
$CH_2=CHCONH\text{-}iPr$ (**3**)	51 ± 6.5	15 ± 0.3
$CH_2=CHCONH\text{-}tBu$ (**4**)	31 ± 5.9	6.1 ± 0.6
$CH_2=CHCO\text{-}Gly\text{-}OMe$ (**5**)	6.3 ± 3.8	28 ± 2.6
$CH_2=CHCO\text{-}D, L\text{-}Ala\text{-}OMe$ (**6**)	Not detected	23 ± 0.3

[a]Hydrogenation reactions catalyzed by **Pd•apo-Fr** were carried out at $7°C$ (pH 7.5) with 30 μm of Pd.
[b]TOFs were calculated on the basis of the ratio of the product to substrate by analyzing the ^1H NMR spectra. Adapted from Reference 40.

channels consisting of negatively charged Asp and Glu residues. Moreover, when the channels in **Pd•apo-Fr** (Fig. 7.2b) were blocked with Tb^{3+} [70, 75], the TOF of (**1**) was decreased to less than 10% of the TOF of (**1**) for **Pd•apo-Fr** without Tb^{3+}. These results suggest that the anionic threefold axis channels are the substrate pathways, as reported for 4-amino-TEMPO (TEMPO: 2,2,6,6-tetramethylpiperidine-N-oxyl) [70]. In comparison with those of the neutral substrates, the TOFs of (**5**) and (**6**) with **Pd•apo-Fr** are smaller than those of (**1**), (**3**), and (**4**), although the electron-deficient alkenes (**1**), (**5**), and (**6**) are expected to be reduced faster than (**3**) and (**4**), if the electronic substituent effect on the hydrogenation is considered. The substrate size is, therefore, discriminated by the threefold axis channels. This is the first example of a catalytic reaction for organic synthesis promoted by a heterogeneous catalyst included in the discrete space of a protein cage.

Next, further catalytic design of the Fr cage was demonstrated by the formation of Pd–Au bimetallic NPs (Pd/Au-NPs) [76]. We prepared apo-rHLFr (recombinant L-chain apo-Fr from horse liver) containing Pd^{2+} or Au^{3+} ions by treatment of apo-rHLFr with 100 equivalent of K_2PdCl_4 or $KAuCl_3$, respectively. The crystal structures of **Pd^{2+}•apo-rHLFr** and **Au^{3+}•apo-rHLFr** were determined using 2.10 and 1.65 Å diffraction data, respectively (Fig. 7.4). The anomalous difference Fourier maps of **Pd^{2+}•apo-rHLFr** indicate the existence of 192 Pd^{2+}-binding sites (Fig. 7.4a). They are divided into two main regions, which are located in the threefold axis channel and an accumulation center (Fig. 7.4c). In the threefold axis channel, three Pd^{2+} ions remain coordinated with His114 and Cys126, which are highly conserved in many kinds of Fr (Fig. 7.4e) [15]. Five Pd^{2+}-binding sites are found in the accumulation center (Fig. 7.4g). The crystal structure of **Au^{3+}•apo-rHLFr** shows that it has two binding regions, similar to those of **Pd^{2+}•apo-rHLFr** (Fig. 7.4). Only three electron densities of Au^{3+} ions were observed at the accumulation center, although a larger number of Pd^{2+} ions were found there, because of the different binding characters of Pd^{2+} and Au^{3+} ions to the carbonyl groups of Asp and Glu residues [77, 78]. The number of Au^{3+} ions deposited at the accumulation site should, therefore, be smaller than the number of Pd^{2+} ions. The results suggest that Pd^{2+} ions are able to bind to the interior surface of apo-rHLFr even after the deposition of Au^{3+} ions in apo-rHLFr.

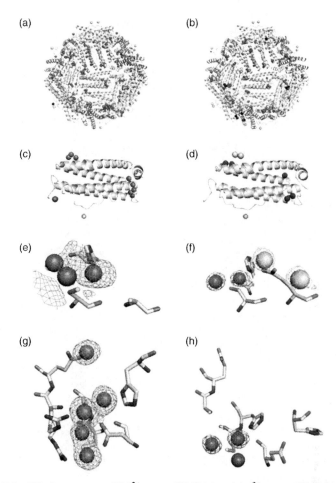

FIGURE 7.4 Whole structure of **Pd^{2+}•apo-rHLF** (a) and **Au^{3+}•apo-rHLF** (b). Monomer unit structures of **Pd^{2+}•apo-rHLF** (c) and **Au^{3+}•apo-rHLF** (d). The threefold channel structures of **Pd^{2+}•apo-rHLF** (e) and **Au^{3+}•apo-rHLF** (f). The accumulation center structures of **Pd^{2+}•apo-rHLF** (g) and **Au^{3+}•apo-rHLF** (h). The Pd, Au, and Cd atoms are indicated as spheres colored dark and light gray, respectively in (a)–(h). Anomalous difference Fourier maps at 4.0 σ are shown as meshes. Adapted from References 63, 76. Reproduced from Reference 76 with permission from The Royal Society of Chemistry.

The different coordination characters of Au^{3+} and Pd^{2+} ions in apo-rHLFr enable us to prepare two types, core–shell and alloy, of Pd/Au-NPs, as shown in Figure 7.5. The alloy Pd/Au-NPs is prepared by co-reduction of a mixture of Au^{3+} and Pd^{2+} ions in apo-rHLFr (Fig. 7.5a). First, Au^{3+} ions are reacted with apo-rHLFr to deposit them at their specific binding sites in the cage, and then Pd^{2+} ions are loaded into the cage to be bound to the binding sites remaining empty after the pretreatment with Au^{3+} ions. Reduction of the mixture of the ions produces Pd/Au-NPs in an alloy

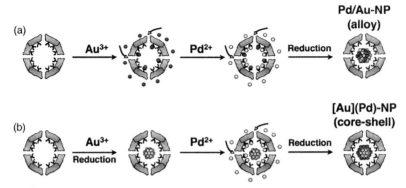

FIGURE 7.5 Schematic drawings of preparation methods of Au/Pd NPs in apo-rHLFr with alloy (a) and core–shell (b) structures. Reproduced from Reference 76 with permission from The Royal Chemical Society.

form. Core–shell NPs are prepared using a sequential reduction method, as shown in Figure 7.5b. In this case, the Au core is first prepared as a monometallic Au-NP in apo-rHLFr, followed by the introduction and reduction of Pd^{2+} ions to form the shell. The sequential reduction method is also used to prepare Pd-core/Au-shell ([Pd](Au)) NPs in apo-rHLFr. All the Pd/Au-NPs were characterized as single-crystal particles without formation of multiple cores. When the hydrogenation of acrylamide was examined using these composites, the TOF value of (**[Au](Pd)-NP)•apo-rHLFr** was 2.5- and 4.2-fold higher than those of **Pd•apo-rHLFr** and (**AuPd-NP)•apo-rHLFr**, respectively, whereas (**[Pd](Au)-NP)•apo-rHLFr** showed no catalytic activity for this reaction. A similar enhancement of the catalytic activity of Pd-NPs by a core(Au)/shell(Pd) bimetallic system has been reported for composites with synthetic polymers or dendrimers [79, 80]. These results, therefore, suggest that [Au](Pd)-NPs can maintain their catalytic reactivity even in apo-rHLFr. Thus, the apo-Fr cage can be used for the preparation of various metal NPs, and acts as a catalytic nanoreactor.

7.3.2 Suzuki–Miyaura Coupling Reaction in Protein Cages

The Suzuki–Miyaura coupling reaction is one of the most important carbon–carbon cross-coupling reactions catalyzed by organometallic complexes. Water-soluble ligands or polymer-bound catalysts have been developed for the use in aqueous-phase organometallic catalysis [71, 72]. Protein cages satisfy the requirements for use in such reactions because the metal complexes accumulating in the cages can promote catalytic reactions in aqueous solutions under ordinary conditions. Moreover, the numbers and reactivities of the complexes can be controlled by altering the coordination environment in the cages. To test the possibility of using protein cages in organometallic catalysis, a Pd(allyl) (allyl = η^3-C_3H_5) complex accumulated in apo-rHLFr was used as a catalyst [81].

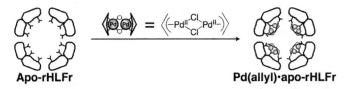

Apo-rHLFr   **Pd(allyl)·apo-rHLFr**

FIGURE 7.6   Schematic drawing of preparation of **Pd(allyl)•apo-rHLFr**. Reproduced from Reference 81 with permission from the American Chemical Society.

Apo-rHLFr-containing Pd(allyl) complexes (**Pd(allyl)•apo-rHLFr**) were constructed by treatment with 100 equivalent of $[Pd(allyl)Cl]_2$ in 50 mM Tris/HCl (pH 8.0) (Fig. 7.6). The crystal structure shows that the binding sites of Pd(allyl) were divided into two regions, that is, the threefold axis channel and the accumulation center, as observed for **Pd^{2+}•apo-rHLFr** (Figs. 7.4 and 7.7) [63]. The Pd(allyl) complexes maintain the same dinuclear structure in each binding region. In the threefold axis channel, Cys126 binds to two Pd(allyl) complexes as a bridging ligand. His114 and a water molecule coordinate with different Pd complexes to retain the dinuclear structure (Fig. 7.7a). The same structure is formed with Cys48, His49, and Glu45, instead of a water molecule, at the accumulation center (Fig. 7.7b). To alter the coordination structure of the Pd(allyl) complexes in the apo-rHLFr cage, two mutants, in which His114 or His49 was replaced by Ala, and their composites with the Pd complexes were prepared. The crystal structure of **Pd(allyl)•apo-H49A-rHLFr** shows that the number of Pd atoms in the cage is 96, which is identical to the value for **Pd(allyl)•apo-rHLFr**. The coordination structures of the dinuclear Pd complex at the accumulation site are exactly conserved because each $O\varepsilon1$ and $O\varepsilon2$ atom of the carboxylate moiety of Glu45 is separately bound to two Pd atoms as a bidentate ligand (Fig. 7.7d). In contrast, **Pd(allyl)•apo-H114A-rHLFr** has thiolato-bridged trinuclear Pd complexes at the threefold axis channel by the deletion of His114 (Fig. 7.7e). The PdB1–PdB1$'$ distance (3.24 Å) indicates that the complexes have no direct bonding interactions among the Pd atoms [82, 83]. The coordination geometry of each Pd atom is a typical square planar structure with an allyl ligand and two Sγ atoms at the threefold axis channel with a six-membered ring structure as a Pd trinuclear complex [84].

A Suzuki–Miyaura coupling reaction of 4-iodoaniline and phenylboronic acid to afford 4-phenylaniline was examined to evaluate the catalytic activities of these composites [85]. The TOFs ([product (mol)] per **Pd(allyl)•apo-rHLFrs** per h) of the coupling reactions were determined by ^{1}H-NMR, based on the consumption of 4-iodoaniline and formation of the product. The activity of **Pd(allyl)•apo-rHLFr** (TOF = 3500) is almost identical to that of **Pd(allyl)•apo-H49A-rHLAFr** (3400). This is a result of the structural conservation of dinuclear $[Pd(allyl)]_2$ with bridging ligation of Asp45, even after the deletion of His49. The activity of **Pd(allyl)•apo-C48A/H49A-rHLFr** (1900) is about half that of **Pd(allyl)•apo-rHLFr** because of the lack of Pd(allyl)-binding sites at the accumulation center. These results suggest that the Pd dinuclear complexes in both the binding areas have similar activities for the catalytic reaction. In contrast, **Pd(allyl)•apo-H114A-rHLFr** (900) and

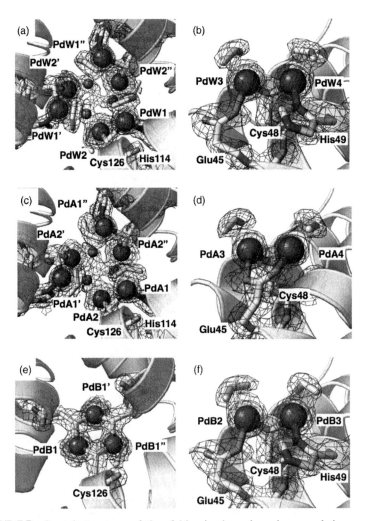

FIGURE 7.7 Crystal structures of threefold axis channels and accumulation centers of **Pd(allyl)•apo-rHLFr** (a and b), **Pd(allyl)•apo-H49A-rHLFr** (c and d), and **Pd(allyl)•apo-H114A-rHLFr** (e and f), respectively. The anomalous difference Fourier maps at 4.0 σ indicate the positions of palladium atoms shown as spherical models. The selected 2|F_o|−|F_c| electron density maps are shown at 1.0 σ. Reproduced from Reference 81 with permission from the American Chemical Society.

Pd(allyl)•apo-C126A-rHLFr (830) show about fourfold lower activity than that of **Pd(allyl)•apo-rHLFr**. The decrease in the activity of **Pd(allyl)•apo-H114A-rHLFr** suggested that the trinuclear Pd cluster at the threefold axis channel interferes with the penetration of substrates or that the geometry of the Pd complexes at the threefold axis channel is different from that of **Pd(allyl)•apo-rHLFr**. At the Pd(allyl)-binding sites of **Pd(allyl)•apo-C126A-rHLFr**, there are no Pd complexes at the threefold axis channel, although the Pd(allyl) dinuclear center is completely maintained at the accumulation center. This composite shows a lower catalytic activity than the expected value (1750). The access of phenylboronic acid into the Fr cage might be inhibited at the channel by electrostatic repulsion with the six negatively charged carboxylic acids of Glu and Asp residues exposed on the channel surface. Thus, the catalytic activities of **Pd(allyl)•apo-rHLFr** could be improved by increasing the number of Pd(allyl) complexes at suitable positions in apo-rHLFr by the introduction of Cys residues.

7.3.3 Polymer Synthesis in Protein Cages

Protein cages play key roles in the biological storage of organic and inorganic polymers such as DNA, RNA, and minerals. These functions have been used to store macromolecular materials such as proteins, and organic and coordination polymers [27, 29, 31, 86–90]. Because most of the incorporations have been performed by dissociation–reassembling reactions of the monomer proteins composing the cages, the recovery yields still remain low. The pH-dependent swelling of virus cages can also be used to accommodate synthetic materials, but there are only a limited number of protein cages with this property. Douglas and coworkers reported that blanched organic polymers were synthesized in heat shock proteins by hetero [3 + 2] cycloaddition [86]. There is remaining control of the molecular weight and the molecular-weight distribution of polymers prepared in the size-limited and discrete interiors of protein cages is necessary. To achieve this, we attempted the *in situ* synthesis of organic polymers in a protein cage involving organometallic catalysts (Fig. 7.8) [91]. It is known that the polymerization of phenylacetylene is catalyzed by Rh(nbd) (nbd = norbornadiene) in aqueous solution [92, 93]. When the complexes are immobilized within the apo-rHLFr cage, it is expected that the reaction occurring within a discrete space could provide polymers with restricted molecular weights and narrow molecular-weight distributions. A composite of apo-rHLFr containing Rh(nbd) complexes (**Rh(nbd)•apo-rHLFr**) was prepared using a procedure similar to that for **Pd(allyl)•apo-rHLFr**. The crystal structure of **Rh(nbd)•apo-rHLFr** shows that three different binding sites are found on the interior surface of the monomer (Fig. 7.9). This means that 72 binding sites of the Rh complexes are located in the 24-mer

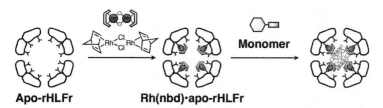

Apo-rHLFr   **Rh(nbd)·apo-rHLFr**

FIGURE 7.8   Schematic drawings of preparation of **Rh(nbd)•apo-rHLFr** and the polymerization reaction. Reproduced from Reference 91 with permission from the American Chemical Society.

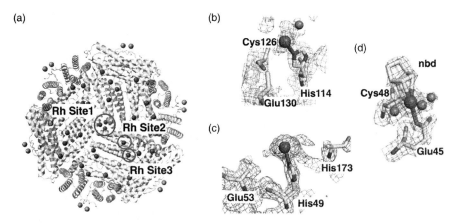

FIGURE 7.9 The whole structure (a), Site 1 (b), Site 2 (c), and Site 3 (d) of **Rh(nbd)•apo-rHLFr** are individually shown [91]. Reproduced from Reference 91 with permission from the American Chemical Society.

Fr cage. One of the three binding sites is at the threefold axis channel with His114 coordination (Fig. 7.9b). The other binding sites are found at the accumulation site bound to His49 or Cys48 (Figs. 7.9c and 7.9d, respectively), with retention of the mononuclear coordination geometry.

The polymerization of phenylacetylene using **Rh(nbd)•apo-rHLFr** was examined. Phenylacetylene monomer (3000 equivalent) was added to a 0.15 M NaCl aqueous solution of **Rh(nbd)•apo-rHLFr** (0.5 μM) in the presence of NaOH (0.3 mM). The color of the reaction mixture changed from colorless to pale yellow as a result of the formation of polyphenylacetylene in the apo-rHLFr cage [94]. When the same reaction was catalyzed by [Rh(nbd)Cl]$_2$ in the absence of apo-rHLFr, water-insoluble polyphenylacetylene precipitated in the buffered solution [91]. The size-exclusion column chromatography (Superdex G-200) elution profile of **Rh(nbd)•apo-Fr** after the polymerization showed co-elution of the protein (280 nm) and the polyphenylacetylene (383 nm) components (Fig. 7.10a). In addition, the elution volume of the peak was the same as those of **Rh(nbd)•apo-rHLFr** and apo-rHLFr (Figs. 7.10b and 7.10c, respectively). These results indicate that the polymerization of phenylacetylene proceeds in the apo-rHLFr cage and that the spherical 24-mer assembly of **Rh(nbd)•apo-rHLFr** is retained during the polymerization reaction. **Rh(nbd)•apo-rHLFr** afforded a *cis*–transoidal polyphenylacetylene main chain, as confirmed by the ^{1}H-NMR signal of the *cis*-olefin proton [95]. The number average molecular weights (M_n) of the polyphenylacetylene polymers produced by catalysis using **Rh(nbd)•apo-rHLFr** and [Rh(nbd)Cl]$_2$ were estimated to be $(13.1 \pm 1.5) \times 10^3$ ($M_w/M_n = 2.6 \pm 0.3$) and $(63.7 \pm 4) \times 10^3$ ($M_w/M_n = 21.4 \pm 0.4$), respectively. The results indicate that the molecular-weight distribution of the polymer prepared in the **Rh(nbd)•apo-rHLFr** cage is narrower than that obtained using [Rh(nbd)Cl]$_2$ in the absence of apo-rHLFr. As can be seen in Figure 7.11, which shows the time

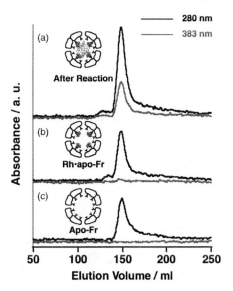

FIGURE 7.10   Elution profiles of size-exclusion column chromatography (Superdex G-200) of the **Rh(nbd)•apo-rHLFr** after the polymerization (a), **Rh(nbd)•apo-rHLFr** (b), and apo-rHLFr (c). Reproduced from Reference 91 with permission from the American Chemical Society.

course of this reaction, the number of phenylacetylene monomers achieved was 811 ± 95 monomers. This value is thought to be the saturated concentration in apo-rHLFr because the number is almost maintained even after 18 h. These results suggest that approximately eight polymer chains are formed in the cage. Most of the Rh complexes in **Rh(nbd)•apo-rHLFr** might, therefore, not be active as polymerization catalysts. This should be clarified with further theoretical investigations [96]. The regulation

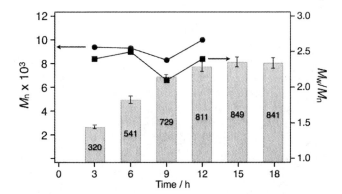

FIGURE 7.11   Time course of monomer increase in **Rh(nbd)•apo-rHLFr** during the polymerization reaction [91]. The time courses of M_n and M_w/M_n are shown as filled circle and square, respectively.

of the polymerization appears to be influenced by the size and environment of the interior space of apo-rHLFr.

7.4   COORDINATION PROCESSES IN FERRITIN

Recently, numerous preparations of inorganic NPs using Fr have been reported [17, 41]. The composites can serve not only as catalytic and magnetic materials but also as bioimaging and drug-delivery systems. Only a little information is available on the mechanism of NP formation because recent progress in this research area has focused on the preparation of new materials in protein cages. It is essential to clarify the mechanism for the synthesis of nanomaterials and for fundamental studies on biomineralization. The formation reaction consists of three processes: (1) penetration, (2) translocation, and (3) accumulation of metal ions. Although investigations of the formation of Fe oxide minerals have been attempted using X-ray crystallography, site-directed mutagenesis, metal substitution, and theoretical calculations [15, 48–61], the detailed processes of metal ion deposition and metal core formation remain obscure. Several research groups have extensively investigated the dynamic coordination mechanism, and elucidated it using advanced techniques, to obtain an in-depth understanding of the fundamental roles of protein–metal-ion interactions in both natural and artificial biomineralization processes [46, 64, 65, 97, 98]. Here, we describe the accumulation mechanisms of Pd ions and complexes with Pd–Fr composite systems [62, 63].

7.4.1   Accumulation of Metal Ions

As described in Section 7.3.1, Pd-NPs with a diameter of approximately 2 nm and a narrow size distribution can be synthesized in Fr cages. It is expected that the size of the Pd-NPs is restricted as a result of the limited number of Pd^{2+} ions bound on the interior surface of apo-Fr. To clarify the accumulation mechanism of metal ions in the apo-rHLFr cage, we attempted to determine the crystal structures of a series of apo-rHLFrs containing different numbers of Pd^{2+} ions, because these composites are expected to be intermediates in the Pd^{2+} ion accumulation reaction occurring in the Fr cage. These structures could provide good models for the binding and nucleation of various metal ions in Fr cages. First, we prepared a series of crystals of apo-rHLFr composites containing low, intermediate, and high contents of Pd^{2+} ions [**X-Pd^{2+}•apo-rHLFrs, X** = L (low), I (intermediate), and H (high)] by mixing with 50, 100, and 200 equivalents of Pd^{2+} ions, respectively.

The crystal structures were determined with resolutions of 1.65, 2.10, and 2.50 Å for **L-Pd^{2+}•apo-rHLFr, I-Pd^{2+}•apo-rHLFr**, and **H-Pd^{2+}•apo-rHLFr**, respectively. The crystal structures of the series of **Pd^{2+}•apo-rHLFrs** containing different amounts of Pd^{2+} ions indicate that there are two deposition areas located in the threefold axis channel and at the accumulation center (Fig. 7.12). There is a large difference between the number of Pd atoms estimated by inductively coupled plasma-optical emission spectrometry (ICP-OES) and bicinchonic acid (BCA) assay

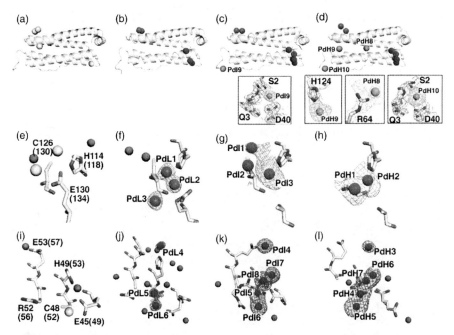

FIGURE 7.12 The binding site structures of **Pd²⁺-free•apo-rHLFr** and **Pd²⁺•apo-rHLFrs**. Monomer unit structures of **Pd²⁺-free•apo-rHLFr, L-, I-,** and **H-Pd²⁺•apo-rHLFr**, respectively (a–d). Threefold channel structures of **Pd²⁺-free•apo-rHLFr, L-, I-,** and **H-Pd²⁺•apo-rHLFr**, respectively (e–h). Accumulation center structures of **Pd²⁺-free•apo-rHLFr, L-, I-,** and **H-Pd²⁺•apo-rHLFr**, respectively (i–l). The Pd²⁺-free•apo-rHLFr structure is taken from PDB ID 1AEW [15]. Pd atoms in **L-, I-,** and **H-Pd²⁺•apo-rHLFrs** are shown as **PdL**s, **PdI**s, and **PdH**s, respectively. The Pd atoms in the threefold axis channels, accumulation centers, and the third site are colored dark gray. Cd atoms are indicated as light gray spheres. Fourier maps at 4.0 σ indicating the positions of individual palladium atoms are shown as purple and blue spheres. The insets of (c) and (d) are close-up views of PdI9, PdH8, PdH9, and PdH10. The original residue numbers of Pd²⁺-free•apo-rHLFr are given in parentheses (e and i) [15]. Reproduced from Reference 63 with permission from the American Chemical Society.

(1.5 Pd atoms/subunit) and that in the crystal structure (six Pd atoms/subunit) of **L-Pd²⁺•apo-Fr** (Table 7.2).

The threefold axis channel, which consists of highly conserved residues, His114, Cys126, Asp127, and Glu130, represents an entrance for metal ions into the Fr cage [54]. The crystal structures of Frs indicate that various metal ions, such as Ca^{2+}, Cd^{2+}, Zn^{2+}, and Tb^{3+}, are able to bind to the channel [15, 55, 58]. The PdL1–PdL3 atoms in **L-Pd²⁺•apo-rHLFr** interact with the three conserved residues, His114, Cys126, and Glu130 (Fig. 7.12f). In particular, the observation of two different conformations of the Cys126 side chain indicates that there are at least two coordination structures for Cys126 to ligate to either PdL1, PdL2, or PdL3. In the channel of **I-Pd²⁺•apo-rHLFr**, three Pd atoms remain in the same region, although their locations are

TABLE 7.2 **Quantitative Analyses of Pd Atoms of Pd^{2+}•apo-rHLFrs and Pd^{2+}•apo-H49A-rHLFrs**

| Reaction conditions ([Pd^{2+}]/[apo-rHLFr]) | Total number of Pd atoms and Pd-binding sites in Pd^{2+}•apo-rHLFr | | |
| | Pd atoms determined by ICP-OES and BCA assay | Pd-binding sites determined by crystal structural analyses | |
		Interior	Exterior
50:1	37 ± 3 (1.5)[a]	144 (6)[a]	0
100:1	88 ± 3 (3.7)[a]	192 (8)[a]	24(1)[a]
200:1	193 ± 10 (8.0)[a]	216 (9)[a]	24(1)[a]
200:1(apo-H49A-rHLFr)	204 ± 13 (8.5)[a]	168 (7)[a]	24(1)[a]
500:1	365 ± 10 (15)[a]	—[b]	—[b]

[a]The number of Pd atoms/subunit is shown in a parenthesis. [b]Failure of crystallization.
Adapted from Reference 63.

somewhat different from those identified in **L-Pd^{2+}•apo-rHLFr**, which has broader anomalous electron densities than those of **L-Pd^{2+}•apo-rHLFr** (Figs. 7.12f and 7.12g). **H-Pd^{2+}•apo-rHLFr** holds two Pd atoms in this region, and electron density is missing from the side chain of Glu130 as a result of conformational flexibility (Fig. 7.12h). The low resolution of the **H-Pd^{2+}•apo-rHLFr** data set might be responsible for the decreased number of Pd atoms observed at the threefold axis channel.

Our results show that Pd^{2+}-binding sites are located at different regions, which include Glu45, Cys48, His49, Arg52, and Glu53 (the accumulation center), close to the ferrihydrite nucleation site (Figs. 7.12i–7.12l) [54]. Pd^{2+} ions generally form square planar complexes whereas Cd^{2+}, Zn^{2+}, and Mn^{2+} prefer octahedral coordination structures, which is similar to the Fe^{2+} ion [2, 99]. Cys48 is expected to be crucial to retaining the square planar geometry of Pd^{2+} ions by the formation of a di- or trinuclear structure with cooperative ligation of Glu45 and water molecules (Fig. 7.12j). In contrast, four well-conserved Glu residues at the ferrihydrite nucleation center have a greater tendency to promote octahedral geometry, as found for Cd^{2+} ions in the crystal structure of mouse L-chain Fr [54]. Moreover, the accumulation center has relatively high B-factors for the side chains of amino acid residues, and the electrostatic potentials of apo-Fr calculated by GRASP show that the threefold axis channel and the ferrihydrite nucleation site are more negatively charged regions than the accumulation center [100]. The results, therefore, suggest that Pd^{2+} ions selectively bind to Cys48 and His49 at the accumulation center with square planar geometry, instead of to the ferrihydrite nucleation center consisting of highly conserved Glu residues to preserve the octahedral coordination geometry. In contrast, it is known that many different metal ions can occupy the apo-Fr cage through the threefold axis channels. These metal ions are expected to first accumulate at the negatively charged interior regions by coordination of the carboxylate groups of Glu residues, and are then mineralized either by oxidation or ligation to inorganic anions [17, 51, 55]. The

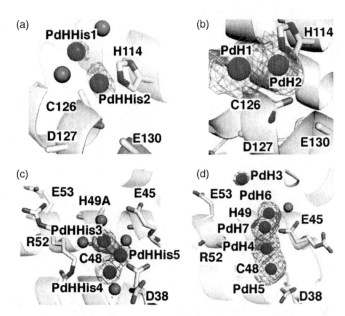

FIGURE 7.13   The crystal structure of **H-Pd²⁺•apo-H49A-rHLFr** [63]. Threefold axis channel structures of **H-Pd²⁺•apo-H49A-rHLFr** (a) and **H-Pd²⁺•apo-rHLFr** (b). Accumulation center structures of **H-Pd²⁺•apo-H49A-rHLFr** (c) and **H-Pd²⁺•apo-rHLFr** (d). Pd atoms in **H-Pd²⁺•apo-H49A-rHLFr** and **H-Pd²⁺•apo-rHLFrs** are shown as **PdHHis**s and **PdH**s, respectively, with the anomalous density maps. Anomalous difference Fourier maps at 4.0 σ indicating the positions of individual palladium atoms shown as dark gray spheres. Reproduced from Reference 63 with permission from the American Chemical Society.

X-ray crystal structures and mutagenesis investigations of apo-Frs show that metal ions such as Cd^{2+}, Zn^{2+}, Mn^{2+}, and Tb^{2+} are located at the ferrihydrite nucleation sites, which consist of Glu53, Glu56, Glu57, and Glu60 [15, 52, 54, 55, 60].

Moreover, His49 gives a weaker electron density with increasing numbers of Pd^{2+} ions at the accumulation centers of **I-** and **H-Pd²⁺•apo-rHLFrs** (Fig. 7.12). To confirm the high-affinity Pd^{2+} sites that play a role in localizing Pd^{2+} ions, we have determined the structure of a mutant (apo-H49A-Fr) with a high content of Pd^{2+} ions. The crystal structure of **H-Pd²⁺•apo-H49A-rHLFr** shows higher occupancy of all the Pd atoms than in **Pd²⁺•apo-rHLFrs** and shows that the number of Pd atoms obtained by quantitative analysis is almost identical to that of the crystal structure (Fig. 7.13 and Table 7.2). At an early stage of the Pd^{2+} accumulation in the apo-rHLFr cage, Pd^{2+} ions are expected to bind to any of the sites in the two areas by several possible binding modes. All of the Pd atoms in apo-rHLFr are, therefore, superimposed on one subunit, with low occupancy of the electron density. The conformational changes of His49 and the surrounding residues, Glu45 and Arg52, stabilize the Pd^{2+} ion bindings and promote an increase in the number of Pd atoms bound to the accumulation center (Figs. 7.12j–7.12l). These results suggest that the cooperation of Cys48 and His49 is a key factor for the formation of a Pd dinuclear coordination structure during the accumulation of Pd^{2+} ions.

When apo-rHLFr was treated with a large excess of Pd^{2+} ions (500 equivalent of K_2PdCl_4), only 365 Pd^{2+} ions were accommodated in apo-rHLFr (Table 7.2). Upon the addition of a large quantity of Pd^{2+} ions, all of the binding sites are fully occupied by Pd^{2+} ions, accompanied by the conformational changes of the amino acid side chains, as shown in a series of coordination structures of **L-**, **I-**, and **H-Pd^{2+}•apo-rHLFrs** (Fig. 7.12). Preparation of 2.0 ± 0.3 nm Pd-NPs with a narrow size distribution in apo-rHLFr is reasonably attributed to the limited number of Pd^{2+} ions accumulated on the internal surface of the apo-Fr, as reported previously [40]. Moreover, on the basis of these crystal structures, we assume that a single Pd^0 NP is formed in apo-rHLFr by the assembly of small Pd^0 seeds generated at each accumulation center within the 24-mer cage upon reduction of Pd^{2+} ions by $NaBH_4$, as proposed for the formation of cobalt oxide NPs in apo-Fr [101].

7.4.2   Accumulation of Metal Complexes

We recently demonstrated the preparation of apo-rHLFr for immobilizing Pd(allyl) complexes for catalyzing a Suzuki–Miyaura coupling reaction (Chapter 7.3.2) [81]. The crystal structure of the apo-rHLFr composite containing 96 Pd(allyl) complexes indicates that the formation of dinuclear Pd(allyl) complexes proceeds via ligation of thiol groups of Cys residues as bridging ligands, as well as ligation of His, Glu, and water molecules at the accumulation center and the threefold axis channel (Figs. 7.7a and 7.7b) [81]. We have determined the crystal structures of apo-rHLFr and selected mutants (apo-C48A-rHLFr, apo-C126A-rHLFr, apo-H49A-rHLFr, and apo-H114A-rHLFr) containing Pd(allyl) complexes to clarify the incorporation and accumulation processes of organometallic Pd complexes at the threefold axis channel and the accumulation center, where the Cys and His residues are expected to play crucial roles in fixing the Pd complexes [62]. The analysis of these structures is expected to provide insights into the design of the interior surface of apo-rHLFr to optimize control of coordination structures of metal complexes for future development of biocatalysts, biosensors, metal–drug systems, and new inorganic materials.

The crystal structure of **Pd(allyl)•apo-C126A-rHLFr** shows that there are no Pd(allyl) complexes bound to the threefold axis channel, as a result of the Cys to Ala mutation (Fig. 7.14a). The quantity of Pd atoms in **Pd(allyl)•apo-C126A-rHLFr** (37 ± 4) determined by quantitative analysis is slightly smaller than the amount observed in the crystal structure [48] (Table 7.3). The apo-C48A-rHLFr mutant contains 62.5 ± 1.5 Pd complexes. This number is greater than the observed quantity of Pd(allyl) complexes in the crystal structure of **Pd(allyl)•apo-C48A-rHLFr** [48]. The different quantity observed for the Cys-deleted mutants suggests that Cys126 is a critical residue for formation of Pd(allyl) dinuclear complexes, and that this residue plays a role in facilitating the uptake of Pd complexes at the threefold axis channel. Cys126 is also expected to influence the uptake of other metal ions such as Pd^{2+}, Cd^{2+}, and Fe^{3+} [53, 63, 102, 103]. When His114, which binds to Pd(allyl) complexes in **Pd(allyl)•apo-rHLFr**, is deleted, a thiolato-bridged trinuclear Pd(allyl) complex

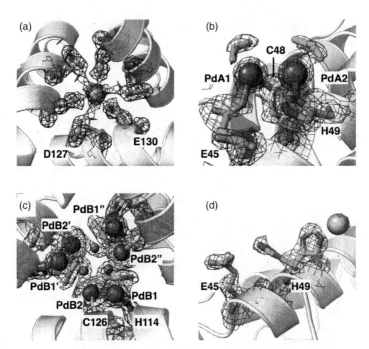

FIGURE 7.14 Crystal structures of threefold axis channels and accumulation centers of **Pd(allyl)•apo-C126A-rHLFr** (a and b), **Pd(allyl)•apo-C48A-rHLFr** (c and d), respectively [62]. The anomalous difference Fourier maps at 4.0 σ indicate the positions of cadmium (a and d) and palladium (b and c) atoms shown as spherical models. The selected $2|F_o| - |F_c|$ electron density maps are shown at 1.0 σ. Reproduced from Reference 62 with permission from the American Chemical Society.

is formed by coordination of three Sγ atoms of Cys126 residues at the threefold axis channel [81] (Fig. 7.7e). The quantity of Pd atoms in the apo-H114AFr, estimated by ICP and BCA measurements (78 ± 6), is essentially identical to the number of Pd atoms observed in the crystal structure [72] (Table 7.3). The results indicate that His114 plays a role in stabilizing the Pd(allyl) dinuclear structure at the threefold channel in apo-rHLFr.

TABLE 7.3   Quantitative Analysis of Pd Atoms in Pd(allyl)•apo-rHLFrs

Reaction conditions [Pd(allyl)]₂/[apo-rHLFr]	Pd atoms determined by ICP-OES and BCA assay	Pd-binding sites determined by crystal structural analysis
100:1 (apo-H114A-rHLFr)	78 ± 6 (3.2)[a,b]	72 (3)[a]
100:1 (apo-C126A-rHLFr)	37 ± 4 (1.5)[a,b]	48 (2)
100:1 (apo-C48A-rHLFr)	62.5 ± 1.5 (2.6)[b]	48 (2)
100:1 (apo-H49A-rHLFr)	99 ± 4 (4.1)[a,b]	96 (4)[a]

[a]The results are taken from Reference 81. [b]The number of Pd atoms/subunit is shown in parentheses.

The crystal structure of **Pd(allyl)•apo-C48A-rHLFr** shows that there are no Pd(allyl) complexes at the accumulation center (Fig. 7.14d). This indicates that Cys48 is a crucial residue for the formation of a dinuclear Pd(allyl) complex. His49 of **Pd(allyl)•apo-C48A-rHLFr** has two conformers, indicating that the side chain of His49 is highly flexible. The imidazole ring of His49 is expected to adopt several conformations, enabling it to adapt to various metal-coordination geometries. When metal ions such as Pd^{2+} and Au^{3+} ions are bound to the accumulation center, the conformation of His49 is dramatically altered by the coordination of the metal ions [63, 76]. The crystal structure of **Pd(allyl)•apo-H49A-rHLFr** shows that the quantity of Pd atoms observed in the cage is 96. This is the same as the quantity observed in **Pd(allyl)•apo-rHLFr**. **Pd(allyl)•apo-H49A-rHLFr** has thiolato–carboxylato dinuclear Pd complexes, which are formed by the ligation of Cys48 and Glu45 at the accumulation center as a result of a different conformation of the side chain of Glu45 (Fig. 7.7d) [81]. These results suggest that the conformational changes of His49 and Glu45 contribute to the stability of the square planar thiolato-bridged dinuclear Pd(allyl) complexes. These substitution experiments with Cys and His show that, (1) Cys126 accelerates the incorporation of metal complexes by inducing a conformational change at the threefold axis channel, and (2) Cys48 is essential for the formation of dinuclear Pd(allyl) complexes at the accumulation center. The quantity of Pd(allyl) complexes bound at the interior surface of apo-rHLFr depends on the number of Cys residues at the appropriate sites. Apo-rHLFr is thus able to bind 96 Pd(allyl) complexes because apo-Fr has a total of 48 Cys residues, which are expected to be needed to maintain the dinuclear Pd(allyl) structures. His and Glu residues, which have flexible side-chain structures in the vicinity of Cys residues, are necessary for stabilization of the dinuclear $[Pd(allyl)]_2(Cys)$ structures. This suggests that introduction of additional Cys residues within the apo-rHLFr cage would promote an increase in the capacity of the cage to accommodate greater quantities of Pd(allyl) complexes. The coordination structures of Pd(allyl) complexes could be controlled by altering the positions of His and Glu residues on the interior surface of apo-rHLFr.

7.5 COORDINATION ARRANGEMENTS IN DESIGNED FERRITIN CAGES

As described in Chapters 1 and 7.4, the accumulation mechanisms of metal ions and complexes in the Fr cage have been explained. These fundamental findings enable us to arrange the coordination of metal complexes existing in the cage. Although several types of metal complexes, such as Gd [32], Pt [33], Rh [91], Pd [81], Prussian blue [31, 104], coordination polymers [27], and phthalocyanine [36] can be incorporated inside protein cages, control of coordination structures has not been established because they are less stably and randomly bound on the interior surfaces. Organometallic complexes are candidates for use in examining the possibility of controlling the coordination structure inside a cage because they have organic ligands stably bound to a metal atom, preventing degradation of the compound in aqueous solutions. We have reported the incorporation of Pd(allyl) complexes into the cage of apo-rHLFr [81]. The coordination structures of the Pd complexes in the apo-rHLFr cage were precisely determined. The crystal structure of **Pd(allyl)•apo-rHLFr** reveals that one subunit of apo-rHLFr has two binding domains for Pd(allyl) complexes centered at Cys48 in the accumulation center and at Cys126 in the threefold axis channel (Fig. 7.7). Moreover, we found that replacing either the Cys or His residues in these two binding sites with Ala dramatically changes the metal-binding structures within the cages of the **Pd(allyl)•apo-rHLFr** mutants [62, 81]. Thus, to rearrange the coordination structures of Pd(allyl) complexes in the apo-rHLFr cage by appending or removing Cys and His residues, two factors should primarily be considered: (1) Cys is essential for the formation of dinuclear Pd(allyl) complexes at the accumulation center, and (2) His residues, which have flexible side-chain structures in the vicinity of the Cys residues, are necessary for controlling the various directions of the dinuclear Pd(allyl) moieties in the cage. On the basis of this hypothesis, new **Pd(allyl)•apo-rHLFr** composites were prepared using the rationally designed mutants apo-E45C/C48A, E45C/R52H, and E45C/H49A/R52C-rHLFrs, in which Pd complexes can be bound with confirmed coordination structures that are different from those in **Pd(allyl)•apo-rHLFr** [105].

Comparison of the accumulation centers of **Pd(allyl)•apo-rHLFr** and **Pd(allyl)•apo-E45C/C48A-rHLFr** shows that the substitution of Cys48 by Ala deletes the [Pd(allyl)]$_2$ moiety bound to the original Cys residue of apo-rHLFr (Figs. 7.15a and 7.15b). Replacement of Glu45 by a Cys brings a new dinuclear center, PdB1–PdB2 (Fig. 7.15b). The reasons for choosing the residue at position 45 as the new location for Cys are that this position avoids potential interference from subunits adjacent to the metal-coordination structure and that the position is at an appropriate distance from both His49 and His173; so these two His residues, His49 and His173, can act as supporting ligands for the PdB1–PdB2 pair. The crystal structure indicates that the positions of PdB1 and PdB2 are fixed (Fig. 7.15b). These mutations can be regarded as operations that shift the only Cys on the interior surface of apo-rHLFr to a new position, where this Cys collaborates with neighboring His residues to integrate one new binding site for a dinuclear Pd complex.

In the second mutant, apo-E45C/R52H-rHLFr, Cys45 is added, while conserving the native Cys48. Thus, the PdC1–PdC2 pair correlated with Cys45, His49, and His173 has an analogous structure to that of PdB1–PdB2 in **Pd(allyl)•apo-E45C/C48A-Fr**, which indicates a certain flexibility of the residues at this binding site (Fig. 7.15c). Another pair, PdC3–PdC4, is stabilized by the thiol group of Cys48, the imidazole group of the mutated His52, and one water molecule. The relatively large shift between PdA3 and PdC3 (1.88 Å) demonstrates that the different location of the coordinating His will cause conformational changes to the Pd atoms. Based on the surrounding residues, allyl ligands belonging to each Pd ion and a water molecule, all the Pd atoms have a typical square planar structure.

Finally, we derived apo-E45C/H49A/R52H-rHLFr from apo-E45C/R52H-rHLFr. A comparison of the crystal structures of **Pd(allyl)•apo-E45C/R52H-rHLFr** and **Pd(allyl)•apo-E45C/H49A/R52H-rHLFr** shows that replacing His49 with Ala does not induce significant structural differences around Cys48 (Fig. 7.15d). The shifts

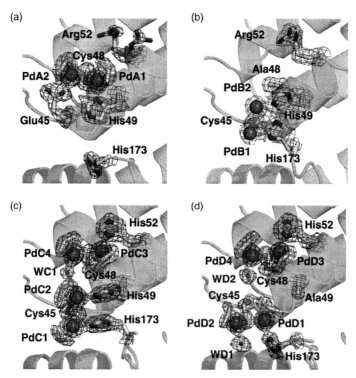

FIGURE 7.15   The accumulation centers of **Pd(allyl)•apo-rHLFr** (a), **Pd(allyl)•apo-E45C/C48A-rHLFr** (b), **Pd(allyl)·apo-E45C/R52H-rHLFr** (c), and **Pd(allyl)•apo-E45C/H49A/R52H-rHLFr** (d) [105]. The Pd atoms are indicated as sphere models. The selected $2|F_o| - |F_c|$ electron density maps at 1.0 σ are shown. Reproduced from Reference 105 with permission from The Royal Chemical Society.

of PdC3–PdD3 and PdC4–PdD4 are just 0.21 and 0.27 Å. However, this mutation significantly alters the geometry of the dinuclear Pd complex bound to Cys45. The orientation of the PdD1–PdD2 pair is twisted to a conformation almost parallel to PdD3–PdD4, with one water molecule coordinated to PdD2, and His173 coordinated to PdD1. This proves the significance of His in modulating the coordination structure of Pd. **Pd(allyl)•apo-E45C/H49A/R52H-rHLFr** contains 144 Pd atoms in each Fr cage, according to ICP/BCA data (154 ± 3 Pd per Fr), suggesting very rigid coordination structures of these four Pd ions.

Composites of Fr with Ru complexes and ferrocene have been prepared [106, 107]. An Ru(*p*-cymene) moiety coordinates with His49, Glu53, and His173 at a position adjacent to the accumulation center by retaining a mononuclear geometry similar to that reported for the Ru complex (Fig. 7.16) [107, 108]. The crystal structures of dimethylaminomethylferrocene/apo-Fr composites were determined and showed the existence of a new anomalous peak at the twofold symmetry axis of the apo-Fr molecule, located between the monomeric subunits of the apo-Fr shell, in which

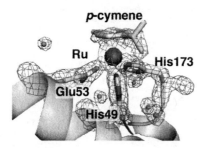

FIGURE 7.16 Close-up view of the coordination structures of a Ru-binding site at the accumulation center of **Ru(*p*-cymene)·apo-rHLFr** [107]. Reproduced from Reference 107 with permission from The Royal Chemical Society.

hydrophobic molecules such as halothane and isoflurane were observed [106, 109]. The data, therefore, show that further coordination arrangements of different types of metal complexes in protein cages can be rationally designed on the basis of the crystal structures.

7.6   SUMMARY AND PERSPECTIVES

It is hoped that this chapter has successfully demonstrated the already established and still growing potential for constructing catalytic systems consisting of inorganic NPs or metal complexes in protein cages. The latest developments have demonstrated that both heterogeneous and homogeneous catalysts are activated in the unique environments of protein cages as catalysts active in organic solvents. They have shown that Fr has great potential as a protein cage for precisely arranging the coordination structures, numbers, and locations of metal complexes. The advantages of Fr are derived from its original function in Fe storage, because the internal surface is inherently more suitable for the accumulation of metal ions and the formation of multinuclear coordination structures without further modification than is the case with other protein cages. We could, therefore, successfully change the amino acid residues for the coordination of metal complexes, to improve their catalytic activity, based on the crystal structures of metal complexes/Fr hybrids. These strategies can be applied to protein assemblies with different morphologies, such as tubes and two-dimensional arrays, to serve as molecular templates for the coordination of metal ions and complexes, rather than for the deposition of metal NPs.

Future developments of catalytic reactions involving coordination compounds in the discrete spaces of protein cages will be expanded to construct cascade reaction systems. One opportunity is to use large or multicomponent natural protein cages. Vault and bacterial microcompartments are expected to have potential for these purposes [10, 25, 110–112]. With respect to the artificially designed assemblies, the construction of periodic protein cages by protein crystallization will be a candidate appropriate for use in coordination chemistry. Only a few groups have published methods for immobilizing and activating metal ions coordinated in cages [113–116]. Finally, the authors hope that the topics discussed in this chapter will contribute to further investigations of protein cages in the next generation.

ACKNOWLEDGMENTS

Parts of this work were supported by Grant-in-Aids for Scientific Research and the Funding Program for Next Generation World-Leading Researchers, from the Japan Society for the Promotion of Science, and the Ministry of Education, Culture, Sports, Science, and Technology, Japan, respectively.

REFERENCES

[1] Bertini, I.; Gray, H. B.; Stiefel, E. I.; Valentine, J. S. *Biological Inorganic Chemistry: Structure & Reactivity*; University Science Books: Sausalito, 2007.

[2] Lippard, S. J.; Berg, J. M. *Principles of Bioinorganic Chemistry*; University Science Books: Mill Valley, 1994.

[3] Qi, D. F.; Tann, C. M.; Haring, D.; Distefano, M. D. *Chem. Rev.* **2001**, *101*, 3081–3111.

[4] Ueno, T.; Abe, S.; Yokoi, N.; Watanabe, Y. *Coord. Chem. Rev.* **2007**, *251*, 2717–2731.

[5] Lu, Y.; Yeung, N.; Sieracki, N.; Marshall, N. M. *Nature* **2009**, *460*, 855–862.

[6] Heinisch, T.; Ward, T. R. *Curr. Opin. Chem. Biol.* **2010**, *14*, 184–199.

[7] Akabori, S.; Sakurai, S.; Izumi, Y.; Fujii, Y. *Nature* **1956**, *178*, 323–324.

[8] Uchida, M.; Klem, M. T.; Allen, M.; Suci, P.; Flenniken, M.; Gillitzer, E.; Varpness, Z.; Liepold, L. O.; Young, M.; Douglas, T. *Adv. Mater.* **2007**, *19*, 1025–1042.

[9] Vriezema, D. M.; Aragones, M. C.; Elemans, J.; Cornelissen, J.; Rowan, A. E.; Nolte, R. J. M. *Chem. Rev.* **2005**, *105*, 1445–1489.

[10] Tanaka, H.; Kato, K.; Yamashita, E.; Sumizawa, T.; Zhou, Y.; Yao, M.; Iwasaki, K.; Yoshimura, M.; Tsukihara, T. *Science* **2009**, *323*, 384–388.

[11] Speir, J. A.; Munshi, S.; Wang, G. J.; Baker, T. S.; Johnson, J. E. *Structure* **1995**, *3*, 63–78.

[12] Xu, Z. H.; Horwich, A. L.; Sigler, P. B. *Nature* **1997**, *388*, 741–750.

[13] Zhang, X. F.; Meining, W.; Fischer, M.; Bacher, A.; Ladenstein, R. *J. Mol. Biol.* **2001**, *306*, 1099–1114.

[14] Kim, K. K.; Kim, R.; Kim, S. H. *Nature* **1998**, *394*, 595–599.

[15] Hempstead, P. D.; Yewdall, S. J.; Fernie, A. R.; Lawson, D. M.; Artymiuk, P. J.; Rice, D. W.; Ford, G. C.; Harrison, P. M. *J. Mol. Biol.* **1997**, *268*, 424–448.

[16] Grant, R. A.; Filman, D. J.; Finkel, S. E.; Kolter, R.; Hogle, J. M. *Nat. Struct. Biol.* **1998**, *5*, 294–303.

[17] Zhang, L.; Swift, J.; Butts, C. A.; Yerubandi, V.; Dmochowski, I. J. *J. Inorg. Biochem.* **2007**, *101*, 1719–1729.

[18] Dickerson, M. B.; Sandhage, K. H.; Naik, R. R. *Chem. Rev.* **2008**, *108*, 4935–4978.

[19] Bode, S. A.; Minten, I. J.; Nolte, R. J. M.; Cornelissen, J. *Nanoscale* **2011**, *3*, 2376–2389.

[20] Douglas, T.; Strable, E.; Willits, D.; Aitouchen, A.; Libera, M.; Young, M. *Adv. Mater.* **2002**, *14*, 415–418.

[21] McMillan, R. A.; Paavola, C. D.; Howard, J.; Chan, S. L.; Zaluzec, N. J.; Trent, J. D. *Nat. Mater.* **2002**, *1*, 247–252.

[22] McMillan, R. A.; Howard, J.; Zaluzec, N. J.; Kagawa, H. K.; Mogul, R.; Li, Y. F.; Paavola, C. D.; Trent, J. D. *J. Am. Chem. Soc.* **2005**, *127*, 2800–2801.

[23] Klem, M. T.; Willits, D.; Solis, D. J.; Belcher, A. M.; Young, M.; Douglas, T. *Adv. Funct. Mater.* **2005**, *15*, 1489–1494.

[24] Prasuhn, D. E.; Kuzelka, J.; Strable, E.; Udit, A. K.; Cho, S. H.; Lander, G. C.; Quispe, J. D.; Diers, J. R.; Bocian, D. F.; Potter, C.; Carragher, B.; Finn, M. G. *Chem. Biol.* **2008**, *15*, 513–519.

[25] Goldsmith, L. E.; Pupols, M.; Kickhoefer, V. A.; Rome, L. H.; Monbouquette, H. G. *Acs Nano* **2009**, *3*, 3175–3183.

[26] Hooker, J. M.; Datta, A.; Botta, M.; Raymond, K. N.; Francis, M. B. *Nano Lett.* **2007**, *7*, 2207–2210.

[27] Lucon, J.; Abedin, M. J.; Uchida, M.; Liepold, L.; Jolley, C. C.; Young, M.; Douglas, T. *Chem. Commun.* **2010**, *46*, 264–266.

[28] Prasuhn, D. E.; Yeh, R. M.; Obenaus, A.; Manchester, M.; Finn, M. G. *Chem. Commun.* **2007**, (*12*), 1269–1271.

[29] Douglas, T.; Young, M. *Nature* **1998**, *393*, 152–155.

[30] Shenton, W.; Mann, S.; Colfen, H.; Bacher, A.; Fischer, M. *Angew. Chem. Int. Ed.* **2001**, *40*, 442–445.

[31] Dominguez-Vera, J. M.; Colacio, E. *Inorg. Chem.* **2003**, *42*, 6983–6985.

[32] Aime, S.; Frullano, L.; Crich, S. G. *Angew. Chem. Int. Ed.* **2002**, *41*, 1017–1019.

[33] Yang, Z.; Wang, X.; Diao, H.; Zhang, J.; Li, H.; Sun, H.; Guo, Z. *Chem. Commun.* **2007**, (*33*), 3453–3455.

[34] Sun, J.; DuFort, C.; Daniel, M. C.; Murali, A.; Chen, C.; Gopinath, K.; Stein, B.; De, M.; Rotello, V. M.; Holzenburg, A.; Kao, C. C.; Dragnea, B. *Proc. Natl. Acad. Sci. USA* **2007**, *104*, 1354–1359.

[35] Ishii, D.; Kinbara, K.; Ishida, Y.; Ishii, N.; Okochi, M.; Yohda, M.; Aida, T. *Nature* **2003**, *423*, 628–632.

[36] Brasch, M.; de la Escosura, A.; Ma, Y. J.; Uetrecht, C.; Heck, A. J. R.; Torres, T.; Cornelissen, J. *J. Am. Chem. Soc.* **2011**, *133*, 6878–6881.

[37] Comellas-Aragones, M.; Engelkamp, H.; Claessen, V. I.; Sommerdijk, N.; Rowan, A. E.; Christianen, P. C. M.; Maan, J. C.; Verduin, B. J. M.; Cornelissen, J.; Nolte, R. J. M. *Nat. Nanotech.* **2007**, *2*, 635–639.

[38] Meldrum, F. C.; Wade, V. J.; Nimmo, D. L.; Heywood, B. R.; Mann, S. *Nature* **1991**, *349*, 684–687.

[39] Meldrum, F. C.; Heywood, B. R.; Mann, S. *Science* **1992**, *257*, 522–523.

[40] Ueno, T.; Suzuki, M.; Goto, T.; Matsumoto, T.; Nagayama, K.; Watanabe, Y. *Angew. Chem. Int. Ed.* **2004**, *43*, 2527–2530.

[41] Yamashita, I.; Iwahori, K.; Kumagai, S. *Biochim. Biophys. Acta.* **2010**, *1800*, 846–857.

[42] Flenniken, M. L.; Willits, D. A.; Brumfield, S.; Young, M. J.; Douglas, T. *Nano Lett.* **2003**, *3*, 1573–1576.

[43] Kramer, R. M.; Li, C.; Carter, D. C.; Stone, M. O.; Naik, R. R. *J. Am. Chem. Soc.* **2004**, *126*, 13282–13286.

[44] Varpness, Z.; Peters, J. W.; Young, M.; Douglas, T. *Nano Lett.* **2005**, *5*, 2306–2309.

[45] Kang, S.; Jolley, C. C.; Liepold, L. O.; Young, M.; Douglas, T. *Angew. Chem. Int. Ed.* **2009**, *48*, 4772–4776.

[46] Pan, Y. H.; Sader, K.; Powell, J. J.; Bleloch, A.; Gass, M.; Trinick, J.; Warley, A.; Li, A.; Brydson, R.; Brown, A. *J. Struct. Biol.* **2009**, *166*, 22–31.

[47] Galvez, N.; Fernandez, B.; Sanchez, P.; Cuesta, R.; Ceolin, M.; Clemente-Leon, M.; Trasobares, S.; Lopez-Haro, M.; Calvino, J. J.; Stephan, O.; Dominguez-Vera, J.M. *J. Am. Chem. Soc.* **2008**, *130*, 8062–8068.

[48] Barnes, C. M.; Petoud, S.; Cohen, S. M.; Raymond, K. N. *J. Biol. Inorg. Chem* **2003**, *8*, 195–205.

[49] Behrens, P.; Baeuerlein, E. *Handbook of Biomineralization*, Wiley-VCH: Weinheim, 2007.

[50] Butts, C. A.; Swift, J.; Kang, S.-G.; Di Costanzo, L.; Christianson, D. W.; Saven, J. G.; Dmochowski, I. J. *Biochemistry* **2008**, *47*, 12729–12739.

[51] Chasteen, N. D.; Harrison, P. M. *J. Struct. Biol.* **1999**, *126*, 182–194.

[52] Crichton, R. R.; Herbas, A.; ChavezAlba, O.; Roland, F. *J. Biol. Inorg. Chem.* **1996**, *1*, 567–574.

[53] Granier, T.; Comberton, G.; Gallois, B.; d'Estaintot, B. L.; Dautant, A.; Crichton, R. R.; Precigoux, G. *Proteins.* **1998**, *31*, 477–485.

[54] Granier, T.; d'Estaintot, B. L.; Gallois, B.; Chevalier, J. M.; Precigoux, G.; Santambrogio, P.; Arosio, P. *J. Biol. Inorg. Chem.* **2003**, *8*, 105–111.

[55] Harrison, P. M.; Artymiuk, P. J.; Ford, G. C.; Lawson, D. M.; Smith, J. M. A.; Treffry, A.; White, J. L. Ferritin: Function and structural design of an iron-storage protein. In *Biomineralization*; Mann, S., Webb, J., Williams, R. J. P., Eds.; VCH: Weinheim, **1989**; pp 257–294.

[56] Hempstead, P. D.; Hudson, A. J.; Artymiuk, P. J.; Andrews, S. C.; Banfield, M. J.; Guest, J. R.; Harrison, P. M. *FEBS Lett.* **1994**, *350*, 258–262.

[57] Kauko, A.; Pulliainen, A. T.; Haataja, S.; Meyer-Klaucke, W.; Finne, J.; Papageorgiou, A. C. *J. Mol. Biol.* **2006**, *364*, 97–109.

[58] Lawson, D. M.; Artymiuk, P. J.; Yewdall, S. J.; Smith, J. M. A.; Livingstone, J. C.; Treffry, A.; Luzzago, A.; Levi, S.; Arosio, P.; Cesareni, G.; Thomas, C. D.; Shaw, W. V.; Harrison, P. M. *Nature* **1991**, *349*, 541–544.

[59] Toussaint, L.; Bertrand, L.; Hue, L.; Crichton, R. R.; Declercq, J. P. *J. Mol. Biol.* **2007**, *365*, 440–452.

[60] Trikha, J.; Theil, E. C.; Allewell, N. M. *J. Mol. Biol.* **1995**, *248*, 949–967.

[61] Wade, V. J.; Levi, S.; Arosio, P.; Treffry, A.; Harrison, P. M.; Mann, S. *J. Mol. Biol.* **1991**, *221*, 1443–1452.

[62] Abe, S.; Hikage, T.; Watanabe, Y.; Kitagawa, S.; Ueno, T. *Inorg. Chem.* **2010**, *49*, 6967–6973.

[63] Ueno, T.; Abe, M.; Hirata, K.; Abe, S.; Suzuki, M.; Shimizu, N.; Yamamoto, M.; Takata, M.; Watanabe, Y. *J. Am.Chem. Soc.* **2009**, *131*, 5094–5100.

[64] Tosha, T.; Ng, H. L.; Bhattasali, O.; Alber, T.; Theil, E. C. *J. Am. Chem. Soc.* **2010**, *132*, 14562–14569.

[65] Kasyutich, O.; Ilari, A.; Fiorillo, A.; Tatchev, D.; Hoell, A.; Ceci, P. *J. Am. Chem. Soc.* **2010**, *132*, 3621–3627.

[66] Theil, E. C. *Annu. Rev. Biochem.* **1987**, *56*, 289–315.

[67] Webb, B.; Frame, J.; Zhao, Z.; Lee, M. L.; Watt, G. D. *Arch. Biochem. Biophys.* **1994**, *309*, 178–183.

[68] Chasteen, N. D. In *Iron Transport and Storage in Microorganisms, Plants, and Animals*; Sigel, A., Sigel, H., Eds.; Marcel Dekker: New York, 1998; pp 498–514.

[69] Yang, D. W.; Nagayama, K. *Biochem. J.* **1995**, *307*, 253–256.

[70] Yang, X. K.; Chasteen, N. D. *Biophys. J.* **1996**, *71*, 1587–1595.

[71] Cornils, B.; Herrmann, W. A. *Aqueous-Phase Organometallic Catalysis*; Wiley-VCH: Weinheim, 2004.

[72] Joo, F. *Aqueous Organometallic Catalysis*; Kluwer Academic Publishers: Dordrecht, 2001.

[73] Wilson, M. E.; Whitesides, G. M. *J. Am. Chem. Soc.* **1978**, *100*, 306–307.

[74] Zhao, M. Q.; Crooks, R. M. *Angew. Chem. Int. Ed.* **1999**, *38*, 364–366.

[75] Harrison, P. M.; Artymuik, P. J.; Ford, G. C.; Lawson, D. M.; Smith, J. M. A.; Treffry, A.; White, J. L. In *Biomineralization: Chemical and Biochemical Perspectives*; Mann, S., Webb, J., Williams, R. J. P., Eds.; Wiley-VCH: Weinheim, 1989; pp 257–294.

[76] Suzuki, M.; Abe, M.; Ueno, T.; Abe, S.; Goto, T.; Toda, Y.; Akita, T.; Yamadae, Y.; Watanabe, Y. *Chem. Commun.* **2009**, 4871–4873.

[77] Freeman, H. C. Metal complexes of amino acids and peptides. In *Inorganic Biochemistry*; Eichhorn, G. L., Ed.; Elsevier: New York, **1973**; pp 121–166.

[78] Sovago, I.; Osz, K. *Dalton Trans.* **2006**, *(32)*, 3841–3854.

[79] Toshima, N.; Harada, M.; Yamazaki, Y.; Asakura, K. *J. Phys. Chem.* **1992**, *96*, 9927–9933.

[80] Scott, R. W. J.; Wilson, O. M.; Oh, S. K.; Kenik, E. A.; Crooks, R. M. *J. Am. Chem. Soc.* **2004**, *126*, 15583–15591.

[81] Abe, S.; Niemeyer, J.; Abe, M.; Takezawa, Y.; Ueno, T.; Hikage, T.; Erker, G.; Watanabe, Y. *J. Am. Chem. Soc.* **2008**, *130*, 10512–10514.

[82] Basato, M.; Grassi, A.; Valle, G. *Organometallics* **1995**, *14*, 4439–4442.

[83] Burrows, A. D.; Mingos, D. M. P. *Transition Met. Chem.* **1993**, *18*, 129–148.

[84] Sellmann, D.; Geipel, F.; Heinemann, F. W. *Eur. J. Inorg. Chem.* **2000**, 271–279.

[85] Viciu, M. S.; Germaneau, R. F.; Navarro-Fernandez, O.; Stevens, E. D.; Nolan, S. P. *Organometallics* **2002**, *21*, 5470–5472.

[86] Abedin, M. J.; Liepold, L.; Suci, P.; Young, M.; Douglas, T. *J. Am. Chem. Soc.* **2009**, *131*, 4346–4354.

[87] Ng, B. C.; Yu, M.; Gopal, A.; Rome, L. H.; Monbouquette, H. G.; Tolbert, S. H. *Nano Lett.* **2008**, *8*, 3503–3509.

[88] Comellas-Aragones, M.; de la Escosura, A.; Dirks, A. J.; van der Ham, A.; Fuste-Cune, A.; Cornelissen, J.; Nolte, R. J. M. *Biomacromolecules* **2009**, *10*, 3141–3147.

[89] Fiedler, J. D.; Brown, S. D.; Lau, J. L.; Finn, M. G. *Angew. Chem. Int. Ed.* **2010**, *49*, 9648–9651.

[90] Worsdorfer, B.; Woycechowsky, K. J.; Hilvert, D. *Science* **2011**, *331*, 589–592.

[91] Abe, S.; Hirata, K.; Ueno, T.; Morino, K.; Shimizu, N.; Yamamoto, M.; Takata, M.; Yashima, E.; Watanabe, Y. *J. Am. Chem. Soc.* **2009**, *131*, 6958-+.

[92] Maeda, K.; Goto, H.; Yashima, E. *Macromolecules* **2001**, *34*, 1160–1164.

[93] Tang, B. Z.; Poon, W. H.; Leung, S. M.; Leung, W. H.; Peng, H. *Macromolecules* **1997**, *30*, 2209–2212.

[94] Cametti, C.; Codastefano, P.; D'Amato, R.; Furlani, A.; Russo, M. V. *Synth. Met.* **2000**, *114*, 173–179.

[95] Simionescu, C. I.; Percec, V. *J. Polym. Sci.:Polym. Chem.Ed.* **1980**, *18*, 147–155.

[96] Ke, Z.; Abe, S.; Ueno, T.; Morokuma, K. *J. Am. Chem. Soc.* in press.

[97] Zeth, K.; Offermann, S.; Essen, L. O.; Oesterhelt, D. *Proc. Natl. Acad. Sci. USA* **2004**, *101*, 13780–13785.

[98] Lopez-Castro, J. D.; Delgado, J. J.; Perez-Omil, J. A.; Galvez, N.; Cuesta, R.; Watt, R. K.; Dominguez-Vera, J.M. *Dalton Trans.* **2012**, *41*, 1320–1324.

[99] Richens, D. T. *The Chemistry of Aqua Ions*, John Wiley & Sons: West Sussex, **1997**.

[100] Gallois, B.; dEstaintot, B. L.; Michaux, M. A.; Dautant, A.; Granier, T.; Precigoux, G.; Soruco, J. A.; Roland, F.; ChavasAlba, O.; Herbas, A.; Crichton, R. R. *J. Biol. Inorg. Chem.* **1997**, *2*, 360–367.

[101] Kim, J. W.; Choi, S. H.; Lillehei, P. T.; Chu, S. H.; King, G. C.; Watt, G. D. *Chem. Commun.* **2005**, 4101–4103.

[102] Levi, S.; Santambrogio, P.; Corsi, B.; Cozzi, A.; Arosio, P. *Biochem. J.* **1996**, *317*, 467–473.

[103] Bou-Abdallah, F.; Zhao, G.; Biasiotto, G.; Poli, M.; Arosio, P.; Chasteen, N. D. *J. Am. Chem. Soc.* **2008**, *130*, 17801–17811.

[104] de la Escosura, A.; Verwegen, M.; Sikkema, F. D.; Comellas-Aragones, M.; Kirilyuk, A.; Rasing, T.; Nolte, R. J. M.; Cornelissen, J. *Chem. Commun.* **2008**, *(13)* 1542–1544.

[105] Wang, Z. Y.; Takezawa, Y.; Aoyagi, H.; Abe, S.; Hikage, T.; Watanabe, Y.; Kitagawa, S.; Ueno, T. *Chem. Commun.* **2011**, *47*, 170–172.

[106] Niemeyer, J.; Abe, S.; Hikage, T.; Ueno, T.; Erker, G.; Watanabe, Y. *Chem. Commun.* **2008**, *(48)*, 6519–6521.

[107] Takezawa, Y.; Bockmann, P.; Sugi, N.; Wang, Z. Y.; Abe, S.; Murakami, T.; Hikage, T.; Erker, G.; Watanabe, Y.; Kitagawa, S.; Ueno, T. *Dalton Trans.* **2011**, *40*, 2190–2195.

[108] Vock, C. A.; Scolaro, C.; Phillips, A. D.; Scopelliti, R.; Sava, G.; Dyson, P. J. *J. Med. Chem.* **2006**, *49*, 5552–5561.

[109] Liu, R. Y.; Loll, P. J.; Eckenhoff, R. G. *FASEB J.* **2005**, *19*, 567–576.

[110] Choudhary, S.; Quin, M. B.; Sanders, M. A.; Johnson, E. T.; Schmidt-Dannert, C. *Plos One* **2012**, *7*, e33342. Epub 2012 Mar 12.

[111] Tanaka, S.; Sawaya, M. R.; Yeates, T. O. *Science* **2010**, *327*, 81–84.

[112] Kang, S.; Douglas, T. *Science* **2010**, *327*, 42–43.

[113] Koshiyama, T.; Shirai, M.; Hikage, T.; Tabe, H.; Tanaka, K.; Kitagawa, S.; Ueno, T. *Angew. Chem. Int. Ed.* **2011**, *50*, 4849–4852.

[114] Guli, M.; Lambert, E. M.; Li, M.; Mann, S. *Angew. Chem. Int. Ed.* **2010**, *49*, 520–523.

[115] Ni, T. W.; Tezcan, F. A. *Angew. Chem. Int. Ed.* **2010**, *49*, 7014–7018.

[116] Wei, H.; Wang, Z. D.; Zhang, J.; House, S.; Gao, Y. G.; Yang, L. M.; Robinson, H.; Tan, L. H.; Xing, H.; Hou, C. J.; Robertson, I. M.; Zuo, J. M.; Lu, Y. *Nat. Nanotech.* **2011**, *6*, 92–96.

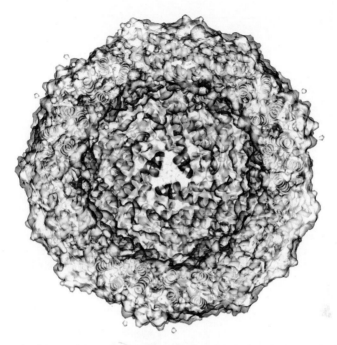

Chapter-1   Ferritin protein cage sliced in half. The view is from inside the ferritin protein cage facing one of the eight ion channels at the 3-fold symmetry axes; ion channels, ~1.5 nm long, form from helical segments (blue-gray) contributed by three of the twenty-four, folded polypeptide subunits; the ion channels connect the environment outside of ferritin cages to the inside, ending in pores at the outer and inner cage surfaces. The mineral growth cavity in the center of the protein cage occupies ~30% of the cage volume with a diameter of ~8 nm, and is surrounded by a layer of protein (green helices and gold contoured surfaces) ~2 nm thick. Graphics produced with Discovery Studio 2.0 from protein structure PDB 3KA3.

Coordination Chemistry in Protein Cages: Principles, Design, and Applications, First Edition.
Edited by Takafumi Ueno and Yoshihito Watanabe.
© 2013 John Wiley & Sons, Inc. Published 2013 by John Wiley & Sons, Inc.

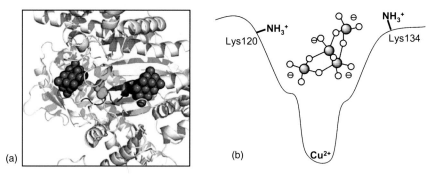

(a)

(b)

FIGURE 2.1 (a) Docking of decavanadate (red spheres) to the "back-door" region of the S1 domain of *Dictyostelium discoideum* myosin. Segments forming the phosphate-binding loop are represented in green (serine), cyan, and orange. Reprinted with permission from Reference 9. © Elsevier. (b) The interaction of tetravanadate ($V_4O_{12}^{4-}$; gray circles, V; empty, O) with two lysines in the pocket of the solvent channel of bovine Cu,Zn superoxide dismutase. One of the possible conformations of tetravanadate is shown (modified from Reference 10).

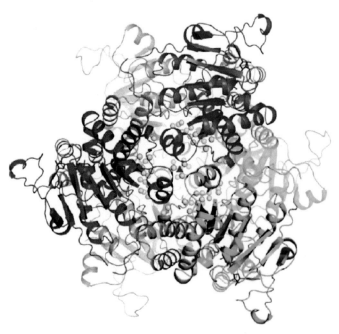

FIGURE 2.2 Structure of the Mo/WSto protein of *A. vinelandii*. The hetero-hexameric $(\alpha\beta)_3$ protein complex has a size of about $100 \times 100 \times 70$ Å^3 and pseudo *32* (D_3) symmetry. The monomers α_1 (red), α_2 (blue), and α_3 (green), and β_1 (light red), β_2 (light blue), and β_3 (light green) assemble as trimers on either side (called interfaces α and β) of the hetero-hexamer. The three monomers of each trimer are related by the crystallographic threefold axis (▲). The ATP molecules are shown as black stick models and the W atoms of the polynuclear tungsten oxide clusters as yellow spheres (see Reference 1).

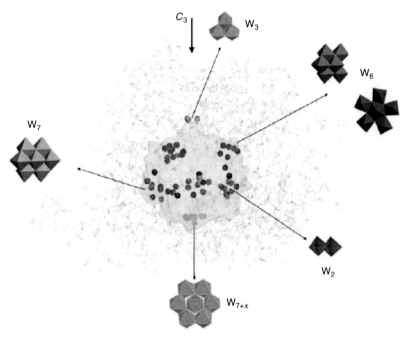

FIGURE 2.3 Surface representation of the Mo/WSto protein. The (roughly) ellipsoidal protein shell (gray) and the nanocavity are shown as a yellow surface. The cavity protein surface forms several well-distributed pockets, which harbor the polynuclear tungsten oxide clusters. The W_3 (yellow) and $W_{7+x} \equiv (W)W_{6+x}$ (green) clusters are positioned on the threefold axis, in front of the interfaces α and β, respectively, and are embedded in a small hydrophilic and a larger hydrophobic pocket, respectively. The binding sites of the three W_6 clusters (red; two idealized structural options are shown) are essentially built up by segments of the α subunit forming a flat and hydrophilic pit. The three W_7 clusters (blue) lie between subunits α and β in a deep pocket, while the W_2 clusters I and II (brown) are located in the vicinity of the W_7 cluster. The clusters are represented as idealized polyhedra constructed by assuming the O positions on the basis of the knowledge of POM chemistry [3a] and the present results; the individual W centers are shown as spheres of the corresponding color in the cavity of the Mo/WSto protein; see Reference 1.

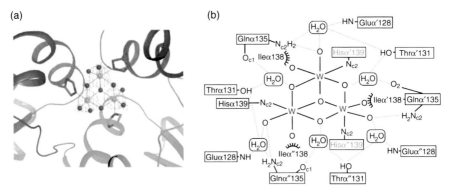

FIGURE 2.4 (a) The structure of the W_3 cluster and its protein surroundings. (b) Amino acid residues and water molecules in the direct environment of the W_3 cluster. The three triangularly arranged W atoms (yellow) are 3.4 Å from each other. Each W atom is linked to five O atoms and the $N_{\varepsilon 2}$ atom of Hisα139 (light green residue in (b); see Reference [1].

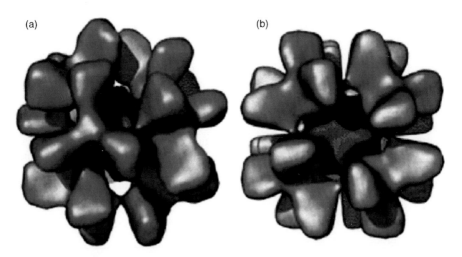

FIGURE 2.7 Reconstruction of an iron-loaded oligomeric (24 subunits) yeast frataxin viewed close to the threefold axis (a) and along the fourfold axis (b). The iron oxide core is shown in red, the protein shell in blue. Reprinted with permission from Reference 55; © American Chemical Society.

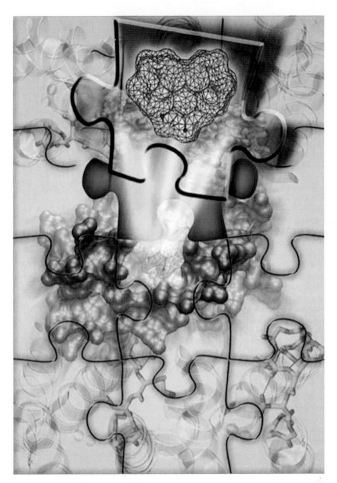

Chapter-3 Cartoon illustrating the design concept in DF3. An artificial functional protein is made up by several secondary structure elements, which, like in a puzzle, should precisely fit each into the other. These secondary structure elements should be carefully designed, in order to provide the protein sufficient stability, and cavity formation, required for activity.

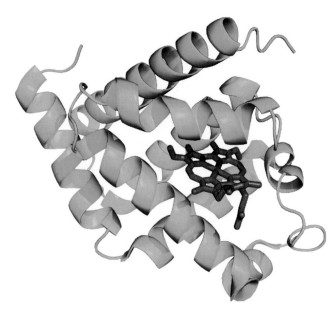

Chapter-4 Crystal structure of sperm whale myoglobin reconstituted with iron porphycene (PDB ID 2D6C). The stick in the heme pocket represents iron porphycene as an artificial prosthetic group.

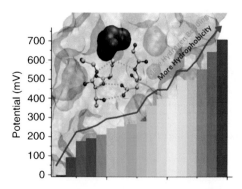

Chapter-5 The wide range of redox potentials of type I copper sites in designed azurin mutants, highlighting the secondary coordination sphere interactions, such as hydrophobicity and hydrogen bonding, important for such control.

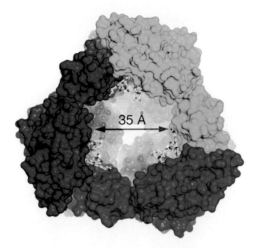

Chapter-6   A Metal-Directed, Tetrahedral Protein Cage Formed in a Crystal Lattice.

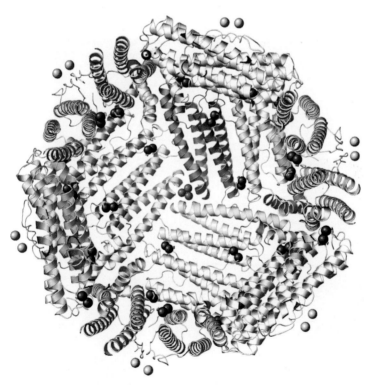

Chapter-7   Artificial organometalloenzymes: Crystal structure of apo-ferritin immobilizing Pd(allyl) complexes. The Pd and Cd atoms are indicated as greenish blue and beige spheres, respectively.

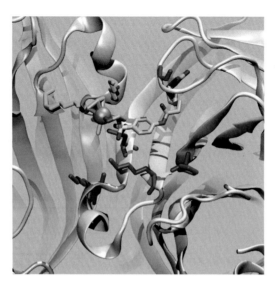

Chapter-8 Metal-catalyzed organic transformations inside a protein scaffold using artificial metalloenzymes.

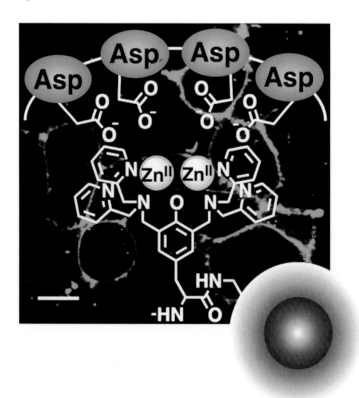

Chapter-9 Fluorescent imaging of GPCRs using Asp-tag/Zn(II)-DpaTyr pair.

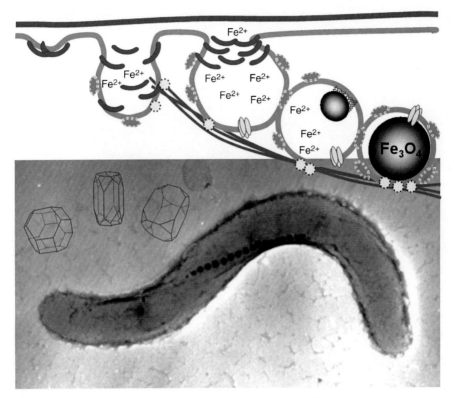

Chapter-10   Magnetosome formation in magnetotactic bacteria: Cytoplasmic membrane is invaginated to form magnetosome membrane, followed by iron accumulation and crystallization.

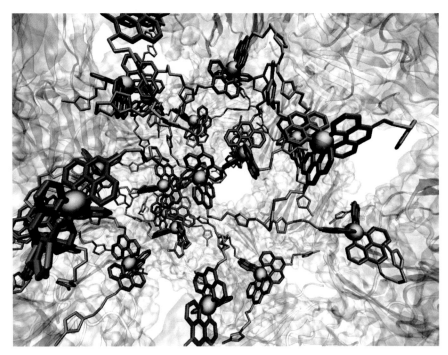

Chapter-11 Modeled view of a coordination polymer inside a protein cage nanoparticle.

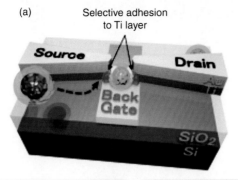

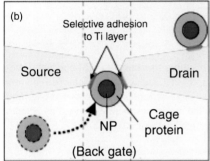
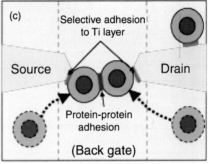

Chapter-12 Schematic illustration of SET device using selective adhesion of ferritin molecule to the surface of Ti layer. (a) Bird's eye view. (b) Nanogap size is less than ferritin molecule. One ferritin molecule is trapped. (c) Nanogap size is larger than one ferritin molecule but less than two. Two ferritin molecules are trapped in the nanogap.

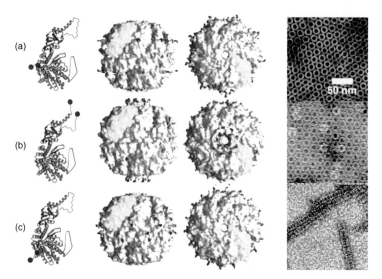

Chapter-13 Circular permutation of TF55B and resulting structure. (a-c) Ribbon and space-filling structures (top and side views) of the (a) 153, (b) 267, and (c) 480 circular permutation mutants with the position of the amino-termini shown as dark dots and the carboxy-termini shown as light dots (left). TEM micrographs show the resulting structures formed by octade-camers of circular permutation mutants (right). Reprinted with permission from Reference 42. Copyright 2006 IOP Publishing Ltd.

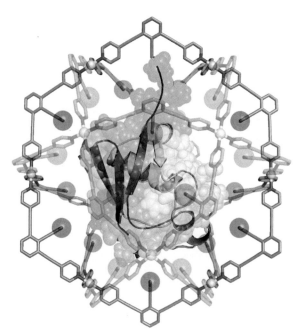

Chapter-14 A protein caged in an artificial metal-organic cage synthesized by mimicking protein cages. An artificial self-assembled cage from 12 palladium ions (shown in yellow) and 24 organic ligands (shown in blue) was developed by the inspiration from protein cages. The size of the hollow cage is sufficiently large to encapsulate a hole ubiquitin. The decollated hydrophilic sugar chains (shown in pink) on the inner surface trapped the protein stably, and the structure was revealed by NMR, ultra centrifugal analysis, and X-ray crystallography.

8

METAL-CATALYZED ORGANIC TRANSFORMATIONS INSIDE A PROTEIN SCAFFOLD USING ARTIFICIAL METALLOENZYMES

V. K. K. Praneeth and Thomas R. Ward

8.1 INTRODUCTION

Enzymes catalyze a wide variety of chemical reactions with high selectivity and activity under mild conditions [1, 2]. However, there are several pitfalls in using enzymes as catalysts compared to homogeneous catalytic systems. For example, enzymatic reactions can often be performed only with a limited range of substrates because of their narrow substrate scope, and most of the reactions need to be carried out in aqueous media. On the other hand, homogeneous systems provide broad substrate scope in nonaqueous conditions. In many cases, homogeneous catalysts and enzymes display complementary features. For these reasons, the topic of *artificial metalloenzymes* has fascinated chemists as a means to create hybrid catalysts. Such hybrid systems are expected to display properties of both enzymatic and homogeneous kingdoms [3]. The construction of artificial metalloenzyme can be carried out by anchoring a metal moiety within a protein scaffold with the help of an anchoring group. This can be done by covalent [2, 4], dative [4], or supramolecular (noncovalent) means [3]. Our research strategy in the construction of artificial metalloenzyme relies on noncovalent attachment of the metal moiety using biotin-(strept)avidin technology (hereafter, (Strep)avidin refers to either avidin (Avi) or streptavidin (Sav)).

The last decade has witnessed an increasing number of publications pertaining to artificial metalloenzymes. This review summarizes our current knowledge on

Coordination Chemistry in Protein Cages: Principles, Design, and Applications, First Edition.
Edited by Takafumi Ueno and Yoshihito Watanabe.
© 2013 John Wiley & Sons, Inc. Published 2013 by John Wiley & Sons, Inc.

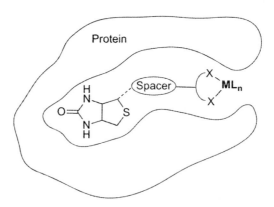

FIGURE 8.1 Construction of artificial metalloenzyme based on the biotin-(strept)avidin technology. An achiral metal complex linked to biotin via a spacer is incorporated within a protein by supramolecular anchoring. Chemogenetic optimization (variation of the spacer or of the chelate, or by point mutations on the host protein) of the hybrid system affords versatile enantioselective hybrid catalysts for a variety of organic transformations.

the catalytic properties of the artificial metalloenzymes in various organic transformations based on our research using the biotin-(strept)avidin technology. Wilson and Whitesides were the first to exploit the nearly irreversible binding of biotin (vitamin H) with avidin ($K_d = \sim 10^{-14}$ M) to generate an artificial metalloenzyme for enantioselective hydrogenation [5]. Avi is a well-characterized homotetrameric protein with four biotin-binding sites. In the late 1970s, their first visionary communication on the asymmetric hydrogenation of alkenes using a biotinylated diphosphine rhodium(I) complex anchored within Avi disclosed moderate enantioselectivity. Initially, we explored the same reaction, but upon combining both chemical and genetic modifications (i.e., chemogenetic optimization), we succeeded in generating artificial metalloenzymes that displayed high conversions and high enantioselectivities. The chemical and genetic modifications refer to the changes made in the ligand structure and/or in the spacer, and site-directed mutagenesis, respectively. These changes are applied to modulate and fine-tune the activity and the selectivity of the corresponding artificial metalloenzyme. This strategy relies on optimizing second coordination sphere interactions between the biotinylated catalyst and the host protein. This chemogenetic approach is illustrated in Figure 8.1. Herein we present the results obtained upon applying this strategy toward the generation of artificial metalloenzymes for various enantioselective transformations.

8.2   ENANTIOSELECTIVE REDUCTION REACTIONS CATALYZED BY ARTIFICIAL METALLOENZYMES

8.2.1   Asymmetric Hydrogenation

As outlined in the introduction, Whitesides's report on the hydrogenation of N-acetamidoacrylic acid to N-acetamidoalanine (N-AcAla) using [Rh(COD)(**biot-1**)]$^+$

TABLE 8.1   Asymmetric Hydrogenation of Alkenes Catalyzed by Artificial Metalloenzymes

Entry	Ligand	Protein	Solvent	% ee (N-AcAla)	% Conv. (N-AcAla)	% ee (N-AcPhe)	% Conv. (N-AcPhe)
1	**Biot-1**	WT Avi	Aqueous	37 (S)	90	_a	_a
2	**Biot-1**	WT Sav	Aqueous	94 (R)	Quant.	93 (R)	84
3	**Biot-2**	WT Sav	Aqueous	20 (S)	55	_a	_a
4	**Biot-3¹-1**	WT Avi	Aqueous	60 (S)	Quant.	_a	_a
5	**Biot-3¹-2**	WT Avi	Aqueous	66 (S)	Quant.	_a	_a
6	**Biot-1**	Sav S112G	Aqueous	96 (R)	Quant.	_a	_a
7	**Biot-1**	Sav S112A	Aqueous	93 (R)	Quant.	94 (R)	94
8	**Biot-4^{meta}-1**	Sav S112K	Aqueous	63 (S)	Quant.	88 (S)	89
9	**Biot-3¹-2**	Sav S112F	Aqueous	64 (S)	Quant.	36 (S)	20
10	**Biot-3¹-2**	Sav S112W	Aqueous	59 (S)	96	33 (S)	8
11	**Biot-(R)-Pro-1**	WT Avi	Aqueous	87 (S)	Quant.	89 (S)	Quant.
12	**Biot-(R)-Pro-1**	Sav S112W	Aqueous	95 (S)	Quant.	95 (S)	Quant.
13	**Biot-(S)-Phe-1**	Sav 112H	Aqueous	87 (S)	Quant.	78 (S)	65
14	**Biot-(S)-Phe-1**	Sav 112M	Aqueous	73 (R)	Quant.	87 (R)	Quant.
15	**Biot-(R)-Pro-1**	WT Sav	EtOAc	83 (S)	90	87 (S)	85
16	**Biot-(R)-Pro-1**	Sav S112W	Immob.	92 (S)	Quant.	94 (S)	89

^{a}Reaction was performed in the absence of N-acetamidocinnamic acid.
Quant., quantitative conversion.

⊂ avidin produced 41% (S)-enantiomer with complete conversion of the substrate (hereafter ⊂ symbolizes the inclusion of an artificial cofactor within a host protein). We speculated that this moderate selectivity may arise from columbic repulsion between avidin (pI = 10.4, i.e., positively charged at pH = 7.0) and the cationic biotinylated complex. Hence, we switched to streptavidin; Sav (pI = 6.2) recombinantly produced in E. coli. Sav, like Avi, is a homotetrameric eight-stranded β-barrel protein, which also displays remarkable affinity for biotin ($K_M = 10^{13}$ M^{-1}) [6]. Using 0.9 mol % Rh catalyst and 0.33 mol % of tetrameric Sav, good enantioselectivity was obtained (94% ee (R)-product) in the reduction of N-acetamidoacrylic acid (Table 8.1, entry 2) [7]. The enantioselectivity was further improved by introducing and modifying the spacer between the biotin anchor and the rhodium moiety. For instance, employing a glycine spacer in conjunction with avidin contributed to improve both the selectivity and the activity (compare entries 4 and 5 with entry 1 in Table 8.1). Further exploration by site-directed mutagenesis (genetic optimization) at position Sav S112, a residue located close to the biotin binding site, produced much improved results. For example, substitution of the serine by a glycine at position 112 (Sav S112G) combined with [Rh(COD)(**Biot−1**)]$^+$ produced the best enantioselectivity (entry 6, Table 8.1). In all cases, however, only (R)-enantiomers could be obtained in >90% ee. The high level of enantioselectivity achieved by [Rh(COD)(**Biot−1**)]$^+$ ⊂ Sav S112G prompted us to prepare the saturation mutagenesis library at position S112X of Sav [8]. In total, 360 artificial metalloenzymes were obtained by the combination of 18 biotinylated diphosphino Rh(I) complexes with 20 Sav mutants (obtained from the saturation mutagenesis library at position Sav S112X). These new artificial metalloenzymes displayed attractive properties in catalytic hydrogenation. The combination of [Rh(COD)(**Biot-4**^{meta}−**1**)]$^+$ with cationic residues (Sav S112K, S112R) afforded N-acetamidoaniline with an enantiomeric excess as high as 88% in favor of the (S)-enantiomer.

Remarkably, these artificial metalloenzymes displayed substrate specificity by discriminating two different substrates varying in shape and size. [Rh(COD)(**Biot-3**^{1}−**2**)]$^+$ showed remarkable discrimination properties in a competition experiment performed using combination of two substrates: N-acetamidoacyrlic acid and N-acetamidocinnamic acid. Entries 9 and 10 in Table 8.1 illustrate the size discrimination provided by the host protein: the smaller substrate N-acetamidoacyrlic acid was converted preferentially [8]. We then developed second-generation artificial metalloenzyme by introducing enantiopure spacers (phenylalanine or proline) between the biotin anchor and the metal chelate [9]. These catalysts displayed pronounced tolerance toward organic solvents and also improved the (S)-selectivities (entries 11–16, Table 8.1). Finally, we developed an immobilized artificial hydrogenase using commercially available biotin–sepharose which, in the presence of the biotinylated Rh complex, yielded >92% ee (entry 16, Table 8.1).

8.2.2 Asymmetric Transfer Hydrogenation of Ketones

Noyori and coworkers have applied d^6-piano-stool complexes of Ru(II), Rh(III), and Ir(III) bearing a Tos-dpen ligand (Tos-dpen: p-toluenesulfonamide-diphenyl-ethylenediamine) to effect asymmetric transfer hydrogenation (ATH) of ketones and imines in both aqueous and nonaqueous media [10]. These catalysts exhibited excellent enantioselectivity and reactivity in these reactions. The ATH reaction mechanism is well understood for ketone reduction compared to the ATH of imines, where a C−H⋅⋅⋅π assisted mechanism has been invoked for the former case. Based on the experimental studies (kinetics and isotope-labeling studies) coupled to density-functional-theory (DFT) calculations, Noyori et al. proposed a six-membered cyclic transition state (TS) leading to enantioselectivity in the ATH of ketones, as depicted in Figure 8.2 [11, 12]. As one can appreciate from the TS, the enantioselection mechanism occurs without direct coordination of the substrate to the metal. This prompted us to explore the ATH of ketones using artificial metalloenzymes. We synthesized d^6-piano-stool complexes (Fig. 8.3) bearing biotinylated N-arylsulfonamide-1,2-ethylenediamine (**Biot-q-LH**, q = ortho, meta, para) [13]. In the ATH of ketones, the activity for the combination of these complexes with (strept)avidin was tested. Better catalytic performance was observed with artificial metalloenzymes having the para-substituted ligand **Biot-p-L**. In general, ruthenium complexes displayed better selectivity and activity compared to Rh and Ir complexes. The promising catalysts were selected and combined with saturation mutagenesis library at position Sav S112X and investigated in the ATH of various ketones. Interestingly, [η^6-(benzene)Ru(**Biot-p-L**)Cl] and [η^6-(p-cymene)Ru(**Biot-p-L**)Cl] produced opposite enantiomers (Table 8.2). This observation illustrates the intriguing role of capping

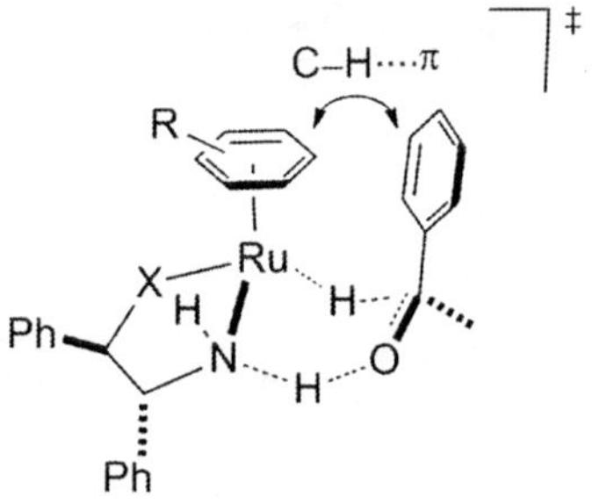

$X = O, NTos$
$R = H, alkyl$

FIGURE 8.2   Postulated TS structure for the ATH catalyzed by d^6-piano-stool complexes. The enantioselective ATH of ketones occurs without direct coordination of the substrate to the central metal ion.

arene in determining the enantioselectivity. Moderate to good enantioselectivities were obtained for various alcohols. The X-ray crystal structure of the (S)-selective [η^6-(benzene)Ru(**Biot-p-L**)Cl] $\subset$ S112K (Fig. 8.4) provided valuable insight regarding the localization of the ruthenium moiety inside the host protein [14]. The absolute configuration around the ruthenium center was determined as (S) by X-ray analysis (correspondingly the absolute configuration of the [η^6-(benzene)Ru(**Biot-p-L**)H] is (R)). The closest lying residues around the Ru center were identified as: K112$_A$,

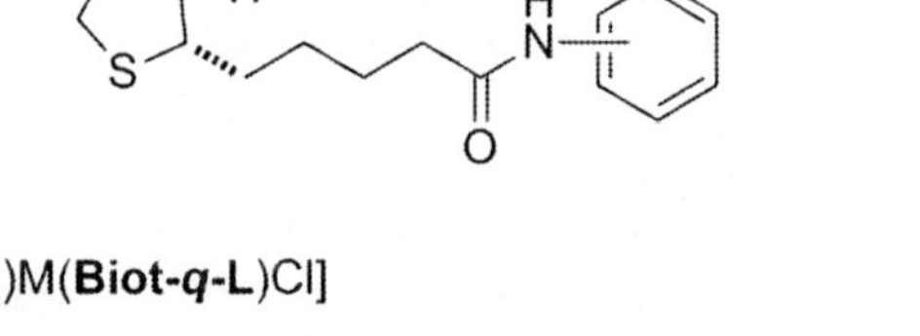

[η^n–(C$_n$R$_n$)M(**Biot-q-L**)Cl]

M = Ru(II), Ir(III), Rh(III)

η^n–C$_n$R$_n$ = cp*, η^6-p-cymene, η^6-benzene, η^6-durene, η^6-mesitylene
q = ortho-, meta-, para-

FIGURE 8.3   Biotinylated d^6-piano-stool complexes for the enantioselective ATH of ketones and imines in the presence of streptavidin.

TABLE 8.2 Selected Results for the ATH of Ketones Using Artifical Metalloenzymes[a]

Entry	Substrate	Protein	Complex	ee (%)	Conv. (%)
1		WT Sav	[η⁶-(p-cymene)Ru(**Biot-p-L**)Cl]	89 (R)	95
2		WT Sav	[η⁶-(benzene)Ru(**Biot-p-L**)Cl]	29 (R)	38
3		S112A	[η⁶-(p-cymene)Ru(**Biot-p-L**)Cl]	91 (R)	98
4		S112A	[η⁶-(benzene)Ru(**Biot-p-L**)Cl]	41 (R)	74
5		P64G	[η⁶-(p-cymene)Ru(**Biot-p-L**)Cl]	94 (R)	92
6		P64G	[η⁶-(benzene)Ru(**Biot-p-L**)Cl]	44 (S)	44
7		L124V	[η⁶-(p-cymene)Ru(**Biot-p-L**)Cl]	96 (R)	97
8		S112Y	[η⁶-(p-cymene)Ru(**Biot-p-L**)Cl]	97 (R)	79
10		S112A	[η⁶-(benzene)Ru(**Biot-p-L**)Cl]	51 (S)	44
11		L124 V	[η⁶-(p-cymene)Ru(**Biot-p-L**)Cl]	87 (R)	20
12		S122AK121N	[η⁶-(benzene)Ru(**biot-p-L**)Cl]	92 (S)	54
13		S112A	[η⁶-(p-cymene)Ru(**Biot-p-L**)Cl]	48 (R)	98
14		S112AK121T	[η⁶-(p-cymene)Ru(**Biot-p-L**)Cl]	88 (R)	Quant.[b]
15		S112AK121W	[η⁶-(benzene)Ru(**Biot-p-L**)Cl]	84 (R)	Quant.[b]
16		S112A	[η⁶-(p-cymene)Ru(**Biot-p-L**)Cl]	69 (R)	97
17		S112AK121T	[η⁶-(p-cymene)Ru(**Biot-p-L**)Cl]	90 (R)	Quant.[b]
18		S112R	[η⁶-(benzene)Ru(**Biot-p-L**)Cl]	70 (S)	95
19		S122AK121N	[η⁶-(benzene)Ru(**Biot-p-L**)Cl]	92 (S)	Quant.[b]
20		S112F	[η⁶-(p-cymene)Ru(**Biot-p-L**)Cl]	76 (R)	95

[a]Reaction conditions: The catalytic runs were performed at 55°C for 64 h using the mixed buffer HCO₂Na (0.48 M) + B(OH)₃ (0.41 M) + MOPS (0.16 M) at pH$_{initial}$ = 6.25. Ru/substrates/formate ratio 1:100:4000.
[b]Quant. = quantitative.

L124$_A$, K112$_B$, and K121$_B$ (A and B refer to two adjacent monomers of the tetrameric Sav, the catalytic moiety being located in monomer A). Among these residues, K112$_B$ seemed ideally positioned to interact with an incoming prochiral substrate. These observations led us to produce a mutagenesis library at position K121X or L124X. The resulting mutants (120 in total) were combined with [η⁶-(benzene)Ru(**Biot-p-L**)] and [η⁶-(p-cymene)Ru(**Biot-p-L**)Cl] and tested in the ATH. Remarkably, the enantioselectivity was further improved yielding up to 97% (R) and 92% (S) for aryl ketones, and 90% (R) for dialkyl ketones (Table 8.2). This latter result clearly demonstrates that the host protein plays a critical role in determining the enantioselectivity. Indeed, the reduction of dialkylketones by homogeneous ATHs is notably problematic as the invoked critical C−H···π interaction cannot operate.

8.2.3 Artificial Transfer Hydrogenation of Cyclic Imines

The catalytic ATH of imines is a versatile method for the synthesis of chiral amines. The practical simplicity of the ATH, avoiding the use of H₂ gas under high pressure,

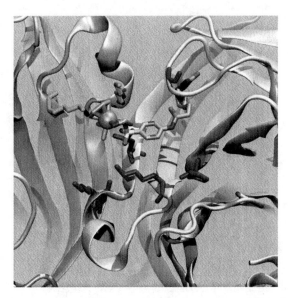

FIGURE 8.4 Closeup view of the X-ray crystal structure of the (S)-selective $[\eta^6$-(benzene)Ru(**biot-p-L**)Cl] $\subset$ S112K. Close-lying residues are depicted as wireframe: G48$_A$, H87$_A$, S112K$_A$, K121$_A$, L124$_A$ in dark gray and S112K$_B$, K121$_B$ in light gray.

is an advantage over hydrogenation reactions. In the past, the ATH of ketones has attracted more attention compared to the ATH of imines. Nevertheless, chiral amines are important compounds in pharmaceuticals and agrochemicals. Following a report by Bäckvall and coworkers [15] on the transfer hydrogenation of imines using Ru complexes, Noyori et al. first reported the ATH of imines employing HCO_2H as the hydrogen source [16].

Our interest in the ATH of imines emerged from our experience with the ATH of ketones using biotinylated metal complexes of type $[\eta^n-(C_nR_n)M(\textbf{Biot-}p\textbf{-L})Cl]$, M = Ru, Ir, or Rh (Fig. 8.3). We tested the ATH of imines on the salsolidine precursor using Biot-p-L in combination with WT Sav (Table 8.3) [17]. Among the metal complexes screened, the [Cp*Ir(Biot-p-L)Cl] complex showed the most promising catalytic activity and selectivity compared to [(p-cymene)Ru(Biot-p-L)Cl] and [Cp*Rh(Biot-p-L)Cl] complexes. This is in sharp contrast to that of the ATH of ketones, where the Ru catalysts outperformed both the Ir and the Rh catalysts. Combining [Cp*Ir(Biot-p-L)Cl] with the saturation mutagenesis library, Sav S112X yielded predominately (R)-salsolidine from the respective imine. Substitution by small amino acids at position S112 (S112G, S112A) resulted in high % ee at pH 6.5 (Table 8.3, entries 3 and 6). Further improvement in the (R)-selectivity was obtained by decreasing the temperature from 55°C down to 5°C, entry 7 in Table 8.3. Importantly, the enantioselectivity was inverted in the presence of cationic residues (S112K or S112R) to give (S)-salsolidine. A preparative scale synthesis was performed by using 0.025 mol % of catalyst to give 86% conversion with 96% ee (entry 9, Table 8.3). Thus, a total

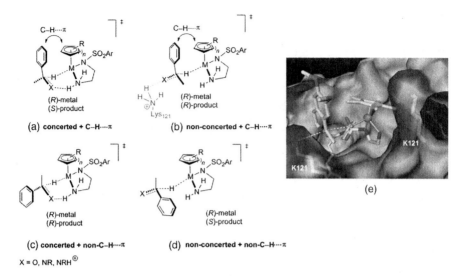

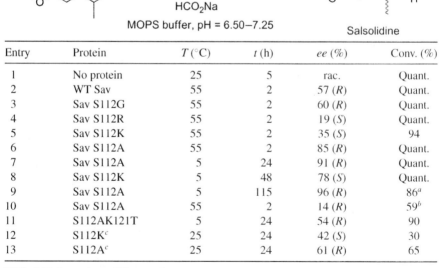

FIGURE 8.5 Possible TSs for the ATH of imines and ketones (a to d). Modeled contact between the imine N and K121 residue (b and e). Reproduced from Reference 17 with permission from Wiley-VCH Verlag GmbH & Co.

TABLE 8.3 Selected Results for the ATH of Cyclic Imines Using Artificial Metalloenzymes

Entry	Protein	T (°C)	t (h)	ee (%)	Conv. (%)
1	No protein	25	5	rac.	Quant.
2	WT Sav	55	2	57 (R)	Quant.
3	Sav S112G	55	2	60 (R)	Quant.
4	Sav S112R	55	2	19 (S)	Quant.
5	Sav S112K	55	2	35 (S)	94
6	Sav S112A	55	2	85 (R)	Quant.
7	Sav S112A	5	24	91 (R)	Quant.
8	Sav S112K	5	48	78 (S)	Quant.
9	Sav S112A	5	115	96 (R)	86[a]
10	Sav S112A	55	2	14 (R)	59[b]
11	S112AK121T	5	24	54 (R)	90
12	S112K[c]	25	24	42 (S)	30
13	S112A[c]	25	24	61 (R)	65

[a]86% yield observed on a 100-mg scale.
[b]4 equiv. biotin added with respect to Sav tetramer.
[c]Precipitated protein from cell extract.
MOPS, 3-morpholinopropanesulfonic acid.

turnover number (TON) up to 4000 can be achieved. Very importantly, this reaction can be performed on unpurified protein samples, recovered by ethanol precipitation method of a cell-free extract, yielding similar selectivity to that of the purified protein (entries 12 and 13, Table 8.3). This is especially attractive for the practical application of these artificial metalloenzymes for parallel screening as this method avoids cumbersome protein purification steps.

Interestingly, [Cp*Ir(Biot-p-L)Cl] $\subset$ S112A produced the same preferred enantiomer for both (R)-1-phenylethanol (from acetophenone reduction) and (R)-salsolidine. This result suggested that, unlike the homogeneous systems, both the ATH of imine and ketone proceed through the same enantioselection mechanism. This result emphasized the importance of second coordination sphere interactions provided by the host protein in dictating the outcome of enantioselectivity. Interestingly, it was observed that an empty biotin-binding site adjacent to Sav may be favorable for selectivity, as shown by the following example. Increasing the ratio of [Cp*Ir(Biot-p-L)Cl] : Sav (tetramer) from 1:4 to 4:4 led to decrease in enantioselectivity.

The X-ray structure of [Cp*Ir(**Biot-p-L**)Cl] $\subset$ S112A provided useful insight into the enantioselection mechanism of the imine reduction. The absolute configuration around the Ir center was assigned as (S), which is also the case in the (S)-selective [η^6-(benzene)Ru(**Biot-p-L**)Cl] $\subset$ S112K]. Based on the assumption that the (S)-configuration at the metal for [Cp*Ir(**Biot-p-L**)Cl] (and thus (R)-[Cp*Ir(**Biot-p-L**)H]) was indeed the competent catalyst, two TSs leading to the observed (R)-products (alcohol or amine) can be envisaged: a nonconcerted + C−H$\cdots\pi$ TS or a concerted + non-C−H$\cdots\pi$ TS (Figs. 8.5b and 8.5c). Modeling studies suggested that, unlike in the case of concerted + non-C−H$\cdots\pi$, no steric clash between the aromatic group of the substrate and the protein is present in a nonconcerted + C−H$\cdots\pi$ TS. In addition, the proximity of the lysine side chain K121$_B$ suggested that the ammonium group may deliver the proton to the product. To test this hypothesis, the double mutant S112AK121T was tested in the ATH of imines. To our delight, a marked decrease in the enantioselectivity was obtained (entry 11, Table 8.3). These results thus suggested that a nonconcerted + C−H$\cdots\pi$ mechanism may be operative in the ATH of both ketones and imines catalyzed by artificial metalloenzymes.

8.3   PALLADIUM-CATALYZED ALLYLIC ALKYLATION

The palladium-catalyzed asymmetric allylic alkylation (AAA) is a powerful tool for the elaboration of enantiopure high-added value compounds [18]. The widely accepted AAA mechanism proceeds via a nucleophilic addition on a palladium-bound η^3-allyl moiety [19], as depicted in Figure 8.6. Hence, the nucleophile does not bind to the metal prior to attack on the prochiral allyl moiety. This feature bears resemblance to enzymatic systems, whereby a substrate does not need to bind an enzyme for a reaction to proceed. Although palladium is absent from the enzymatic world, it is unrivalled in the homogeneous kingdom when it comes to C–C bond-forming reactions. Typical homogeneous ligands for such reactions include of diphosphine chelates [19].

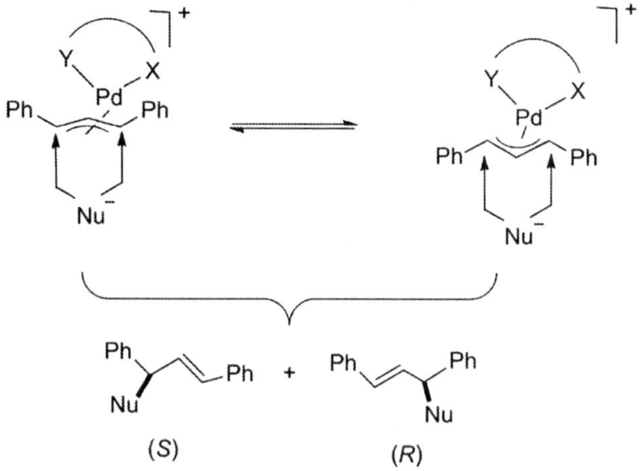

FIGURE 8.6 The postulated nucleophile-assisted enantiodiscrimination mechanism in the palladium-catalyzed AAA reaction.

We speculated that the library of biotinylated diphosphine ligands used for the creation of artificial hydrogenases could prove useful to implement artificial allylic alkylases based on the biotin-(strept)avidin technology. We thus proceeded to screen a library of biotinylated catalysts prepared *in situ* by combining $[Pd(\eta^3-R_2allyl)Cl]_2$ (R = H, Ph) and the biotinylated ligands, displayed in Scheme 8.1, in the presence of (strept)avidin [20]. In the AAA of 1,3-diphenylallylacetate, all the metal–ligand combinations yielded only the hydrolysis product, 1,3-diphenylallyl alcohol. The AAA efficiency was significantly improved by adding the surfactant, didodecyldimethylammonium bromide (DMB), which suppressed the hydrolysis of the substrate to yield the corresponding alcohol. Various biotinylated ligands were screened in the presence of (strept)avidin and DMB for AAA. Among the ligands screened, **Biot-4**^{ortho}**-1** behaved very well, affording >90% *ee* (*R*), with high conversion. When aqueous solutions containing up to 45% dimethyl sulfoxide (DMSO) were used instead of DMB, the *ee* could be further improved to 95% (entry 3, Table 8.4). Achieving good (*S*)-selectivities proved problematic however. To overcome this challenge, we had to resort to a double mutant used in conjunction with **Biot-(*R*)-Pro-1** to afford the product in 82% *ee* (*S*) (entry 7, Table 8.4). Again here, the AAA results highlight the critical importance of the chemogenetic optimization strategy to achieve good activities and enantioselectivities.

8.4   OXIDATION REACTION CATALYZED BY ARTIFICIAL METALLOENZYMES

8.4.1   Artificial Sulfoxidase

The asymmetric sulfoxidation is attracting increased interest mainly due to the importance of enantiopure sulfoxides present in several drugs [21]. Jacobsen, Katsuki, and

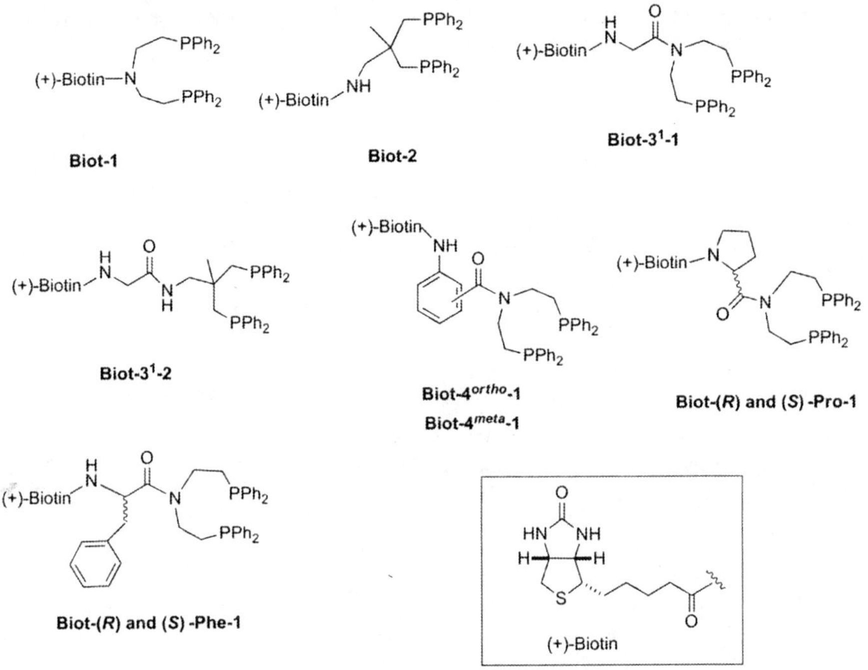

SCHEME 8.1   Biotinylated ligands used in the asymmetric hydrogenation and allylic alkylation reactions.

their coworkers developed versatile homogeneous catalysts based on Schiff base ligands for sulfoxidation and epoxidation reactions [22–25]. In the field of artificial metalloenzymes, the enantioselective sulfoxidation is the transformation that has attracted most interest (Scheme 8.2) [26–33]. Inspired by Jacobsen's systems, our initial effort in the catalytic asymmetric sulfoxidation reaction was based on biotinylated Mn-salen complexes incorporated into streptavidin, Mn-**Sal-x** ⊂ Sav (**x** = **a**, **b**, or **c**) [34]. Using thioanisole as substrate and H_2O_2 as the oxidant, the Mn-**Sal-x** ⊂ Sav systems displayed moderate conversion and low selectivity. We hypothesized that the Schiff base ligand may be hydrolyzed during catalysis and thus explored the possibility of introducing the "naked" metal moiety into Sav, abbreviated [VO]$^{2+}$ ⊂ WT sav [35]. Upon combining 2 mol % VOSO$_4$ with 1 mol % tetrameric Sav, an artificial sulfoxidase resulted. In the presence of 5 equivalents t-BuOOH at pH 2.2, a variety of aryl-alkyl- and dialkyl sulfides were oxidized selectively to the corresponding sulfoxides in moderate to good enantioselectivities (Table 8.5). Catalysis in the presence of 4 equivalents of biotin yielded racemic sulfoxide product. Similarly, the low biotin affinity Sav mutant (Sav D128A) showed no enantioselectivity. These results provided strong evidence that the asymmetric sulfoxidation was indeed taking place inside the biotin-binding pocket, that is, the catalytically active vanadyl moiety was located inside this pocket. Additionally, electron paramagnetic resonance (EPR)

TABLE 8.4 Selected Results for the AAA Using Artificial Metalloenzymes Based on the Biotin–Streptavidin Technology

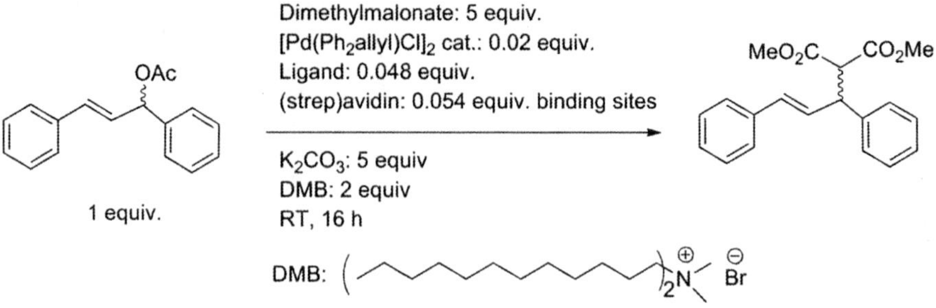
OAc
1 equiv.
Dimethylmalonate: 5 equiv.
[Pd(Ph2allyl)Cl]2 cat.: 0.02 equiv.
Ligand: 0.048 equiv.
(strep)avidin: 0.054 equiv. binding sites
K2CO3: 5 equiv
DMB: 2 equiv
RT, 16 h
MeO2C
CO2Me
DMB:
N
Br

Entry	Ligand	Protein	*ee* (%)	Conv. (%)
1	**Biot-4^{ortho}-1**	Sav S112A	90 (R)	95
2	**Biot-4^{ortho}-1**	Sav S112A[a]	93 (R)	20
3	**Biot-4^{ortho}-1**	Sav S112A[b]	95 (R)	90
4	**Biot-4^{ortho}-1**	Sav S112Q	31 (S)	96
5	**Biot-(R)-Pro-1**	Sav S112Y	80 (R)	87
6	**Biot-(R)-Pro-1**	Sav S112G	54 (S)	96
7	**Biot-(R)-Pro-1**	Sav S112G-V47G	82 (S)	92

[a] No DMB added.
[b] 45% DMSO was added instead of DMB.

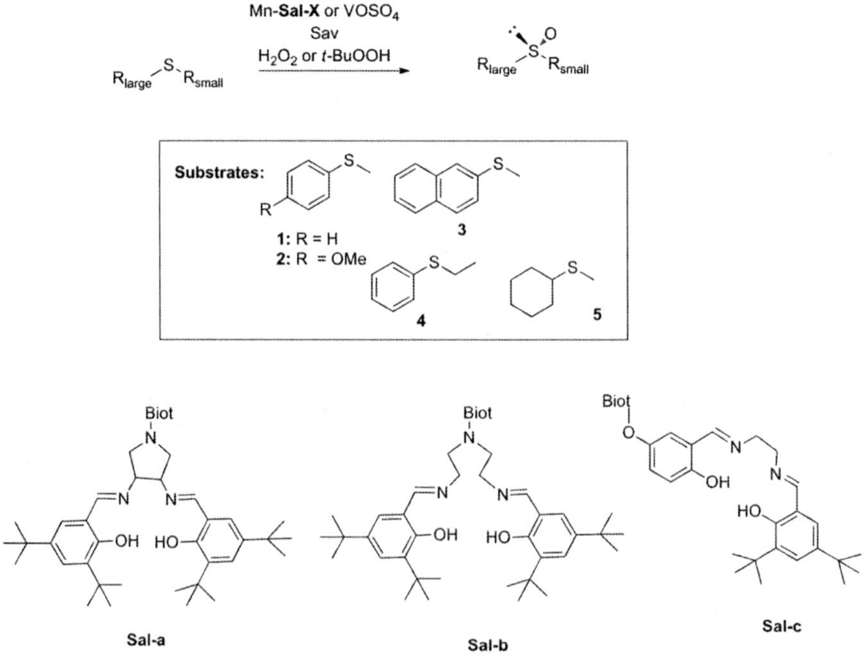
Mn-Sal-X or VOSO4
Sav
H2O2 or t-BuOOH
Rlarge
Rsmall
S
O
Substrates:
1: R = H
2: R = OMe
3
4
5
Biot
Sal-a
Sal-b
Sal-c
OH
HO

SCHEME 8.2 Artificial metalloenzymes for the enantioselective sulfoxidation.

TABLE 8.5   Selected Results for the Enantioselective Sulfoxidation Catalyzed by Artificial Metalloenzymes with Streptavidin as Host Protein

Entry	Catalyst	Substrate	Protein	*ee* (%)	Conv. (%)
1	No metal	1	WT Sav	4 (*R*)	7
2[a]	Mn-**Sal-c**	1	Sav S112D	13 (*R*)	32
3[b]	VOSO$_4$	1	WT Sav	46 (*R*)	94
4[b]	VOSO$_4$	2	WT Sav	90 (*R*)	Quant.
5[b]	VOSO$_4$	2	WT Sav	0	96
6[b]	VOSO$_4$	2	Sav D128A	0	97
7[b]	VOSO$_4$	3	WT Sav	93 (*R*)	53
8[b]	VOSO$_4$	4	WT Sav	90 (*R*)	96
9[b]	VOSO$_4$	4	WT Sav	86 (*R*)	61

[a]0.1 mM Mn-**Sal-c**; 0.033 mM Sav; 5mM sulfide; 5 mM H$_2$O$_2$; RT.
[b]0.2 mM VOSO$_4$; 0.1 mM Sav; 10 mM sulfide; 50 mM *t*-BuOOH; RT.

analysis of [VO]$^{2+}$ $\subset$ WT sav displayed a spectrum identical to that of [VO(H$_2$O)$_5$]$^{2+}$, suggesting the vanadium ion interacts only via second coordination sphere contacts with the host protein.

8.4.2   Asymmetric *cis*-Dihydroxylation

Sharpless's method for asymmetric *cis*-dihydroxylation (AD) of olefins is perhaps the most versatile enantioselective oxidation reaction used in industry [36]. The Sharpless AD system is based on catalytic amount of OsO$_4$ combined with cinchona alkaloid ligands derived from a quinidine or a quinine framework. Among the substrates screened, most classes of prochiral olefins show excellent enantioselectivities exceeding 90% *ee*; however, *cis*-olefins often yield moderate enantioselectivities. In nature, AD of arenes is achieved by a family of nonheme iron-dependent Rieske dioxygenase enzymes for bioremediation purpose [37]. Interestingly, the proposed reaction mechanism for both the enzyme- and synthetic Os-catalyzed AD involves second coordination sphere interaction between the substrate and the active site, whereby the substrate is not directly bound to the metal center in the TS (Fig. 8.7) [38, 39].

In search for an artificial metalloenzyme that catalyze AD reaction, we screened the catalytic activity of OsO$_4$ incorporated within various protein scaffolds [40]. Using α-methyl styrene as the substrate, wild-type streptavidin showed better selectivity and TON as compared to other proteins (Table 8.6). Interestingly, crystal-soaking experiments followed by X-ray analysis, revealed anomalous electron density (assigned to osmium) at positions H127, H87, K80, and K131. Removal of these side chains by site-directed mutagenesis led to only a marginal decrease in the enantioselectivity. These results highlighted that the thermodynamic products, revealed by X-ray crystallography, were not catalytically competent, merely acting as osmium sink. Competition experiments performed in the presence of biotin led to a drastic decrease in enantioselectivity. These results suggested that the catalytically active Os moiety is

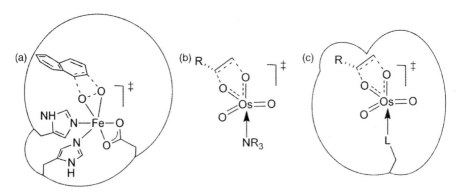

FIGURE 8.7 (a) Proposed TS structures for naphthalene dioxygenase. (b) Sharpless's osmium-catalyzed asymmetric dihydroxylation. Upon incorporation within a host protein, OsO_4 may act as an artificial dihydroxylase. Reproduced from Reference 40 with permission from Wiley-VCH Verlag GmbH & Co.

located in the proximity of the biotin-binding pocket. To gain further indirect insight on the Os-binding site and to improve the catalytic performance of the artificial dihydroxylases, screening with various Sav mutants at positions Sav S112, K121, and L124 was carried out. These results highlighted the critical role of lysine 121 as well as the potential for optimization by mutagenesis at position S112. In general, these hybrid systems showed broad substrate scope with good selectivity in many cases. Excellent selectivity was observed for allyl phenyl sulfide (Table 8.6, entries 12 and 13) and *cis*-β-methyl styrene (Table 8.6, entry 20) in the presence of SAV mutants D128A, S112Y, and S122T, respectively. However, a drawback was the modest TON obtained. Removal of the Os-binding sites identified by X-ray crystallography may allow to address this issue.

8.5 SUMMARY AND PERSPECTIVES

Thus far, seven reactions have been implemented with artificial metalloenzymes based on the biotin–avidin technology. It thus appears that any organometallic reaction compatible with aqueous solutions may be amenable to the creation of artificial metalloenzymes. In order to give a competitive advantage over homogeneous catalysts, the next step requires directed evolution protocols for the optimization of these hybrid catalysts. For this purpose, cell-debris-tolerant organometallic catalysts need to be identified. Indeed, as directed evolution relies on *en-masse* screening, it will not be possible to purify the protein mutants before proceeding to the screening. A first major advance in this direction was achieved with the identification of [Cp*Ir(**Biot-p-L**)Cl] which was shown to be tolerant to cell debris. Further improvements in this direction are necessary to ultimately allow screening thousands of mutants expressed on Petri dishes.

TABLE 8.6   OsO$_4$·Streptavidin Catalyzed Asymmetric *cis*-Dihydroxylation of Olefins

Entry	Olefin	Protein	*ee* (%)	TON[a]
1		WT Sav	95 (*R*)	27
2		Avidin	2 (*R*)	16
3		BSA	77 (*S*)	4
4		Sav S112M	97	16
5		Sav D128E	80 (*R*)	21
6		WT Sav + biotin[b]	9 (*R*)	13
7		WT Sav	40 (*S*)	13
8		S112Y	82 (*S*)	14
9		Sav D128A	77 (*R*)	21
10		Sav D128E	0	9
11		WT Sav	2 (*R*)	4
12		Sav S112Y	71 (*S*)	7
13		Sav D128A	71 (*R*)	10
14		Sav D128E	12 (*R*)	5
15		WT Sav	30 (1*R*,2*S*)	13
16		Sav S112Y	7 (1*S*,2*R*)	12
17		Sav S112M	41 (1*R*,2*S*)	12
18		WT Sav	90 (1*R*,2*S*)	26
19		Sav S112Y	45 (1*R*,2*S*)	12
20		Sav S112T	92 (1*R*,2*S*)	16
21		WT Sav	62 (3*R*,4*R*)	6
22		Sav S112Y	26 (3*R*,4*R*)	5
23		Sav S112T	68 (3*R*,4*R*)	6

[a] TON = mol product/mol K$_2$[OsO$_2$(OH)$_4$].
[b] 1.05 equivalents of (+)-biotin were added per protein monomer.

Compared to other protein scaffolds for the implementation of artificial metalloenzymes, (strept)avidin seems like a "privileged" scaffold. The reasons for this versatility remain mysterious. Our current hypothesis is that proteins with a given catalytic function are difficult to use as host for the creation of artificial metalloenzymes. Proteins which merely act as transporters (myoglobin, serum albumins, (strept)avidin, etc.) may be more suited for the creation of artificial metalloenzymes. We are currently actively pursuing both these avenues.

REFERENCES

[1] Lu, Y.; Berry, S. M.; Pfister, T. D. *Chem. Rev.* **2001**, *101*, 3047–3080.

[2] Qi, D.; Tann, C.-M.; Haring, D.; Distefano, M. D. *Chem. Rev.* **2001**, *101*, 3081–3111.

[3] Ward, T.R. *Acc. Chem. Res.* **2011**, *44*, 47–57.

[4] Yamamura, K.; Kaiser, E. T. *J. Chem. Soc., Chem. Commun.* **1976**, 830–831.

[5] Wilson, M. E.; Whitesides, G. M. *J. Am. Chem. Soc.* **1978**, *100*, 306–307.

[6] Pazy, Y.; Kulik, T.; Bayer, E. A.; Wilchek, M.; Livnah, O. *J. Biol. Chem.* **2002**, *277*, 30892–30900.

[7] Collot, J.; Gradinaru, J.; Humbert, N.; Skander, M.; Zocchi, A.; Ward, T. R. *J. Am. Chem. Soc.* **2003**, *125*, 9030–9031.

[8] Klein, G.; Humbert, N.; Gradinaru, J.; Ivanova, A.; Gilardoni, F.; Rusbandi, U. E.; Ward, T. R. *Angew. Chem. Int. Ed.* **2005**, *44*, 7764–7767.

[9] Rusbandi, U. E.; Lo, C.; Skander, M.; Ivanova, A.; Creus, M.; Humbert, N.; Ward, T. R. *Adv. Syn. Catal.* **2007**, *349*, 1923–1930.

[10] Ikariya, T.; Murata, K.; Noyori, R. *Org. Biomol. Chem.* **2006**, *4*, 393–406.

[11] Yamakawa, M.; Yamada, I.; Noyori, R. *Angew. Chem. Int. Ed.* **2001**, *40*, 2818–2821.

[12] Yamakawa, M.; Ito, H.; Noyori, R. *J. Am. Chem. Soc.* **2000**, *122*, 1466–1478.

[13] Letondor, C.; Humbert, N.; Ward, T. R. *Proc. Natl. Acad. Sci. U. S. A.* **2005**, *102*, 4683–4687.

[14] Creus, M.; Pordea, A.; Rossel, T.; Sardo, A.; Letondor, C.; Ivanova, A.; Letrong, I.; Stenkamp, R.E.; Ward, T. R. *Angew. Chem. Int. Ed.* **2008**, *47*, 1400–1404.

[15] Wang, G. Z.; Bäckvall, J. E. *J. Chem. Soc., Chem. Commun.* **1992**, 980–982.

[16] Uematsu, N.; Fujii, A.; Hashiguchi, S.; Ikariya, T.; Noyori, R. *J. Am. Chem. Soc.* **1996**, *118*, 4916–4917.

[17] Dürrenberger, M.; Heinisch, T.; Wilson, Y. M.; Rossel, T.; Nogueira, E.; Knörr, L.; Mutschler, A.; Kersten, K.; Zimbron, M. J.; Pierron, J.; Schirmer, T.; Ward, T. R. *Angew. Chem. Int. Ed.* **2011**, *50*, 3026–3029.

[18] Trost, B. M.; Crawley, M. L. *Chem. Rev.* **2003**, *103*, 2921–2944.

[19] Trost, B. M.; Machacek, M. R.; Aponick, A. *Acc. Chem. Res.* **2006**, *39*, 747–760.

[20] Pierron, J.; Malan, C.; Creus, M.; Gradinaru, J.; Hafner, I.; Ivanova, A.; Sardo, A.; Ward, T. R. *Angew. Chem. Int. Ed.* **2008**, *47*, 701–705.

[21] Carreno, M. C. *Chem. Rev.* **1995**, *95*, 1717–1760.

[22] Zhang, W.; Loebach, J. L.; Wilson, S. R.; Jacobsen, E. N. *J. Am. Chem. Soc.* **1990**, *112*, 2801–2803.

[23] Jacobsen, E. N.; Zhang, W.; Muci, A. R.; Ecker, J. R.; Deng, L. *J. Am. Chem. Soc.* **1991**, *113*, 7063–7064.

[24] Tanaka, H.; Nishikawa, H.; Uchida, T.; Katsuki, T. *J. Am. Chem. Soc.* **2010**, *132*, 12034–12041.

[25] Matsumoto, K.; Yamaguchi, T.; Fujisaki, J.; Saito, B.; Katsuki, T. *Chem. Asian. J.* **2008**, *3*, 351–358.

[26] van de Velde, F.; Könemann, L.; van Rantwijk, F.; Sheldon, R. A. *Chem. Commun.* **1998**, 1891–1892.

[27] Ueno, T.; Koshiyama, T.; Ohashi, M.; Kondo, K.; Kono, M.; Suzuki, A.; Yamane, T.; Watanabe, Y. *J. Am. Chem. Soc.* **2005**, *127*, 6556–6562.

[28] Ohashi, M.; Koshiyama, T.; Ueno, T.; Yanase, M.; Fujii, H.; Watanabe, Y. *Angew. Chem. Int. Ed.* **2003**, *42*, 1005–1008.

[29] Carey, J. R.; Ma, S. K.; Pfister, T. D.; Garner, D. K.; Kim, H. K.; Abramite, J. A.; Wang, Z.; Guo, Z.; Lu, Y. *J. Am. Chem. Soc.* **2004**, *126*, 10812–10813.

[30] Mahammed, A.; Gross, Z. *J. Am. Chem. Soc.* **2005**, *127*, 2883–2887.

[31] Rousselot-Pailley, P.; Bochot, C.; Marchi-Delapierre, C.; Jorge-Robin, A.; Martin, L.; Fontecilla-Camps, J. C.; Cavazza, C.; Menage, S. *ChemBioChem* **2009**, *10*, 545–552.

[32] Ricoux, R.; Allard, M.; Dubuc, R.; Dupont, C.; Marechal, J. D.; Mahy, J. P. *Org. Biomol. Chem.* **2009**, *7*, 3208–3211.

[33] Sansiaume, E.; Ricoux, R.; Gori, D.; Mahy, J.-P. *Tetrahedron Asymm.* **2010**, *21*, 1593–1600.

[34] Pordea, A.; Mathis, D.; Ward, T. R. *J. Organomet. Chem.* **2009**, *694*, 930–936.

[35] Pordea, A.; Creus, M.; Panek, J.; Duboc, C.; Mathis, D.; Novic, M.; Ward, T. R. *J. Am. Chem. Soc.* **2008**, *130*, 8085–8088.

[36] Kolb, H.C.; Vannieuwenhze, M. S.; Sharpless, K. B. *Chem. Rev.* **1994**, *94*, 2483–2547.

[37] Costas, M.; Mehn, M. P.; Jensen, M. P.; Que, L., Jr. *Chem. Rev.* **2004**, *104*, 939–986.

[38] Lu, Y.; Yeung, N.; Sieracki, N.; Marshall, N. M. *Nature* **2009**, *460*, 855–862.

[39] Karlsson, A.; Parales, J. V.; Parales, R. E.; Gibson, D. T.; Eklund, H.; Ramaswamy, S. *Science* **2003**, *299*, 1039–1042.

[40] Köhler, V.; Mao, J.; Heinisch, T.; Pordea, A.; Sardo, A.; Wilson, Y. M.; Knörr, L.; Creus, M.; Prost, J. C.; Schirmer, T.; Ward, T. R. *Angew. Chem. Int. Ed.* **2011**, *50*, 10863–10866.

PART IV

APPLICATIONS IN BIOLOGY

9

SELECTIVE LABELING AND IMAGING OF PROTEIN USING METAL COMPLEX

YASUTAKA KURISHITA AND ITARU HAMACHI

9.1 INTRODUCTION

Protein is one of the most important molecules in biological systems, as well as nucleic acid, lipid, and carbohydrate. Proteins are composed of only 20 kinds of amino acids and folded into individual three-dimensional structures, to exhibit sophisticated functions such as molecular recognition, enzymatic catalysis, signal transduction, mass transportation, and so on. Biological systems maintain their homeostasis by regulating protein functions, so that it is not too much to emphasize that proteins primarily regulate most of the complicated biological systems. In various aspects of protein researches, protein labeling is one of the most powerful methods for their functional analysis. For example, when a protein of interest (POI) is labeled with a fluorescent dye, it can be visualized by fluorescence, so that we can monitor and evaluate its expression, localization, and function in rather easy and direct manner. Therefore, how to selectively label POI with probes such as fluorescent molecules under various conditions is crucial for the subsequent analysis.

Many of the protein labeling methods have been reported so far, which are roughly divided into two categories. One is the so-called genetically encodable labeling method, the other is based on posttranslational modification. A representative example of the former method employs fluorescent proteins such as green fluorescent protein (GFP) [1, 2]. Fluorescent proteins are fused to POI by genetic engineering and the

Coordination Chemistry in Protein Cages: Principles, Design, and Applications, First Edition.
Edited by Takafumi Ueno and Yoshihito Watanabe.
© 2013 John Wiley & Sons, Inc. Published 2013 by John Wiley & Sons, Inc.

fused protein is expressed in live cells or organisms. Due to the high specificity and the established protocol for expression in live cells, the GFP-fusion method is now recognized as a standard for fluorescence imaging. Being a very powerful method, it still has some drawbacks, that is the rather large size of fluorescent proteins sometimes interferes/disturbs the structure or function of the fused proteins. An example of the latter methods, on the other hand, is based on chemical bioconjugation [3–6]. In general, the bioconjugation often suffers from its low selectivity in terms of protein and its site to be labeled. Thus, this chemical labeling was conducted toward purified proteins in test tube, not under crude cellular conditions in most cases. Many efforts have been made to improve such low selectivity, and some of the approaches based on the affinity-based protein labeling seem promising in the future [7, 8].

Tag–probe pair method uses both a genetically encoded tag and a synthetic compound for protein labeling, that is a combination of the selective expression of tag using genetic method and a flexible usage of synthetic compounds such as fluorescent, MRI active, affinity, or cross-linker reagents. Thus, the tag–probe pair strategy may be categorized into the middle of the above mentioned two methods. This strategy can be further divided into three different methods in detail, which depend on the incorporated tag fragment, that is, enzyme tag, enzyme-substrate tag, and metal-chelating peptide tag (Fig. 9.1).

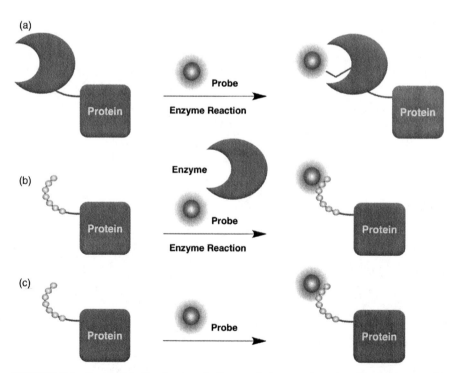

FIGURE 9.1 Protein labeling by genetically encoded tag–probe pairs. (a) enzyme tag, (b) substrate-peptide tag and (c) metal-chelating tag.

The first one uses whole enzyme as a tag, where a suicide substrate modified with a probe can be irreversibly attached to the tag enzyme by the enzymatic reaction [9–13]. The method such as SNAP-tag [9] or Halo-tag [10] is demonstrated to be applicable to intracellular protein labeling. However, the size of the enzyme tag may be problematic, as in the case of the GFP-fusion method. In the second one, a small peptide that can become a suitable substrate of a specific enzyme is used as a tag, which may suppress serious interference to the fused protein functions [14–19]. This method is, however, too hard to be used in cells due to the serious requirements both of the limited conditions for the enzymatic reaction and of the sufficient expression level of the corresponding enzyme. The last one is based on a metal-chelation system, in which a short peptide is selectively bound to a metal complex using coordination chemistry. This method has several advantages, including that the artificial peptide is small enough to minimize the interference to the fused protein, and no additional expression of other factors such as enzymes is needed. In this chapter, we focus on this metal-chelating method and describe the representative examples as a newly emerged strategy for protein analysis.

9.2   TAG–PROBE PAIR METHOD USING METAL-CHELATION SYSTEM

9.2.1   Tetracysteine Motif/Arsenical Compounds Pair

Combination of tetracysteine motif (tetraCys) and arsenical compounds was reported by Tsien et al., as a pioneering example of tag–probe pair methods for protein labeling (Fig. 9.2a) [20]. They found that tetraCys motif CCXXCC, where X is noncysteine amino acid, was recognized by biarsenical compounds having a fluorescein scaffold, named FlAsH, with high affinity and specificity. They clearly demonstrated the fluorescent imaging of the tetraCys-fused protein by labeling with FlAsH in live cell systems. FlAsH, which was virtually nonfluorescent, showed a dramatically large fluorescence increment upon the reversible complex formation with tetraCys motif. Following FlAsH, they developed analogs of FlAsH, such as ReAsH, a resorufin derivative which is fluorescent in the red region, and HoXAsH, a xanthone derivative which is fluorescent in the blue region (Fig. 9.2b). They were nonfluorescent by themselves and rapidly formed a fluorescent complex with tetraCys motif, similar to FlAsH [21].

In addition to varying fluorescent arsenical probes, the peptide tag part was also optimized in order to extend the utility of tetraCys motif/arsenical compounds pair. Tsien et al. quantitatively evaluated the affinity of FlAsH for a variety of tetra-Cys motifs. They initially placed four cysteins at the i, $i + 1$, $i + 4$, and $i + 5$ positions of an α-helix, so that the four thiol groups would form a parallelogram on one side of the helix. After a meanwhile, it was clear that alternation of the amino acid residues between the cysteins to Pro-Gly showed the tightest binding to FlAsH, suggesting that the more preferable peptide conformation is a hairpin rather than α-helix [21]. To develop more sophisticated peptide sequences, they subsequently attempted to optimize the peptide sequence using a retrovirally

(a)

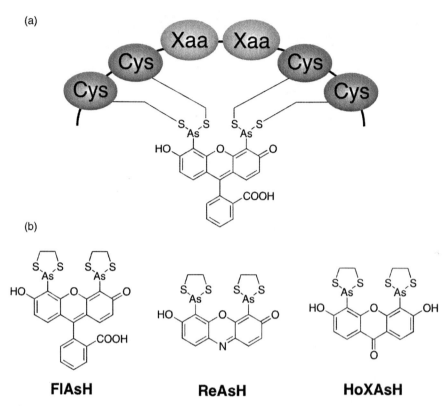

(b)

FlAsH   **ReAsH**   **HoXAsH**

FIGURE 9.2   (a) TetraCys/arsenical compounds pair for protein labeling. (b) Molecular structures of biarsenical conjugated fluorophores.

transduced mammalian cell-based library approach, by which they randomized the sequence surrounding the tetraCys motif [22]. They finally found that HRWCCPGC-CKTF and FLNCCPGCCMEP exhibited the fluorescence more strongly than the original sequence.

As aforementioned, POI fused with tetraCys motif can be selectively labeled by arsenical compounds. Various utilities of this method have been demonstrated by the following examples. Tsien et al. used this method for bioimaging of a selective protein, revealing that newly synthesized connexin43 in live cell was transported predominantly in 100–150 nm vesicles to the plasma membrane and incorporated at the periphery of existing gap junctions [23]. Bishop et al. reported that tetraCys/arsenical compound pair could allosterically modulate the enzymatic activity of the tethered protein tyrosine phosphatase (PTP) [24]. The catalytic activity of tetraCys-fused PTP strongly decreased when incubated with FlAsH. In addition, Kodadek et al. demonstrated that tetraCys/arsenical compound pair was useful to attach a cross-linker reagent for protein–protein interaction analysis [25].

9.2.2   Oligo-Histidine Tag/Ni(II)-NTA Pair

The complementary recognition pair of oligo-histidine tag (His-tag) and Ni(II) complex nitrilotriacetic acid (Ni(II)-NTA) is an indispensable tool for biotechnology (Fig. 9.3) [26]. A variety of proteins, indeed, have been expressed as the fused ones with His-tag for purification and immobilization [27, 28].

Ebright et al. reported for the first time that the His-tag-fused proteins could be labeled by fluorophore having one or two Ni(II)-NTA (**1**, **2**-2Ni(II), **3**, **4**-Ni(II)) with high affinity and specificity (Fig. 9.4) [29]. After a meanwhile, Vogel et al. reported that the His-tag-fused membrane protein could be labeled using the corresponding Ni(II)-NTA probes (**5**, **6**-Ni(II)) in live cell (Fig. 9.4) [30]. They labeled Ni(II)-NTA conjugated chromophore to a ligand-gated ion channel, ionitropic $5HT_3$ serotonin receptor ($5HT_3R$), and monitored the ligand-receptor interactions by fluorescence measurement.

In order to improve the relatively low affinity between His-tag and Ni(II)-NTA for application in cellular systems, Piehler et al. attempted to enhance the affinity between His-tag and Ni(II)-NTA by using multivalent strategy [31]. They demonstrated that the high selectivity of tris Ni(II)-NTA (**7**, **8**, **9**-3Ni(II)) toward cumulated histidines (His10) enabled selective labeling of proteins in cell lysates and on live cell surfaces (Fig. 9.5).

However, the use of paramagnetic Ni(II) decreases the inherent fluorescence quantum yield of the tethered fluorophore, leading to the limitation of the utility of the fluorescent labeling based on the Ni(II)-chelation technique for bioimaging. Lippard et al. reported that the photophysical properties of the fluorophore are not significantly influenced by Ni(II)-NTA functionality that is installed at the 6-position of the fluorescein ring (**10**-Ni(II)) (Fig. 9.6a) [32]. Tsien et al., on the other hand, attempted

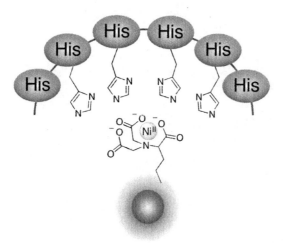

FIGURE 9.3   His-tag/Ni(II)-NTA pair for protein labeling.

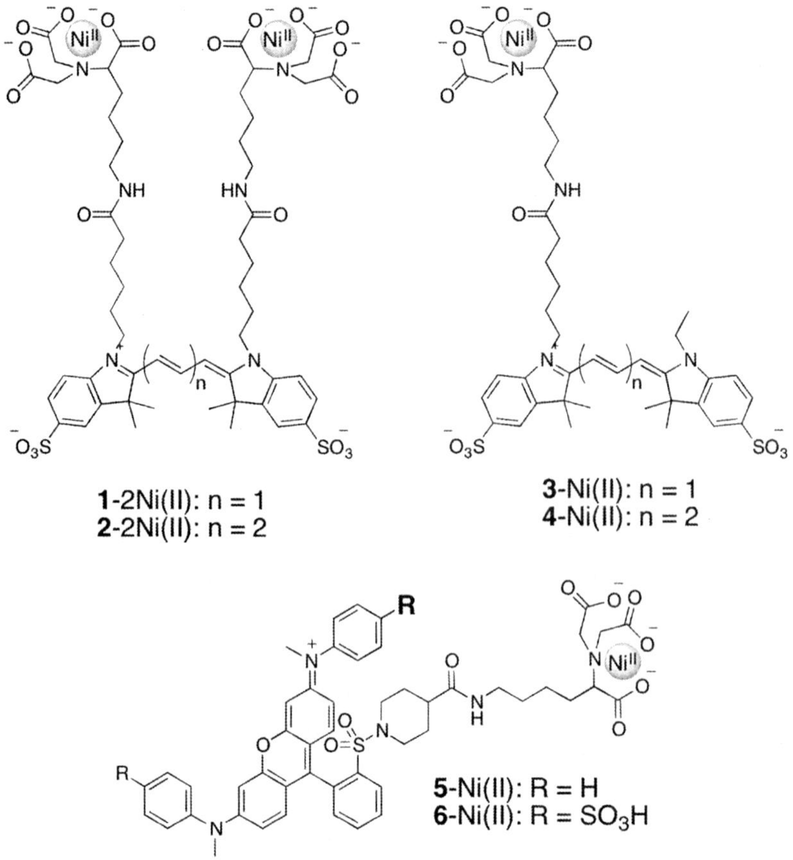

1-2Ni(II): n = 1
2-2Ni(II): n = 2

3-Ni(II): n = 1
4-Ni(II): n = 2

5-Ni(II): R = H
6-Ni(II): R = SO$_3$H

FIGURE 9.4   Molecular structures of chromophore-conjugated Ni(II)-NTA.

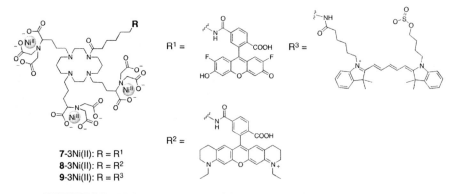

R^1 =

R^2 =

R^3 =

7-3Ni(II): R = R^1
8-3Ni(II): R = R^2
9-3Ni(II): R = R^3

FIGURE 9.5   Molecular structures of fluorophore-conjugated tris Ni(II)-NTA.

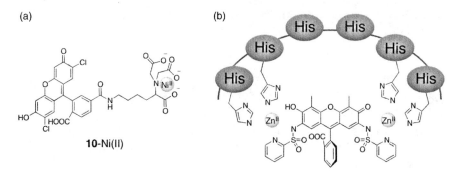

FIGURE 9.6   (a) Molecular structure of fluorophore-conjugated Ni(II)-NTA. (b) His-tag/ HisZiFit pair for protein labeling.

to overcome this drawback by replacing Ni(II) with Zn(II) complex, named HisZiFiT [33]. Because Zn(II) is not a fluorescent quencher, the fluorophore of the probe can be directly participated in metal chelator. HisZiFiT was contributed to resolve a current controversial issue concerting externalization of the stromal interaction molecule STIM1 upon depletion of Ca(II) from ER (Fig. 9.6b).

Although His-tag/Ni(II)-NTA pair is one of the most useful techniques for protein labeling, the reversibility sometimes limits its utility, particularly in the postlabeling analyses such as SDS-PAGE or Western blotting. In the recent years, methods for the covalent labeling of the His-tag-fused proteins have been reported (Fig. 9.7). Auer et al. reported a strategy that used Ni(II)-NTA probe with a photoreactive moiety, generating a covalent linkage between the His-tag and the probe by UV light irradiation [34]. Hamachi et al. reported a strategy based on chemical conjugation, which used a nucleophilic reaction between a histidine residue of the tag and the

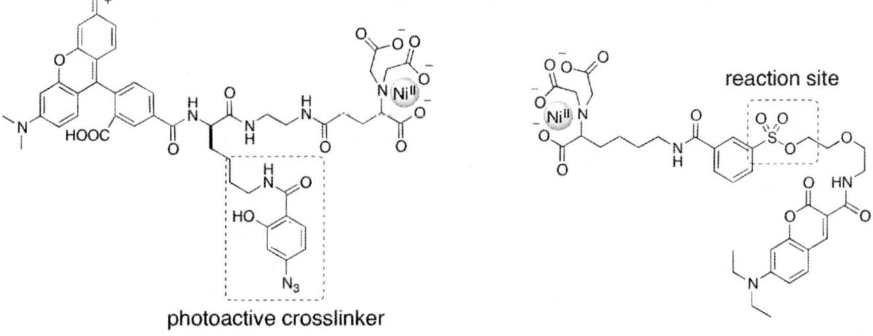

photoactive crosslinker

reaction site

11-Ni(II)   **12**-Ni(II)

FIGURE 9.7   Molecular structures of fluorophore-conjugated Ni(II)-NTA for covalent labeling.

electrophilic tosylate group of the Ni(II)-NTA probe accelerated by the proximity effect [35].

9.2.3   Oligo-Aspartate Tag/Zn(II)-DpaTyr Pair

Hamachi et al. developed another tag–probe pair which utilized the multivalent coordination chemistry between oligo-aspartate sequence (Asp-tag), which is a combination of a rare peptide sequence among the naturally occurring sequence, and a dinuclear Zn(II) complex based on the L-tyrosine scaffold (Zn(II)-DpaTyr) (Fig. 9.8) [36].

In the early stage of this study, the binding affinity of **1**-2Zn(II), a binuclear Zn(II)-DpaTyr complex, to the sequential oligo-aspartate peptides (D2-D5) was evaluated by isothermal titration calorimetry (ITC) experiments (Fig. 9.9). The binding affinity increased by almost 10-fold for a one-unit increase in Asp from D2 to D4, and this trend seems to be saturated at the D4 peptide. The binding affinity between D4 peptide and **1**-2Zn(II), was $(6.9 \pm 0.2) \times 10^5$ M^{-1}. For further enhancement of the binding affinity for Asp-tag, they utilized multivalent strategy. The binding affinity between D4-G-D4 peptide and **2**-4Zn(II), a tetranuclear Zn(II)-DpaTyr complex was determined to be $(1.8 \pm 0.3) \times 10^7$ M^{-1}, which is strong enough for biological applications.

Actually, the muscarinic acetylcholine receptor (mAChR) tethering Asp-Tag, which is expressed in CHO cell, was successfully labeled and fluorescently visualized by **3**-4Zn(II), a tetranuclear Zn(II)-DpaTyr complex having a cyanine dye (Fig. 9.10).

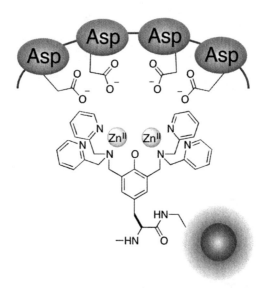

FIGURE 9.8   Asp-tag/Zn(II)-DpaTyr pair for protein labeling.

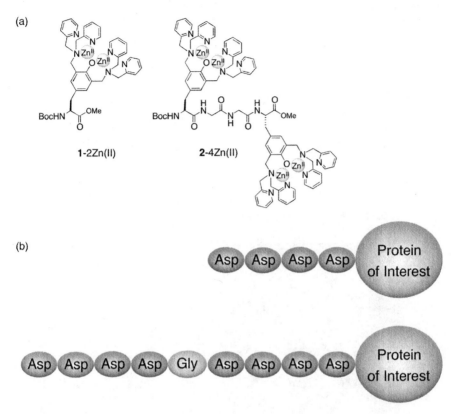

FIGURE 9.9   (a) Molecular structures of Zn(ii)-DpaTyr. (b) Sequences of the D4 and D4-G-D4 tethered to a protein of interest.

Subsequently, they reported two additional methods for ratiometric fluorescence detection of the Asp-tag-fused proteins (Fig. 9.11). One employed a pH sensitive fluorophore [37]. It is conceivable that the local pH in the proximity to the D4 peptide is rather acidic than the bulk pH due to its accumulated carboxylate residues. Therefore, they designed a new probe **4**-2Zn(ii), which possessed seminaphthorhodafluor (SNARF) as a pH-responsive dual-emission fluorophore. As expected, a typical ratiometric fluorescence change immediately occurred in response to the D4 peptide addition, in which the emission at 628 nm due to the basic phenolate form of SNARF decreased and the emission at 586 nm due to the acidic phenol form increased. Another method utilized the wavelength change by the equilibrium shift of the pyrene momomer/dimer fluorescence emission, where they designed **5**-2Zn(ii) having a pyrene as a fluorophore [38]. When **5**-2Zn(ii) bound to the repeated Asp peptide, the pyrene excimer emission was observed due to the pyrene assembly on the peptide. Excess addition of the peptide, in contrast, induced the pyrene monomer emission due to the destruction of the pyrene assembly on the peptide.

(a)

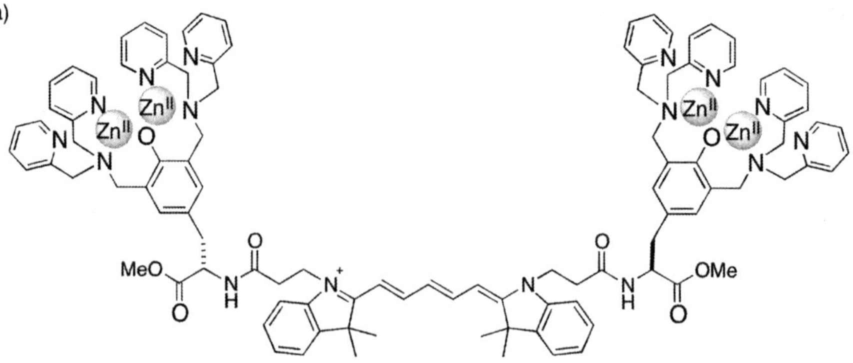

(b)

FIGURE 9.10   (a) Molecular structure of fluorophore-conjugated Zn(II)-DpaTyr. (b) Fluorescence imaging of m1AChR on the surface of CHO cells labeled with **3**-4Zn(II).

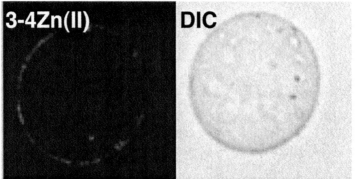

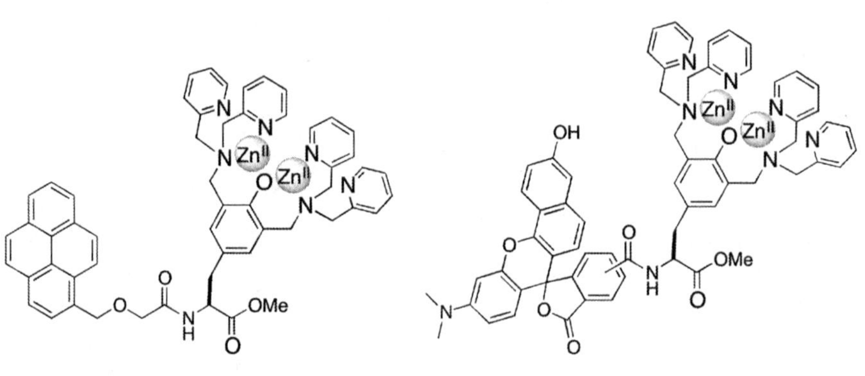

4-2Zn(II)   **5**-2Zn(II)

FIGURE 9.11   Molecular structures of fluorophore-conjugated Zn(II)-DpaTyr for ratiometric fluorescence detection.

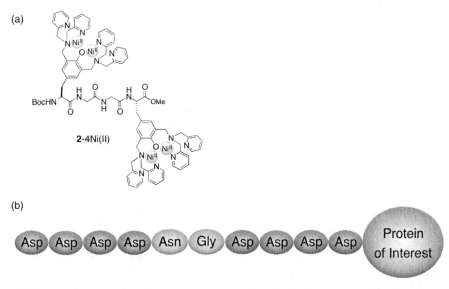

FIGURE 9.12   (a) Molecular structures of Ni(II)-DpaTyr. (b) Sequences of the D4-NG-D4 tethered to a protein of interest.

They also reported that Ni(II)-DpaTyr has strong affinity for Asp-tag (Fig. 9.12) [39]. Indeed, **2**-4Ni(II) showed a remarkably strong affinity ($K_{app} = 2.0 \times 10^9$ M^{-1}) for D3-NG-D3 peptide, the value of which is approximately equal to 100-fold larger than that of the above mentioned **2**-4Zn(II) and D4-G-D4 peptide. Taking advantage of the strong affinity, they successfully applied this novel pair for the deactivation of D3-NG-D3 tag-fused β-galactosidase by using the chromophore-assisted light inactivation (CALI) technique.

Asp-tag/Zn(II)-DpaTyr pair was recently extended to the covalent protein labeling system, named as the reactive tag system (Fig. 9.13a). The covalent labeling has several advantages, including that various artificial molecules can be permanently incorporated and the postlabeling analyses such as SDS-PAGE and Western blotting can be conveniently performed. For covalent bond formation between D4 tag and Zn(II)-DpaTyr, a nucleophilic reaction of α-haloketone with a cystein thiol group was employed (Figs. 9.13b and 9.13c) [40]. At first, they evaluated the reactivity of **6**-2Zn(II) as a thiol-reactive probe with a series of the D4 peptides Cys-Ala$n$-Asp4 ($n = 0, 2, 4, 6, 8$) possessing one Cys group at various distal positions from D4 sequence. Remarkably, it was found that Cys-Ala6-Asp4 peptide showed the dramatically rapid reaction, which completed within 15 min.

This reactive tag system was then applied for labeling of tag-fused G-protein coupled receptor (GPCR) proteins expressed on live cell surfaces [41]. They utilized a peptide tag and a probe that is composed of cystein-containing Cys-Ala6-Asp4 × 2 tag and tetranuclear Zn(II)-DpaTyr probes containing a reactive α-chloroacetyl moiety (**7**, **8**, **9**-4Zn(II)) (Fig. 9.14). The utility of this system was demonstrated

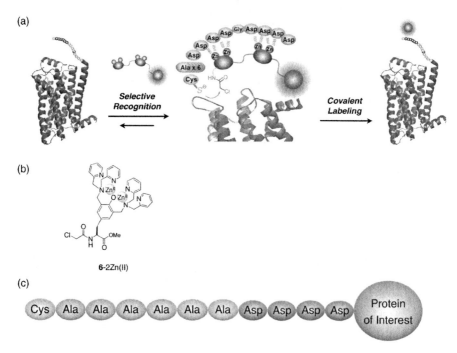

FIGURE 9.13 (a) Covalent protein labeling of Asp-tag-fused protein using reactive tag system. (b) Molecular structures of the labeling probe. (c) Sequences of the C-A6-D4 tethered to a protein of interest.

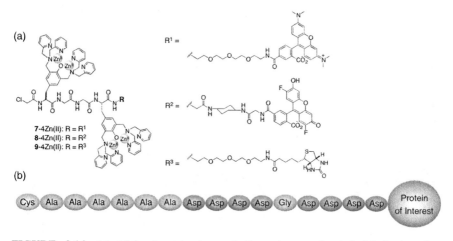

FIGURE 9.14 (a) Molecular structure of fluorophore-conjugated labeling probes. (b) Sequences of the C-A6-D4 × 2 tethered to a protein of interest.

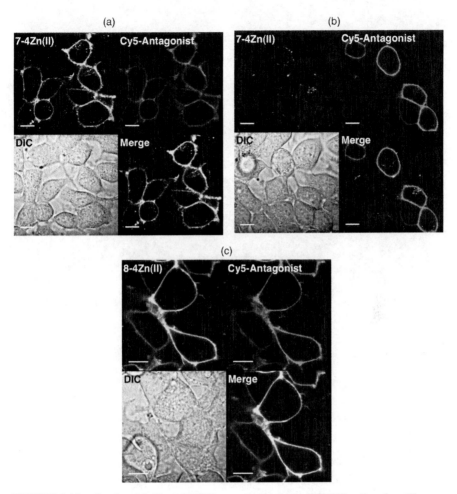

FIGURE 9.15 Covalent labeling of GPCRs expression on the HEK293 cell surface. Fluorescence imaging of (a) C-A6-D4 × 2 tag-fused B2R with **7**-4Zn(II), (b) B2R lacking C-A6-D4 × 2 tag with **7**-4Zn(II), (c) C-A6-D4 × 2 tag-fused B2R with **8**-4Zn(II).

by several functional analyses of GPCRs, such as Western blotting, fluorescence visualization of the stimuli-responsive internalization of GPCRs, and pH change in endosomes containing the internalized GPCRs (Fig. 9.15).

9.2.4 Lanthanide-binding Tag

Lanthanides, such as Tb(III) and Eu(III), possess several attractive properties for application in biotechnology (Fig. 9.16). For example, given the appropriate coordination environment and the presence of a sensitizing chromophore, lanthanides

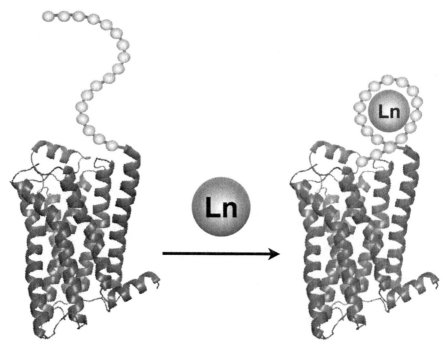

FIGURE 9.16 LBT that selectively binds with lanthanides for protein functional analysis.

display long-lived luminescence emission that is ideal for time-resolved experiments and lanthanide-based resonance energy transfer (LRET). Imperiali et al. developed lanthanide-binding tags (LBTs), which can be incorporated into proteins for various functional analyses. In the early stage, they discovered the peptide sequence (ACADYNKCGWYEELECAA) with high affinity for Tb(III) ($K_D = 0.220 \pm 0.03$ μM), which showed substantially strong luminescence intensity upon binding with Tb(III) [42]. Thereafter, using the strategy of solid-phase split-and-pool combinatorial peptide synthesis, they found the more sophisticated peptide (YIDTNNDG-WYEGDELLA) that binds Tb(III) with the higher affinity ($K_D = 57$ nM) [43].

In this technique, it is predictable that increasing the number of bound lanthanides would improve the capabilities of these tags. Imperiali et al. designed a double LBT (dLBT) which concatenates two lanthanide-binding motifs using a structurally well-characterized single-LBT sequence as a starting point. The dLBT (GPGYODTNNDGWIEGDELYIDTNNDGWIEGDELLA)-fused ubiquitin showed threefold greater luminescence intensity upon Tb(III) binding than the single LBT [44].

It was also clear that the paramagnetic lanthanides are useful in NMR study because these induce residual dipolar couplings (RDC) and pseudo-contact shifts (PCS) that yield valuable distance constraints for protein structural analysis [44–46]. Furthermore, LBTs offer a heavy-atom site for solving the phase problem in X-ray crystallographic study of proteins [46, 47]. Imperiali et al. reported that phasing of the single-wavelength X-ray data (at 2.6 Å resolution) using only the anomalous signal from the two tightly bound Tb(III) ions in the dLBT led to clear electron density maps [47]. In addition, lanthanide luminescence-based techniques are valuable for investigating the functions and dynamics of proteins [48–50]. Imperiali et al. monitored the interaction between Src homology domain (SH2) and phosphopeptides by LRET [49]. The LRET between the LBT-fused SH2-bound Tb(III) and the peptide tethering organic fluorophore was elaborately used for evaluating the specific recognition of the SH2 domain and the peptide-binding partner.

9.3 SUMMARY AND PERSPECTIVES

In this chapter, protein labeling methods based on the metal-chelation systems are described. Recently, development of protein labeling methods is considered as fast-growing research area, so that many methods have been reported. Among various methods, the methods based on coordination chemistry present with several advantages, including no need to incorporate a large enzyme/protein domain into the target proteins, availability of a wide range of labeling reagents, and flexible setting of the labeling conditions, which would be expected to contribute the rapidly expanding studies for protein functional analysis in living systems.

REFERENCES

[1] Lippincott-Schwartz, J.; Patterson, G. H. *Science* **2003**, *300*, 87–91.

[2] Shaner, N. C.; Steinbach, P. A.; Tsien, R. Y. *Nat. Methods* **2005**, *2*, 905–909.

[3] Hamachi, I.; Nagase, T.; Shinkai, S. *J. Am. Chem. Soc.* **2000**, *127*, 12065–12067.

[4] Chen, G.; Heim, A.; Riether, D.; Yee, D.; Migrom, Y.; Gawinowicz, M. A.; Sames, D. *J. Am. Chem. Soc.* **2003**, *128*, 8130–8133.

[5] Takaoka, Y.; Tsutsumi, H.; Kasagi, N.; Nakata, E.; Hamachi, I. *J. Am. Chem. Soc.* **2006**, *128*, 3273.

[6] Koshi, Y.; Nakata, E.; Miyagawa, M.; Tsukiji, S.; Ogawa, T.; Hamachi, I. *J. Am. Chem. Soc.* **2008**, *130*, 245.

[7] Tsukiji, S.; Miyagawa, M.; Takaoka, Y.; Tamura, T.; Hamachi, I. *Nat. Chem. Biol.* **2009**, *5*, 341–343.

[8] Wang, H.; Koshi, Y.; Minato, D.; Nonaka, H.; Kiyonaka, S.; Mori, Y.; Tsukiji, S.; Hamachi, I. *J. Am. Chem. Soc.* **2011**, *133*, 12220–12228.

[9] Keppler, A.; Gendreizig, S.; Gronemeyer, T.; Pick, H.; Vogel, H.; Johnsson, K. *Nat. Biotechnol.* **2003**, *21*, 86–89.

[10] Los, G. V.; Darzins, A.; Karassina, N.; Zimprich, C.; Learish, R.; McDougall, M. G.; Encell, L. P.; Friedman-Ohana, R.; Wood, M.; Vidugirls, G.; Zimmerman, K.; Otto, P.; Klaubern, D. H.; Wood, K. V. *Promega Cell Notes* **2005**, *11*, 2–6.

[11] Miller, L. W.; Cai, Y.; Sheetz, M. P.; Cornish, V. W. *Nat. Methods* **2005**, *2*, 255–257.

[12] Mizukami, S.; Watanabe, S.; Hori, Y.; Kikuchi, K. *J. Am. Chem. Soc.* **2009**, *131*, 5016–5017.

[13] Hori, Y.; Ueno, H.; Mizukami, S.; Kikuchi, K. *J. Am. Chem. Soc.* **2009**, *131*, 16610–16611.

[14] George, N.; Pick, H.; Vogel, H.; Johnsson, N.; Johnsson, K. *J. Am. Chem. Soc.* **2004**, *126*, 8896–8897.

[15] Chen, I.; Howarth, M.; Lin, W.; Ting, A. Y. *Nat. Methods* **2005**, *2*, 99–104.

[16] Lin, C. W.; Ting, A. Y. *J. Am. Chem. Soc.* **2006**, *128*, 4542–4543.

[17] Fernandez-Suarez, M.; Baruah, H.; Martinez-Herbandez, L.; Xie, K. T.; Baskin, J. M.; Bertozzi, C. R.; Ting, A. Y. *Nat. Biotechnol.* **2007**, *25*, 1483–1487.

[18] Popp, M. W.; Antos, J. M.; Grotenberg, G. M.; Spooner, E.; Ploegh, H. L. *Nat. Chem. Biol.* **2007**, *3*, 707–708.

[19] Tanaka, T.; Yamamoto, T.; Tsukiji, S.; Nagamune, T. *Chembiochem.* **2008**, *9*, 802–807.

[20] Griffin, B. A.; Adams, S. R.; Tsien, R. Y. *Science* **1998**, *281*, 269–272.

[21] Adams, S. R.; Campbell, R. E.; Gross, L. A.; Martin, B. R.; Walkup, G. K.; Yao, Y.; Liolis, J.; Tsien, R. Y. *J. Am. Chem. Soc.* **2002**, *124*, 6063–6076.

[22] Martin, B. R.; Giepmans, B. N. G.; Adams, S. R.; Tsien, R. Y. *Nat. Biotechnol.* **2005**, *23*, 1308–1314.

[23] Gaietta, G.; Deerinck, T. J.; Adams, S. R.; Bouwer, J.; Tour, O.; Laird, D. W.; Sosinsky, G. E.; Tsien, R. Y.; Ellisman, M. H. *Science* **2002**, *296*, 503–507.

[24] Zhang, X.-Y.; Bishop, A. C. *J. Am. Chem. Soc.* **2007**, *129*, 3812–3813.

[25] Chase, B. L.; Archer, T.; Burdine, L.; Gillette, T. G.; Kodadek, T. *J. Am. Chem. Soc.* **2007**, *129*, 12348–12349.

[26] Houchuli, E.; Dobeli, H.; Schacher, A. *J. Chromatogr.* **1987**, *411*, 177–184.

[27] Arnau, J.; lauritzen, C.; Petersen, G. E.; Petersen, J. *Protein Expr. Purif.* **2006**, *48*, 1–13.

[28] Tomizaki, K.; Usui, K.; Mihara, H. *Chembiochem.* **2005**, *6*, 782–789.

[29] Kpanidis, A. N.; Ebright, Y. W.; Ebright, R. E. *J. Am. Chem. Soc.* **2001**, *123*, 12123–12125.

[30] Guignet, E. G.; Hovius, R.; Vogel, H. *Nat. Biotechnol.* **2004**, *22*, 440–444.

[31] Lata, S.; Gavutis, M.; Tampe, R.; Piehler, J. *J. Am. Chem. Soc.* **2006**, *128*, 2365–2372.

[32] Goldsmith, C. R.; Jaworski, J.; Sheng, M.; Lippard, S. J. *J. Am. Chem. Soc.* **2006**, *128*, 418–419.

[33] Hauser, C. T.; Tsien, R. Y. *Proc. Nat. Acad. Sci. USA* **2007**, *104*, 3693–3697.

[34] Hintersteiner, M.; Weidemann, T.; Kimmerlin, T.; Filiz, N.; Buehler, C.; Auer, M. *Chembiochem.* **2008**, *9*, 1391–1395.

[35] Uchinomiya, S.-H.; Nonaka, H.; Fujishima, S.-H.; Tsukiji, S.; Ojida, A.; Hamachi, I. *Chem. Commun.* **2009**, 5880–5882.

[36] Ojida, A.; Honda, K.; Shinmi, D.; Kiyonaka, S.; Mori, Y.; Hamachi, I. *J. Am. Chem. Soc.* **2006**, *128*, 10452–10459.

[37] Honda, K.; Nakata, E.; Ojida, A.; Hamachi, I. *Chem. Commun.* **2006**, 4024–4026.

[38] Honda, K.; Fujishima, S.-H.; Ojida, A.; Hamachi, I. *Chembiochem.* **2007**, *8*, 1370–1372.

[39] Ojida, A.; Fujishima, S.-H.; Honda, K.; Nonaka, H.; Uchinomiya, S.-H.; Hamachi, I. *Chem. Asian J.* **2010**, *5*, 877–886.

[40] Nonaka, H.; Tsukiji, S.; Ojida, A.; Hamachi, I. *J. Am. Chem. Soc.* **2007**, *129*, 15777–15779.

[41] Nonaka, H.; Fujishima, S.-H.; Uchinomiya, S.-H.; Ojida, A.; Hamachi, I. *J. Am. Chem. Soc.* **2010**, *132*, 9301–9309.

[42] Franz, K. J.; Nitz, M.; Imperiali, B. *Chembiochem.* **2003**, *4*, 265–271.

[43] Nitz, M.; Franz, K. J.; Maglathlin, R. L.; Imperiali, B. *Chembiochem.* **2003**, *4*, 272–276.

[44] Martin, L. J.; Hahnke, M. J.; Nitz, M.; Wohnert, J.; Silvaggi, N. R.; Allen, K. N.; Schwalbe, H.; Imperiali, B. *J. Am. Chem. Soc.* **2007**, *129*, 7106–7113.

[45] Wohnert, J.; franz, K. J.; Nitz, M.; Imperiali, B.; Schwalbe, H. *J. Am. Chem. Soc.* **2003**, *125*, 13338–13339.

[46] Barthelmes, K.; Reynolds, A. M.; Peisach, E.; Jonker, H. R. A.; DeNunzio, N. J.; Allen, K. N.; Imperiali, B.; Schwalbe, H. *J. Am. Chem. Soc.* **2011**, *133*, 808–819.

[47] Sculimbrene, B. R.; Imperiali, B. *J. Am. Chem. Soc.* **2006**, *128*, 7346–7352.

[48] Goda, N.; Tenno, T.; Inomata, K.; Iwaya, N.; Sasaki, Y.; Shirakawa, M.; Hiroaki, H. *Biochimi. Biophys. Acta* **2007**, *1773*, 141–146.

[49] Silvaggi, N. R.; Martin, L. J.; Schwalbe, H.; Imperiali, B.; Allen, K. N. *J. Am. Chem. Soc.* **2007**, *129*, 7114–7120.

[50] Hartley, M. D.; Larkin, A.; Imperiali, B. *Bioorg. Med. Chem.* **2008**, *16*, 5149–5156.

10

MOLECULAR BIOENGINEERING OF MAGNETOSOMES FOR BIOTECHNOLOGICAL APPLICATIONS

ATSUSHI ARAKAKI, MICHIKO NEMOTO, AND TADASHI MATSUNAGA

10.1 INTRODUCTION

Organisms produce inorganic/organic composite materials, called biominerals, show amazing structures and exhibit excellent physical/chemical properties that often outperform artificial materials [1]. The complex architectures are formed under conditions that are incredibly mild, given an idea to use the process for material synthesis. For the biomineralization process to occur, specific subunit compartments or microenvironments need to be created, in order to stimulate crystal formation at certain functional site and inhibition or prevention of the process at all other sites [2, 3]. The highly specific control of morphology, composition, location, orientation, and crystallographic phase all indicate the existence of functional molecules that attribute for these regulations in each biominerals [4–6], although the mechanisms remain largely unknown thus far.

One of the best understood examples of biomineralization studies is in magnetotactic bacteria. Magnetotactic bacteria have an ability to synthesize nano-sized single-domain magnetite crystals (Fe_3O_4) which are aligned in chain, enabling the cells swim or migrate along with magnetic field lines [7]. For the synthesis of magnetite crystals, the cells use a specialized intracellular membranous compartment, also referred to as magnetosomes. The biomineralization process in magnetosomes is

Coordination Chemistry in Protein Cages: Principles, Design, and Applications, First Edition.
Edited by Takafumi Ueno and Yoshihito Watanabe.
© 2013 John Wiley & Sons, Inc. Published 2013 by John Wiley & Sons, Inc.

strictly genetically controlled, leading to highly specific crystal size, shape, number, and their assembly for a given bacterial strain [8]. Recent studies on molecular biology of magnetotactic bacteria much improved our understandings of the formation mechanisms. The knowledge gives us an idea to use magnetosomes, their proteins, and the bacterial cells as magnetic materials for biotechnological applications.

Magnetic particles are widely used in the development of medical and diagnostic applications such as magnetic resonance imaging (MRI), cell separation, environmental inspections, drug delivery, and hyperthermia [9–14]. The major advantage of magnetic particles to use in such applications is that they can be easily manipulated by magnetic force. Magnetic particles enable rapid and easy separation of target molecules bound to the magnetic particles from reaction mixtures. Therefore, magnetic particles are beneficial to use in fully automated system, providing minimal manual labor and superior reproducibility.

Magnetosomes are individually covered with thin organic membrane, which confers high and even dispersion in aqueous solutions compared with artificial magnetites. The size ranges of magnetosomes are typically 20–100 nm in diameter, which makes them magnetically stable single domain. From the basic studies on magnetotactic bacteria and their magnetosome formation mechanisms, methods for functional design of magnetosomes and magnetotactic bacterial cells were established through molecular bioengineering. In the following section, we introduce recent progress on the basic studies of magnetosome formation mechanism and the design of functional magnetic particles and their applications to biotechnology based on their unique properties.

10.2 MAGNETITE BIOMINERALIZATION MECHANISM IN MAGNETOSOME

Over the past few years, the genomes of several magnetotactic bacteria have been sequenced. Based on the comprehensive studies including proteome and transcriptome analyses of magnetotactic bacteria, an outline of the molecular mechanism of magnetosome formation has been illustrated. Clarification of magnetosome formation mechanism contributes to the molecular design of magnetosome surface and the biotechnological application in a broad range of research disciplines. Recent progresses on the molecular studies in magnetotactic bacteria were introduced.

10.2.1 Diversity of Magnetotactic Bacteria

Magnetotactic bacteria have been identified and observed to inhabit various aquatic environments including pond, river, and ocean by date. Since the first peer-reviewed report of magnetotactic bacteria in 1975 [7], various morphological types including cocci, spirilla, vibrios, ovoid bacteria, rod-shaped bacteria, and multicellular bacteria have been characterized [8, 15]. The *Magnetospirillum magnetotacticum* strain MS-1 was the first isolate of magnetotactic bacteria [16]. After the first isolation of strain MS-1, a number of *Magnetospirillum* species including the well-studied

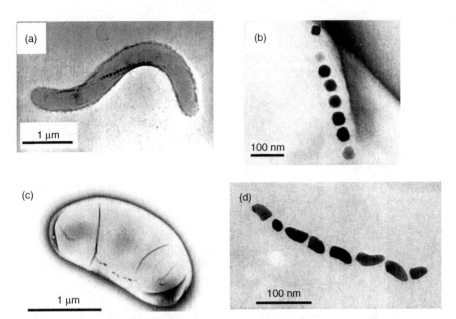

FIGURE 10.1   Transmission electron micrographs of *M. magneticum* strain AMB-1 (a) and *D. magneticus* strain RS-1 (c). Magnetosomes of strain AMB-1 (b) and strain RS-1 (d).

strains, *Magnetospirillum magneticum* AMB-1 [17] (Fig. 10.1a), and *Magnetospirillum gryphiswaldense* strain MSR-1 [18] were isolated and characterized. *Magnetospirillum* sp. is a nitrate-reducing bacteria that synthesize cubo-octahedral-shaped magnetosomes (Fig. 10.1b). They are capable of growing under both microaerobic and aerobic conditions in liquid or solid media, which makes them ideal candidates for genetic manipulation. The coccus *Candidatus* Magnetococcus marinus strain MC-1 and the vibrio *Candidatus* Magnetovibrio blakemorei strain MV-1 producing elongated pseudohexagonal prismatic magnetosomes were also isolated [19, 20]. The magneto-ovoid strain MO-1, which is phylogenetically related to strain MC-1 and possessing unique flagellar sheath, was isolated from the Mediterranean Sea [21]. Most of the isolates of magnetotactic bacterial strains are identified in the group of α-Proteobacteria. *Desulfovibrio magneticus* is the isolate of magnetotactic bacterium classified under the δ-Proteobacteria (Fig. 10.1c) [22, 23]. This bacterium produces irregular, bullet-shaped magnetosomes (Fig. 10.1d). In addition to its unique magnetosome morphology, the distinct physiological and biochemical characteristics of strict anaerobic, sulfate-reducing bacterium, strain RS-1 among other isolated magnetotactic bacteria was reported [24]. Recently, extremophilic magnetotactic bacteria were isolated from highly alkaline aquatic environments ($\sim$pH 9.5). Those bacteria form a bullet-shaped magnetite magnetosome, and closely relate to nonmagnetotactic, alkaliphilic bacterium, *Desulfonatronum* [25]. A greigite (Fe_3S_4)-producing magnetotactic bacterium strain BW-1 forming a clade within the sulfate-reducing bacterium in δ-Proteobacteria has also been isolated [26]. Genomic analysis of this strain revealed the presence of two different magnetosome gene clusters that might be responsible for greigite biomineralization and magnetite biomineralization.

Microscopic analysis of magnetically collected environmental sample has revealed the existence of various magnetotactic bacteria that could not be isolated and cultured in the laboratory [27]. Magnetotactic bacteria categorized in the subphyla γ-Proteobacteria and the phylum *Nitrospirae* were confirmed by 16S rDNA sequence analyses of environmental sample [28]. *Candidatus* Magnetobacterium bavaricum, which forms numerous (∼1000) bullet-shaped magnetosomes in one cell is classified into phylum *Nitrospirae*. In addition to strain RS-1, two δ-Proteobacterial magnetotactic bacteria were also identified from environmental sample. Large spherical magnetotactic bacteria, also called multicellular magnetotactic prokaryotes (MMPs), collected from sulfidic environments, produce greigite (Fe_3S_4) magnetosomes [29, 30]. Barbell-shaped magnetotactic bacteria forming chains of two to five cocci were identified as a member of the *Desulfobulbaceae* family [31]. Further molecular biological analysis using these yet unisolated or uncharacterized species might provide new leads to the better understanding of magnetosome formation by magnetotactic bacteria.

10.2.2   Genome and Proteome Analyses of Magnetotactic Bacteria

The first complete genome sequence of a magnetotactic bacteria was obtained for *M. magneticum* AMB-1 [32]. The genome comprised a single circular 5-Mbp chromosome and 4559 predicted open reading frames (ORFs). Within the genome sequence, a remarkable number of sensor and response domains, numerous vestiges of past exogenous gene transfers, such as insertion sequence (IS) elements, integrases, and large regions containing phage-coding genes were identified. The IS-concentrated regions had lower GC contents than the average GC content of the entire genome, which suggests gene transfer from other bacteria or phages. During the genome analysis, we obtained a spontaneous nonmagnetic mutant that lacks a 98 kb genomic region, which exhibited the characteristics of a genomic island (also referred to as magnetosome island): it has a low GC content; is located between two 1.1 kb repetitive sequences; and contained an integrase in the flanking region of the first repetitive sequence. The spontaneous deletion of this 98 kb genomic region before and after excision from the chromosome was detected by polymerase chain reaction (PCR) amplification [33]. A spontaneous nonmagnetic mutant that lacks the approximately 80 kb magnetosome island was also obtained with the *M. gryphiswaldense* MSR-1 strain [34]. Subsequently, various phenotypes of mutants showing altered numbers, sizes, and alignment of magnetosomes were isolated and analyzed [35]. The genotypic analysis of mutants revealed several gene deletions in different regions of magnetosome island, and suggesting the occurrence of frequent transposition events in the cell. Considering the typical structure and the dynamics after excision from the chromosome, the magnetosome island, which might be essential for magnetosome synthesis, was considered to be derived from lateral gene transfer.

Besides the first whole genome sequencing of magnetotactic bacteria in strain AMB-1 [32], the whole genome sequence of strain MC-1 [36] and the draft genome assembly of strain MS-1 and MSR-1 have become available. Comparative genome analysis of these α-Proteobacterial magnetotactic bacteria revealed the common gene sets located within and outside of the magnetosome island [37]. Recently, partial genome sequence including magnetosome island region of marine magnetotactic bacterium, strain MV-1 was sequenced and comparative analysis of magnetosome islands among *Magnetospirillum* species, strains MC-1 and MV-1, was conducted [38]. The results revealed the distinct variations in gene order and sequence similarity of magnetosome island genes in these microorganisms. Numbers of α-Proteobacterial magnetotactic bacteria-specific genes encode unknown proteins as well as a putative sensor protein that might be related to magnetotaxis. The results indicated that magnetosome island encoded not all the proteins required for magnetosome formation and magnetotaxis. This is consistent with the results previously obtained by the analysis of the gene located outside of the magnetosome island that related to magnetosome formation [39, 40]. The functions of these gene products are described in Section 10.2.3.

The whole genome sequence analysis of *Desulfovibrio magneticus* strain RS-1, the isolate of magnetotactic bacteria outside of the α-Proteobacteria, was reported [41]. Comparative genomics between strain RS-1 and the four α-Proteobacterial strains revealed the presence of three separate gene regions (*nuo-* and *mamAB*-like gene clusters, and gene region of a cryptic plasmid) conserved in all magnetotactic bacteria. The *nuo* gene cluster, encoding NADH dehydrogenase (complex I), was also common to the genomes of three iron-reducing bacteria exhibiting uncontrolled extracellular and/or intracellular magnetite synthesis. A cryptic plasmid, pDMC1, encodes three homologous genes that exhibit high similarities with those of other magnetotactic bacterial strains. In addition, the *mamAB*-like gene cluster, encoding the key components for magnetosome formation such as iron transport and magnetosome alignment, was conserved only in the genomes of magnetotactic bacteria as a similar genomic island-like structure. The findings suggested the presence of core genetic components for magnetosome biosynthesis; these genes may have been acquired into the magnetotactic bacterial genomes by multiple gene transfer events during proteobacterial evolution.

A new strategy to prepare genomic DNA from a single species of uncultured magnetotactic bacterium was also developed [42]. The strategy enables to obtain genomic information from yet unisolated magnetotactic bacteria. Metagenomic approach was also conducted and magnetosome island clusters were identified from the metagenome library of uncultured magnetotactic bacteria [43]. By using micromanipulation and whole genome amplification, the partial genome analyses of uncultured magnetotactic bacteria, *Candidatus* Magnetobacterium bavaricum of *Nitrospira* phylum was reported [44]. The previously identified proteobacterial magnetotactic bacteria-specific gene sets that might be essential for magnetosome formation were partially present in the putative magnetosome island of *Candidatus* Magnetobacterium bavaricum. Based on the phylogenetic analysis, the presence of common magnetotactic ancestor of *Nitrospira* and Proteobacteria was implied.

The genome analyses accelerated the molecular understanding of magnetosome formation using comprehensive proteome and transcriptome analysis. Magnetotactic bacteria uptake a larger amount of iron compared to other well-characterized bacteria. To elucidate the iron uptake system for magnetosome formation, global gene expression of strain AMB-1 grown under iron-rich and iron-deficient conditions was analyzed [45]. The results indicate that despite the unusual high iron requirement of strain AMB-1, it utilizes robust but simple iron uptake systems similar to those of other gram-negative bacteria.

Based on the genome sequences, proteome analyses of magnetosome membrane proteins have been conducted in *M. magneticum* AMB-1, *M. gryphiswaldense* MSR-1, and *D. magneticus* RS-1, from which numerous novel proteins with potentially crucial roles in magnetosome biomineralization have been discovered [46–48]. In the analysis of magnetosome membrane proteins from *M. gryphiswaldense* MSR-1, approximately 30 proteins were identified based on an incomplete genomic sequence [46]. A considerable number of identified proteins were found to be assigned in gene clusters located within the magnetosome island. The comprehensive proteome analysis of magnetosome membrane based on the whole genome sequence was conducted with *M. magneticum* AMB-1, wherein a total of 78 proteins were identified from the surface of its magnetosome [47]. The identified proteins included several homologues of magnetosome membrane proteins in strain MSR-1, which was also encoded in magnetosome island in strain AMB-1. Furthermore, comparative proteome analysis between the protein fractions acquired from the outer membrane, cytoplasmic membrane, magnetosome membrane, and cytoplasmic–periplasmic fractions was conducted. A high degree of similarity was observed between the protein profiles of the magnetosome and cytoplasmic membranes of the AMB-1 strain. Fatty acid comparative analysis also indicated that both these fractions showed similar profiles. These results suggest that the magnetosome membrane could have been derived from the cytoplasmic membrane. Direct evidence of this phenomenon was also reported by Komeili et al. [49]. Invagination of the cytoplasmic membrane was clearly observed by electron cryotomography. Proteome analysis of *D. magneticus* RS-1 revealed a presence of magnetosome membrane proteins commonly exist in strain RS-1 and *Magnetospirillum* sp. [48]. Those include actin-like protein, TPR containing protein, cation diffusion facilitator, and several oxidation–reduction proteins. Some of them were identified from the gene region of cryptic plasmid pDMC1, in which similar region is also found from strain MC-1. The results indicate that δ-Proteobacterial magnetotactic bacterium strain RS-1 produces magnetosome using similar process as described in α-Proteobacterial magnetotactic bacteria.

10.2.3   Magnetosome Formation Mechanism

Since the discovery of magnetotactic bacteria, magnetosome membrane proteins have been believed to play key roles in their biosynthesis. A number of proteins were isolated from magnetosome, and characterized. One of the major proteins of magnetosome membrane Mms16 is 16 kDa protein of a small GTPase [40]. This protein is speculated to prime the invagination of the cytoplasmic membrane for vesicle formation in *M. magneticum* AMB-1, though the subsequent study in *M. gryphiswaldense* MSR-1 proposes that the function is poly(3-hydroxybutyrate) depolymerization [50]. MamA protein (corresponding to Mms24 and Mam22), which has tetratricopeptide repeat (TPR) domain, a protein–protein interaction module, was also a dominant magnetosome membrane protein in *Magnetospirillum* sp. The crystal structure revealed MamA folds as a sequential TPR protein with a unique hook-like shape [51]. The MamA-deficient mutant formed less number of magnetosome in the cells [52]. Furthermore, GFP-fused MamA displayed dynamic localization during cell growth. These analyses suggested that MamA acts as a scaffold protein for magnetosome formation by interacting with other magnetosome formation—proteins. The MpsA showing homology with acetyl-CoA carboxylase (transferase) and the acyl-CoA-binding motif was also identified from magnetosome membrane. This protein is considered to function as a mediator of cytoplasmic membrane invagination [53]. MamY protein was isolated from the small magnetite crystals (average particle size was approximately 28 nm), which were separated from regular-sized magnetite crystals (average particle size was approximately 59 nm) using a novel size fractionation technique [54]. Homology search and secondary structure prediction suggested that MamY shares the similarity with eukaryotic membrane deformation proteins, such as bin/amphiphysin/Rvs (BAR) protein. The BAR proteins have been known to involve in the eukaryotic membrane compartmentalization process by constricting the membrane. *In vitro* functional characterization using recombinant protein showed that MamY protein bound directly to the liposomes, causing them to form long tubules like the CentaurinBAR protein, used as a positive control (Fig. 10.2). In addition, *mamY*-deficient mutant showed to have expanded magnetosome vesicles

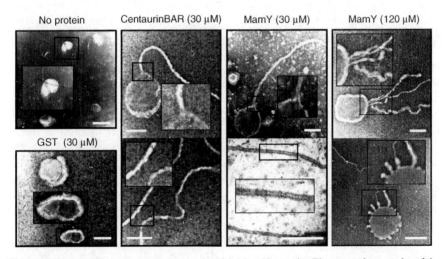

FIGURE 10.2   Liposome tubulation assay with MamY protein. Electron micrographs of the liver lipid liposomes incubated with the proteins. MamY protein interacted with the liposome surfaces, leading to the formation of tubules with outer diameters of 20 nm. CentaurinBAR and GST proteins were used as the positive and negative controls, respectively. The inset shows liposomes at a higher magnification. Scale bar = 200 nm. Reprinted with permission from Reference 54. Copyright 2010 Molecular Microbiology.

and a greater number of small magnetite crystals relative to the wild-type strain. Based on the above results, the function of MamY protein is considered to constrict the magnetosome membrane during magnetosome vesicle formation. MamE is a bifunctional protein with a protease-independent role in magnetosome protein localization and a protease-dependent role in maturation of small magnetite crystals [55]. MamM, a cation diffusion facilitator family protein, is a multifunctional protein that is involved in the crystallization initiation and regulation of proper localization of other magnetosome proteins [56]. MamB, which also exhibits similarity with cation diffusion facilitator family, was found to interact with several other proteins including the PDZ1 domain of MamE.

The protein regulating magnetite shape was also identified from magnetosome in *M. magneticum* AMB-1. The Mms6 protein was identified from the protein fraction tightly bound to the magnetite surface with other homologous protein designated Mms5, Mms7 (identical to MamD), and Mms13 (identical to MamC) [57]. These proteins show the common amphiphilic features containing hydrophobic *N*-terminal and hydrophilic *C*-terminal regions. The *N*-terminal regions in Mms5, Mms6, and MamD possess a common leucine and glycine repetitive sequence— LGLGLGLGAWGPX(L/I)(L/V) GX(V/A)GXAGA. *In vitro* binding assay using recombinant protein revealed that Mms6 is an iron-binding protein. Furthermore, the magnetite particles synthesized in the presence of Mms6 *in vitro* showed cuboidal morphology similar to the AMB-1 strain magnetosomes with sizes ranging from 20 to 30 nm in diameter. These results suggest that Mms6 binds iron ions to initiate magnetite crystal formation and/or regulates the morphology by producing a self-assembled framework structure. Recent study using *mms6*-deficient mutant demonstrated that Mms6 controls morphology of the cubo-octahedral magnetite crystals. The detailed function of Mms6 protein is described in Section (10.2.4).

In addition to the protein analysis, mutant analysis also revealed a number of magnetosome formation-related proteins. The *magA*, the gene encoding a proton-driving H^+/Fe(II) antiporter protein, was isolated from a magnetosome depleted Tn5 mutant [39]. MagA appears to play a role in iron efflux in the former and iron influx in the latter. A tungsten-containing aldehyde ferredoxin oxidoreductase (AOR), which plays a role in aldehyde oxidation, was also isolated from transposon mutant. AOR was found to be expressed under the microaerobic conditions and localized in the cytoplasm of AMB-1. Transmission electron microscopy (TEM) of the mutant revealed that no magnetosomes were completely synthesized. Furthermore, the iron uptake and growth of AOR-deficient mutant was lower than the wild type under microaerobic condition [58]. The results indicate that AOR may contribute to ferric iron reduction during magnetosome synthesis under microaerobic respiration. A nonmagnetic transposon mutant of cytoplasmic ATPase showed defective activity of iron uptake [59]. Even though the detailed action of cytoplasmic ATPase is unclear, it might promote an iron transport through the cytoplasmic membrane by energizing ATP-driven iron uptake transporters for magnetosome synthesis.

With the development of genetic manipulation techniques, comprehensive mutant analyses of the magnetotactic bacteria-specific genes were conducted to clarify the magnetosome formation-related proteins. MamK is a homologue of the bacterial actin-like protein, and is revealed to form filaments to establish the chain-like structure of magnetosomes through appropriate subcellular targeting [49]. MamK nucleates at multiple sites to form long filaments that were confirmed via protein expression in *Escherichia coli* [60]. *In vitro* polymerization of recombinant MamK was examined, and formation of long filamentous bundles was observed [61]. The MamJ deletion mutant also showed the lack of magnetosome chain structure. MamJ is an acidic protein associated with the filamentous structure, and it directs the assembly of the magnetosome chain [62]. LimJ, which is a paralogue of MamJ, is also required for the proper assembly of the magnetosome chain [63]. Direct interaction between MamJ and MamK has been revealed, and it is proposed that MamJ plays a role in chain assembly and maintenance [64]. The *mamGFDC* operon was knocked out by Cre–loxP system in strain MSR-1. The magnetite crystals produced by the mutant strain were 75% of the wild-type size and had less regular morphology and chain organization [65]. The MamGFDC are homologues of a magnetite shape regulating protein, Mms6. In contrast to the drastic shape defect of Δ*mms6* strain, morphological change of crystal in Δ*mamGFDC* was only a little. These results showed that Mms6 might be mainly responsible for the magnetite shape regulation in *Magnetospirillum* sp. Recent comprehensive genetic analyses using a number of magnetosome island genes deletion mutants revealed that a number of genes previously unrecognized are also involved in the magnetosome formation [66, 67]. These results indicated that the *mamAB* gene cluster encodes the factors essential for magnetosome membrane biogenesis. The results also suggested that magnetosomes are assembled in a step-wise manner in which membrane biogenesis, magnetosome protein localization, and biomineralization are placed under discrete genetic control.

On the basis of the results obtained from the above studies, it has been described that the mechanism underlying magnetosome formation involves multiple processes including ferric iron reduction, invagination of the cytoplasmic membrane followed by vesicle formation and alignment, vesicular iron accumulation, and iron oxide crystallization (Fig. 10.3).

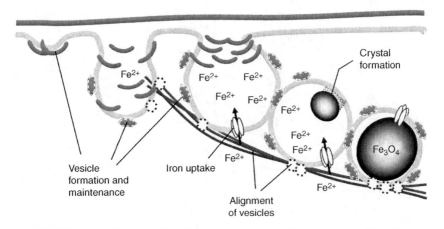

FIGURE 10.3 Scheme of hypothesized mechanism of magnetosome formation.

10.2.4   Morphological Control of Magnetite Crystal in Magnetosomes

The magnetite crystal produced by magnetotactic bacteria is characteristic in its narrow size distribution and uniform shape. The crystals of magnetosomes are in the size range for single-domain magnetite, which are from 20 to 100 nm. From the observation of already isolated and yet unisolated environmental magnetotactic bacteria, magnetosomes with species-specific various morphologies were observed.

The Mms6 protein regulating the cubo-octahedral magnetite crystals was isolated from *M. magneticum* AMB-1. The Mms6 protein has a hydrophilic domain containing a series of aspartic acid and glutamic acid, and hydrophobic domain rich in leucine and glycine as described in Section 10.2.3. Based on the previous studies using other biomineral proteins containing the amphiphilic structure, isolated from enamel [68] or eggshell [69], hydrophobic domain of Mms6 is considered to involve in the protein assembly to form a functional unit for the crystal shape regulation. *In vitro* characterization also indicated the self-aggregation property of Mms6 protein [70]. The acidic domain has been considered as an iron-binding domain providing a nucleation site for iron mineralization and/or inhibits the crystal growth by blocking specific faces. The gene deletion of *mms6* in strain AMB-1 revealed that the deletion mutant contained significantly smaller crystals ((minor axis + major axis) / 2 = 27.4 ± 8.9) than the wild type (48.3 ± 12.5) [71]. The shape factor (0.74 ± 0.23) of *mms6* deletion mutant was also lower than that of the wild-type strain (0.92 ± 0.16) (Fig. 10.4). The magnetite crystals in *mms6* deletion mutant were further evaluated

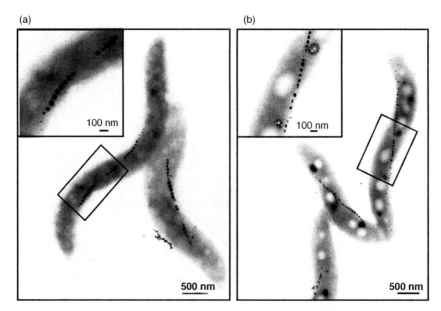

FIGURE 10.4   Transmission electron micrographs of magnetites formed in wild-type strain (a) and *mms6* deletion mutant strain (b). The inset images are magnifications of the square area in the whole cell images.

with HRTEM. In contrast to the cubo-octahedral magnetite crystals in the wild-type and complementation strains, which have (111) and (100) crystal faces, uncommon crystal faces, such as (210), (211), and (311), were detected from the magnetite crystals in *mms6* deletion mutant. In addition, the amounts of proteins tightly bound to the magnetite crystals (Mms5, Mms7, and Mms13) decreases in the deletion mutant. The lack of Mms6 might influence the homeostasis of these proteins tightly bound to the magnetite surface. The detailed functional mechanism of Mms6 for magnetite shape regulation is under investigation.

The presence of *mms6*-homologous genes in the genome of magnetotactic bacteria was further discussed. The genome of strain AMB-1, MSR-1, MC-1, and MV-1 possess *mms6* and *mms6*-homologous gene sets [36–38]. These strains all produce the magnetite crystal consisting of (111) and (100) faces. In contrast, those *mms6*-homologous genes are absent in the genome of strain RS-1, which produces irregular, bullet-shaped magnetite crystals [41]. Although it was the partial information, the magnetosome island gene region in *Candidatus* Magnetobacterium bavaricum also does not contain *mms6*-homologous genes [44]. This observation indicated that *mms6*-homologous involve in the formation of magnetite crystal with (111) and (100) faces. Strain RS-1 and *Candidatus* Magnetobacterium bavaricum may have the functional counterparts of *mms6*-homologous required to control the bullet-shaped magnetite crystals.

As described in Section 10.2.2, recently the comprehensive proteomic analysis of magnetosomes in strain RS-1 was conducted, and a number of RS-1 specific proteins including a collagen-like protein and a putative iron-binding protein were identified [48]. These RS-1 specific novel proteins are likely to control the bullet-shaped morphology of magnetite crystal in strain RS-1. Further investigation of these proteins might uncover the mechanism underlying the formation of the bullet-shaped magnetite crystals in magnetotactic bacteria.

10.3   FUNCTIONAL DESIGN OF MAGNETOSOMES

The magnetic particles have been widely used in the development of medical and diagnostic applications such as MRI [12], cell separation [11, 14], hyperthermia [9], and site-specific chemotherapy [13]. The magnetosomes produced by magnetotactic bacteria can offer substantial advantages over other commercialized magnetic particles used for various applications for the following reasons: (1) the superior magnetic property derived from single-domain magnetite with narrow size distribution and uniform shape, (2) high dispersion property in aqueous solution because of lipid bilayer membrane surrounding each magnetite crystals, and (3) functionalized magnetic particles for various application can be easily generated by using already developed genetic modification or chemical modification techniques. In addition, mass cultivation technique developed with *M. magneticum* AMB-1 allowing for a steady supply of magnetosomes for the industrial applications.

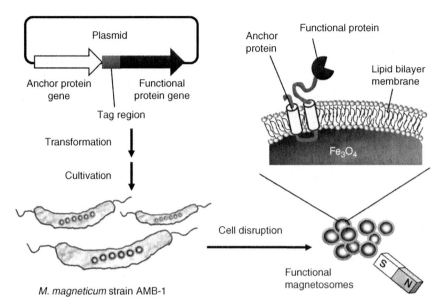

FIGURE 10.5   Schematic for preparation of protein-magnetosome by gene fusion technique.

10.3.1   Protein Display on Magnetosome by Gene Fusion Technique

Genetic transformation techniques developed with *M. magneticum* AMB-1 enables the protein display onto the magnetosome as well as the generation of the mutant strain for the functional analysis of gene products. By the development of a fusion technique involving anchor proteins isolated from magnetotactic bacteria, functionally "active" protein molecule has been successfully displayed on the magnetosome (Fig. 10.5). In order to improve the efficiency of displayed protein onto magnetosomes, transformation techniques, anchor proteins, and promoters for foreign gene expression have been investigated.

For efficient protein display onto magnetosomes, stable genetic transformation techniques of magnetotactic bacteria are essential. The gene transformation of *M. magneticum* AMB-1 was first achieved by using the broad-host-range vector pRK415 and conjugative gene transfer system [72]. However, a low copy number of pRK415 resulted in the low level expression of target genes. To improve the transformation efficiency, a new shuttle vector was constructed based on the cryptic plasmid of *M. magneticum* MGT-1 [73]. Using the newly constructed shuttle vector, pUMG, a high copy number (39 ± 10 copies/cell) followed by a high level expression of the reporter gene was achieved. At the same time, highly efficient electrotransformation technique for *M. magneticum* AMB-1 was developed using pUMG.

The magnetosome membrane proteins used as the anchor molecules for the efficient and stable protein display were also investigated. Previously, the MagA (46.8 kDa) and Mms16 (16 kDa) proteins have been used as the anchor molecules for displaying luciferase [74], acetate kinase [75], protein A [76], the estrogen receptor hormone-binding domain [77], and G protein-coupled receptors [78], respectively. However, with these anchor proteins, protein expression level and stability were limited. After the discovery of the protein sets tightly bound to magnetite crystals [57], the anchoring properties of one of these proteins, Mms13 (13 kDa), were investigated. By using Mms13, stable localization and higher expression level of a large protein, luciferase (61 kDa) compared to the MagA and Mms16, were confirmed [79]. After this work, various proteins were displayed on magnetosome using Mms13 as a stable anchor molecule [80–84].

The strong promoters for the expression of recombinant genes in strain AMB-1 were also investigated based on the genome and proteome analysis [85]. From the proteome analysis, several proteins highly expressed in strain AMB-1 were identified. Then, the promoter activities of the putative promoter regions of the selected proteins were evaluated. As a result, efficient protein expression onto magnetosomes was achieved by using the promoter region of *msp3*, which showed 400 times higher activity of Mms13-luciferase than the *magA* promoter.

Through the improvement of genetic transformation techniques, anchor proteins and promoters, the efficiency of protein display on magnetosome has been dramatically increased. However, the display of transmembrane proteins on magnetosome membranes remains a technical challenge due to the cytotoxic effects of the proteins when they are overexpressed in bacterial cells. Since more than half of all drug targets have been shown to be transmembrane proteins, and the analysis of the interactions of transmembrane proteins and their ligands is the most promising method of the drug screening, a new strategy for the heterologous expression of transmembrane proteins on magnetosomes has been desired. Recently, a tetracycline-inducible expression vector was developed with strain AMB-1 for the efficient expression of transmembrane proteins on magnetosomes [86]. The tetracycline-inducible expression system is based on the tetracycline resistance operon of *E. coli* transposon Tn10 [87]. In this system, the tetracycline analog, anhydrotetracycline is used as an inducer of gene expression. By using GFP and luciferase as reporter proteins, successful induction of target protein expression in strain AMB-1 was confirmed. Furthermore, the protein expression level could be controlled by modulating the concentration of the anhydrotetracycline. The truncated form of transmembrane protein, tetraspanin CD81 was successfully displayed on magnetosomes by adding anhydrotetracycline at mid-log phase whereas growth inhibition of the cell by the toxic effect of CD81 was observed when anhydrotetracycline was added at the time of inoculation. These results suggest that the inducible expression system will be a useful tool for the expression and display of transmembrane and other potentially cytotoxic proteins on the magnetosome.

In addition to the protein display technique, functional small molecule, biotin has been successfully displayed on magnetosomes using posttranslational modification system in the cell [80–82]. In the system, biotin acceptor peptide (BAP) or biotin carboxyl carrier protein (BCCP) was displayed on magnetosome using the already developed recombinant gene expression system with strain AMB-1. The biotin was immobilized to BAP or BCCP expressed on magnetosomes by *in vitro* biotynilation using biotin and *E. coli* biotin ligase [82]. The biotin ligase is one of the posttranslational enzymes conserved in all organisms which label the biotin to specific lysine residues of biotin enzyme including BCCP. Furthermore, *in vivo* biotynilation of BCCP expressed on magnetosomes using endogenously expressed biotin ligase was successfully achieved [80]. The newly developed system has greater advantage than chemical immobilization techniques of biotin in terms of following points: (1) biotin can be site-selectively immobilized to the target protein, and (2) biotynilated magnetosomes can be acquired by simply cultivating transformant cells. Further studies will make it possible to display diverse small molecules on magnetosomes using the posttranslational modification reaction.

Among the various applications of functionalized magnetosomes, cell separation is one of the major applications [11, 14, 88, 89]. For the cell-associated applications including cell separation, minimization of the nonspecific binding of nanoparticles to the cell surface is required. Recently, a polypeptide that is polar and uncharged and functions to minimize nonspecific adsorption of nanoparticles to cell surfaces was developed [90]. The designed polypeptide is composed of multiple units consisting of four asparagines (N) and one serine (S) residue and is referred to as the NS polypeptide. To verify the function of NS polypeptide, the interaction between the magnetosomes displaying NS polypeptides and cells was analyzed. As a result, NS polypeptides on magnetosomes functioned as a barrier to block the particle aggregation and minimize the nonspecific adsorption of cells to the nanoparticles. This technology, incorporating a functional polypeptide, may represent a completely new strategy for surface modification of magnetosomes for use in a variety of cell-associated applications.

Mass production of magnetosomes was also investigated in order to obtain functional magnetosomes for biotechnological use [91–94]. To enhance the productivity, the culture conditions including iron source, nutrients, and reducing agents in the medium were investigated [92]. Fed-batch culture of AMB-1 harboring a stably maintained expression plasmid in the cells for recombinant protein was carried out in a 10-l fermenter under microaerobic conditions. The addition of fresh nutrients was feedback controlled as a function of the pH of the culture. The yield of magnetosomes was optimized by adjusting the rate of ferric iron addition. Feeding ferric quinate at 15.4 μg/min yielded 7.5 μg/L magnetosomes [94]. Furthermore, stable expression of the recombinant gene in strain AMB-1 was indicated. Enrichment of the growth medium with L-cysteine, yeast extract, and polypeptone enhanced both bacterial growth and magnetosome production [93]. The presence of L-cysteine in the medium was useful for the induction of cell growth. Strict anaerobic conditions led to a prolonged lag phase and limited the final cell density. With regard to the iron sources, ferrous sulfate and ferric gallate dramatically enhanced the magnetosome yield as compared to ferric quinate, an iron chelate that is conventionally used. The optimized conditions increased the cell density to 0.59 ± 0.03 gram cells (dry weight) per liter, and the magnetosome production to 14.8 ± 0.5 gram (dry weight) per liter in a fed-batch culture for 4 days [93]. The optimized culture condition is applicable for the production of other functional magnetosomes.

10.3.2   Magnetosome Surface Modification by *In Vitro* System

In addition to the gene fusion technique, magnetosome surface can be functionalized by *in vitro* modifications with biomolecules (enzymes, antibodies, or oligonucleotides) [95–99]. Several techniques based on different principles have been developed. The chemical modification is the most conventional method to immobilize functional biomolecules onto magnetosome surfaces. Homofunctional cross-linkers, which contain 2 aldehydes or *N*-hydroxysuccinimide (NHS) esters reacted with amines on magnetosomes and functional biomolecules. A heterofunctional cross-linker, *N*-succinimidyl 3-(2-pyridyldithio)propionate (SPDP), which contains an NHS ester group, was used for the thiolation of magnetosomes. Heterofunctional cross-linkers were used to define the direction of functional biomolecules after chemical conjugation. Sulfosuccinimidyl 4-(*N*-maleimidomethyl)-cyclohexane-1-carboxylate (SMCC), which contains heterofunctional reaction sites (the NHS ester and maleimide), was reacted with biomolecules [96, 97]. Sulfo-NHS-LC-LC-biotin was used to modify the magnetic particles with biotin [95, 98]. The biotin was then utilized for immobilization of streptavidin. The streptavidin-modified magnetosomes were designed to immobilize biotin-modified antibodies or oligonucleotides [95]. These chemical modification techniques are relatively easy to introduce functional molecules on magnetosome surface, and will be used for the preparation of magnetosomes functionalized by various biomolecules.

As an alternative method, *in vitro* modification of magnetosome surface with proteins was developed. Artificial integration of useful proteins is a powerful method for the high efficient assembly of proteins on magnetosomes. Electrostatic and hydrophobic interactions are considered to be the driving force for spontaneous insertion of proteins and peptides into lipid membranes. MagA-luciferase fusion proteins prepared from recombinant *E. coli* membranes were integrated by sonication [100]. Maximum luminescence was obtained, which was 18 times higher than that of the recombinant luciferase-MagA displayed on magnetosomes generated using the gene fusion techniques. Furthermore, an antimicrobial peptide, temporin L, and its derivative were also employed as anchor peptides and immobilized streptavidin on magnetosomes. Temporin L is a cationic linear peptide with varied structure, length, and orientation. It forms an amphiphilic structure with the hydrophobic part that is organized in a helix when associated with a membrane. This spontaneous integration mechanism was applied for protein assembly onto magnetosomes [101]. The *C*-terminal of temporin L was incorporated into a magnetosome membrane, and the *N*-terminal was located on the magnetosome membrane surface. The conjugation of biotin to the *N*-terminal of temporin L leads to the binding of streptavidin on a magnetosome membrane. Improved efficiency of the integration of functional proteins is obtained by selecting the most suitable anchor molecule under optimized conditions.

As described above, the membrane is useful to immobilize functional molecules onto the magnetosome surface. The magnetosome membrane extracted from bacterial cells consists of 98% lipid and 2% other compounds including proteins and lipopolysaccharides (LPS) [89]. The LPS was acquired from the outer membrane of cells during the magnetosome extraction and purification processes. In the biotechnological applications of magnetosomes, such unneeded cellular components might cause nonspecific adsorption, leading to the inhibition of specific reaction and/or reduction of signal-to-noise ratio in detection. In order to eliminate such components, a membrane reconstruction technique was developed to create a nontoxic magnetic nanoparticle, enveloped by a stable phospholipid bilayer, displaying only the target proteins [89]. Lipids and proteins were removed from magnetosome surface using surfactants (Fig. 10.6a). Only Mms13-protein A was remaining on the surface after the treatment. The magnetosome surface was then reconstructed with phosphatidylcholine. The protein A-magnetosomes with reconstructed magnetosome membrane showed high dispersibility and lower antigenicity without the loss of ability to immobilize antibodies, leading to the successful separation of specific target cells.

The reconstruction strategy of magnetosome surface was applied for the modification with aminosilane compounds. Organosilane compounds can be covalently bound to magnetites introducing a positive ionic charge beneficial for restoring dispersion and facilitating ionic interactions between DNA molecules. The amino groups introduced onto magnetites are also useful for chemical covalent immobilization of proteins [102], biotin, and oligonucleotide [103]. After the removal of membrane lipid, bacterial magnetite crystals were treated with the aminosilane compounds, such as 3-aminopropyltriethoxysilane (APTES), N(trimethoxysilylpropyl)isothiouronium chloride (NTIC), and 3-[2-(2-aminoethyl)-ethylamino]-propyltrimethoxysilane) (AEEA), that can be covalently bound to surface –OH groups of magnetites [104]. Moreover, polyamidoamine (PAMAM) dendrimer was constructed to use AEEA-modified bacterial magnetite as a core (Fig. 10.6b) [105]. A PAMAM dendrimer can introduce a dense outer amine shell

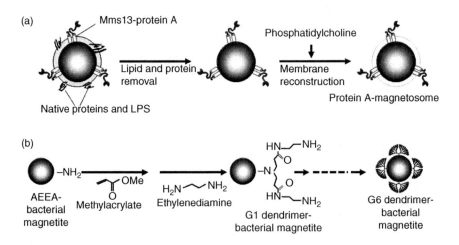

FIGURE 10.6 Schematic diagrams of magnetosome surface reconstruction by *in vitro* systems. Reconstruction of magnetosome membrane with phosphatidylcholine (a), and PAMAM dendrimer modification of bacterial magnetite (b).

through a cascade type generation. The outer amine concentration doubles with every layer generated and is limited only by steric interferences. A PAMAM coating may be used to reduce the magnetite agglomeration, and the terminal groups on the periphery can be tailored to control composite solubility. The increased cationic contribution will be useful for generating a colloidal suspension with an increased surface area for DNA extraction (see Section 10.4.3).

10.3.3 Protein-mediated Morphological Control of Magnetite Particles

For the practical use of magnetic iron oxides in medical and diagnostic applications, the particle size and morphology are of great importance because these factors strongly affect their magnetic properties. The development of reliable and reproducible methods that enable controlled synthesis of magnetic particles in size and morphology has, therefore, been a major challenge in this field over the years.

We have examined magnetite synthesis by a partial oxidation of ferrous hydroxide in the presence and absence of Mms6 [106]. The ferrous solution containing Mms6 was incubated under ambient conditions to produce Mms6-iron complexes prior to magnetite crystal formation. During the reaction, the color of the ferrous solution changed from transparent blue to bluish green. The observed nonmagnetic green precipitate is considered to be mainly composed of iron hydroxide ($Fe(OH)_2$ and $Fe(OH)_3$). The coexistence of Mms6 in the iron solution seems to either accelerate or stabilize the iron hydroxide formation process. Magnetite formation was then evident due to the progressive conversion of the initially nonmagnetic green precipitate into a strongly black magnetic precipitate when the solution was heated to 90°C. The reaction at room temperature yields primary small iron hydroxides that aggregate to form a larger iron hydroxide core. Mms6 in the iron solution are considered to be involved in the acceleration or stabilization of the iron hydroxide-formation process. The size distribution and circularity of the synthesized magnetite particles in the presence of various polypeptides were statistically analyzed from the TEM images. Mms6-mediated synthesis of magnetite by this method produced crystals of a uniformed size and narrow size distribution with a cubo-octahedral morphology, similar to those of magnetosomes observed in *M. magneticum* strain AMB-1 (Fig. 10.7a). The crystals formed in the absence of Mms6 were octahedral (Fig. 10.7b), larger with an increased size distribution. This type of control over the size and morphology of the particles has previously only been observed in the case of high temperature synthetic methods.

In addition, we designed short peptides by mimicking characteristic amino acid sequences of Mms6, and utilized them for the *in vitro* magnetite synthesis. The magnetite synthesis using the peptides containing *C*-terminal acidic region of Mms6 resulted in the formation of a uniformed size and narrow size distribution with a cubo-octahedral morphology [106]. Particles that were synthesized in the presence of short peptides harboring the *C*-terminal acidic region of Mms6 exhibited a spherical morphology with circularities of 0.70–0.90 similar to those of magnetosomes and particles formed in the presence of the Mms6 protein. In contrast, a rectangular morphology with circularities of 0.60–0.85 were obtained when other peptides were

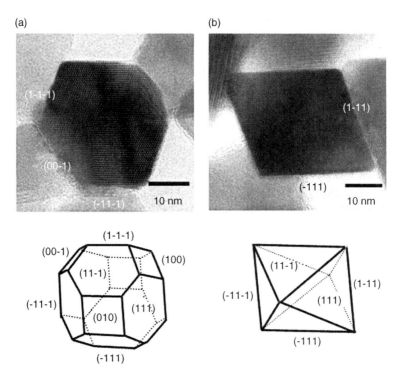

FIGURE 10.7 Transmission electron microscopic analysis of magnetite crystals prepared by partial oxidation of ferrous hydroxide. Magnetites synthesized in the presence (a) and absence of Mms6 protein (b). Reprinted with permission from Reference 106. Copyright 2007 Biomaterials.

used for synthesis. These results indicated that the magnetite particles synthesized by the *in vitro* chemical synthetic method with the Mms6 peptides revealed similar features that of biogenic magnetites and the method presents an alternative route for controlling the size and shape of magnetite crystals without the use of organic solvent and high temperatures. Recently, the method was further developed for the synthesis of shape-regulated cobalt ferrite ($CoFe_2O_4$) [107] or cobalt-doped magnetite nanoparticles [108]. This novel method using protein in mild condition might also enable enzyme fixation at ambient temperature as well as establish environment-friendly novel magnetite synthesis method.

10.4 APPLICATION

Nanoparticles are necessary for the growing of a variety of cell-associated applications, such as cell separation, *in vivo/in vitro* imaging, and drug or gene delivery. Nanoparticles have a high surface area to volume ratio, and the ability and property of nanoparticles are affected seriously by surface modification with various molecules.

In order to develop superior nanoparticles which are useful for cell-associated applications, it is needed to use suitable surface modification to the nanoparticles.

Magnetosomes purified from *M. magneticum* strain AMB-1 are 40–100 nm in size, covered with a lipid bilayer membrane. The lipid bilayer imparts magnetosomes with better dispersion qualities in solutions compared with artificial magnetic particles. Superior dispersion permits various applications of magnetosomes. Antibody, enzyme, and receptor protein have been immobilized on magnetosomes and applied to immunoassay, enzymatic reactions, cell separation, and DNA isolation techniques.

10.4.1 Enzymatic Bioassays

Proteins immobilized on magnetosomes can be applied for various bioassays. The bioassays using magnetosomes enable rapid and conventional detection by magnetic collection. In addition, the size of magnetosomes provides large surface reaction area to volume ratio, leading to high efficient reaction between ligand and receptor. Furthermore, various proteins were stably embedded within the lipid bilayer membrane or attached on the surface as native forms with their specific enzymatic activities. The first idea to use magnetosome for bioassay was proposed in 1987 [109]. In this examination, magnetosomes were directly extracted from environmental magnetotactic bacteria harvested from pond sediment. The activity of glucose oxidase immobilized on magnetosomes was determined to be 40 times higher than that immobilized on artificial magnetites.

Based on this investigation, magnetosomes were then used for immunoassays. Fluorescence immunoassay was constructed to use fluorescein isothiocyanate (FITC) conjugated with anti-IgG antibody-magnetosomes [110]. The fluorescence quenching caused by agglutination of FITC-anti-IgG antibody-magnetosome conjugates was measured by using a fluorescence spectrophotometer. The relative fluorescence intensity correlated linearly with a concentration of IgG in the range of 0.5–100 ng/mL. Alternatively, chemiluminescent enzyme immunoassay was developed to use anti-IgG antibody-magnetosomes [96]. A good relationship was obtained between the luminescence intensity and mouse IgG concentration in the range of $1-10^5$ fg/mL. The chemiluminescence enzyme immunoassay was further investigated to use the Z-domain of protein A-magnetosomes [11] and protein G-magnetosomes [88]. The antibody was accurately oriented on the magnetosome due to its association with protein A and protein G, unlike that in the immobilization by chemical conjugation. When a chemiluminescence sandwich immunoassay was performed using antibody–protein A-magnetosome complexes and magnetosomes chemically immobilized with antibodies, the antigen-binding activity per microgram of antibodies for antibody–protein A-magnetosome complexes was two times higher than that for the antibody–magnetosome conjugates prepared by chemical immobilization [111]. Human insulin concentrations in blood serum were measured by a fully automated sandwich immunoassay using antibody–protein A-magnetosome complexes and alkaline phosphatase-conjugated antibodies as primary and secondary immunoreactants, respectively. Dose–response curves were obtained from the luminescence intensity for human insulin concentrations. A detection limit of 2 μU/mL and a linear correlation between the signal and the concentration was apparent over the range of 19–254 μU/mL [111]. On the other hand, Wacker et al. developed the magneto immuno-PCR method by using antibodies immobilized onto magnetosomes as the capture phase [112]. Using this method, a detection limit of 320 pg/mL for the hepatitis B surface antigen was reported.

Magnetosomes displaying receptors were developed and utilized for receptor-binding assays. GPCR represents one of the most predominant families of transmembrane proteins and is a prime target for drug discovery. Various challenges with regard to the ligand screening of GPCRs have been undertaken; however, these proteins are generally expressed at very low levels in the cell and are extremely hydrophobic, rendering the analysis of ligand interaction very difficult. Moreover, purification of GPCRs from cells is frequently time-consuming and typically results in the loss of native conformation. In magnetosomes displaying D1 dopamine receptor (D1R), GPCR was used as a model for a ligand-binding assay [78]. Efficient assembly of D1R into the lipid membrane of nano-sized magnetosomes was accomplished, and this was used for competitive dopamine-binding assays. This system conveniently refines the native conformation of GPCRs without the need for detergent solubilization, purification, and reconstitution after cell disruption. This novel system provides advantages for studying various membrane proteins, which are usually difficult to assay.

Pyrosequencing is a bioluminometric DNA sequencing method based on the sequencing-by-synthesis principle by employing a cascade of several enzymatic reactions. A highly thermostable enzyme, pyruvate phosphate dikinase (PPDK) which converts pyrophosphate pyrophosphoric acid (PPi) to ATP, was expressed on magnetosomes and used for pyrosequencing [113]. The pyrosequencing relies on the incorporation of nucleotides by DNA polymerase, which results in the release of PPi. The production of ATP by PPDK-displaying magnetosomes can be utilized by luciferase in a luminescent reaction. PPDK-displaying magnetosomes were applied in pyrosequencing and a target oligonucleotide was successfully sequenced [114]. The PPDK enzyme was recyclable in each sequence reaction as it was immobilized onto magnetosomes that could be manipulating a magnet. These results illustrate the advantages of using enzyme-displaying magnetosomes as biocatalysts for repeat usage. The PPDK-displaying magnetosomes also has a benefit for the scale-down of pyrosequencing reaction volumes, thus, permitting high throughput data acquisition.

10.4.2 Cell Separation

Direct separation of target cells from mixed population, such as peripheral blood, umbilical cord blood, and bone marrow, is an essential technique for various therapeutic or diagnosis applications. Immunomagnetic particles have been used preferentially in target cell separation because the methodology is more rapid and simple compared with cell sorting using flow cytometry. The magnetic bead technology facilitates simple, rapid, and efficient enrichment of target cell populations. In general, nano-sized magnetic particles rather than micro-sized beads are preferred for cell separation because the cells that are separated using the latter are subsequently

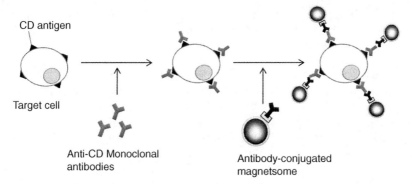

CD antigen

Target cell

Anti-CD Monoclonal
antibodies

Antibody-conjugated
magnetsome

FIGURE 10.8   Schematic diagram of magnetic cell separation using antibody-conjugated magnetosomes.

useful for flow cytometric analysis. Micro-sized magnetic particles on the other hand have inhibitory effects on cell growth and differentiation after magnetic separation. Because magnetosomes consist of ferromagnetic iron oxide, they are easily separated from cell suspensions by using a permanent magnet with no special column (Fig. 10.8).

Antibody-conjugated magnetosomes were utilized for cell separation from peripheral blood. To enrich for target cells, cell surface antigens, such as cluster of different (CD) antigens, were used as markers. The procedure for positive selection involves incubation of mononuclear cells and mouse monoclonal antibodies against different cell surface antigens. CD8-, CD14-, CD19-, CD20-, and CD34-positive cells were efficiently enriched from peripheral blood [11, 88]. The average purities of the separated mononuclear cells of CD19$^+$ and CD20$^+$ were 97.5% and 97.6%, respectively. More than 95% of the recovery ratio was obtained for other cells. The separated CD34$^+$ cells separated using magnetosomes maintained their capability of colony formation as hematopoietic stem cells without inhibition of their proliferation and differentiation abilities. Furthermore, the cell separation efficiency was improved to use protein G which is an IgG-binding protein that interacts with a wide range of antibody types. To use protein G-magnetosomes binding with anti-CD monoclonal antibodies, B lymphocytes (CD19$^+$ cells) or T lymphocytes (CD3$^+$ cells), which represent less than 0.05% in whole blood cells, were successfully separated at a purity level of more than 96% [83]. This level was superior to that from previous reports using other magnetic separation approaches. Specific separation of target cells using magnetosomes was achieved by simple magnetic separation from cell suspensions by using a permanent magnet with no special magnetic column.

An alternative approach to increase cell separation efficiency using a novel functional polypeptide which minimizes nonspecific adsorption of magnetic particles to cells was introduced. A NS polypeptide, which is 100 amino acids composed of repeated units of four asparagines and one serine residue (NS), as a molecule for magnetosome surface modification was designed [90]. The polypeptide designed was composed of multiple units consisting of four asparagines (N) and one serine (S) residue and is referred to as the NS polypeptide. Modification of the magnetosome surface with the NS polypeptide results in reduction of magnetosome–magnetosome and magnetosome-cell interactions. When the NS polypeptide is used in a single fusion protein as a linker to display protein G on magnetic particles, the particle acquires the capacity to specifically bind target cells and to avoid nonspecific adsorption of nontarget cells. $CD19^+$ cells represent 4.1% of leukocytes, and in peripheral blood were determined to be less than 0.004% of the total cells. $CD19^+$ cells were separated directly from peripheral blood with over 95% purity using protein G-displaying magnetosomes bound to anti-CD19 monoclonal antibodies with the NS polypeptide. Display of protein G-NS polypeptide fusion proteins on magnetosomes significantly improved the recognition and binding to target cells and minimized adsorption of nontarget cells. This technology, incorporating a functional polypeptide, may represent a completely new strategy for surface modification of nanoparticles for use in a variety of cell-associated applications.

10.4.3 DNA Extraction

DNA extraction is an important technique for DNA sequencing, PCR, and gene recombination. Solid-phase extraction methods using microbeads or membrane filters have been widely used for DNA extraction. Magnetic particles offer advantages over conventional methods due to quick processing times, reduced chemical requirements, and ease of magnetic separation. Moreover, DNA extraction using magnetic particles is commonly accepted as an automation-friendly procedure. Many automated instruments have been developed for extraction of DNA from blood samples. We developed a DNA extraction method using amine-modified magnetic particles introduced in Section 10.3.2.

Bare magnetite particles have capability to adsorb DNA; however, they form aggregates due to magnetic attractive forces reducing usable surface area for adsorption. In order to overcome this problem, we developed magnetite particles coupled to silane compounds introducing a positive ionic charge that is beneficial for restoring dispersion and may facilitate an ionic interaction with DNA [104]. Bare magnetosomes were coated with 3-aminopropyltriethoxysilane, N-(trimethoxy-silylpropyl) isothiouronium chloride or 3-[2-(2-aminoethyl)-ethylamino]-propyltrimethoxysilane (AEEA). The DNA-binding efficiency increased with the number of amino groups present on the silane compounds and was 14-fold higher than with untreated bacterial magnetite. Addition of AEEA to aqueous solutions containing coated bacterial magnetite increased efficiency due to co-condensation of DNA. From 10^8 E. coli cells, 7.1 μg of high purity DNA was recovered using 100 μg of bacterial magnetites. E. coli DNA extracted with modified bacterial magnetite was suitable for restriction enzyme digestion and PCR. Over 99.5% of DNA in solution was bound to AEEA-modified bacterial magnetites [103], and recovery of approximately 85% of adsorbed DNA was attained by using dNTP in place of phosphate buffer [115]. The desorption mechanism was considered to be the replacement of DNA with phosphate ion or dNTP (Fig. 10.9a).

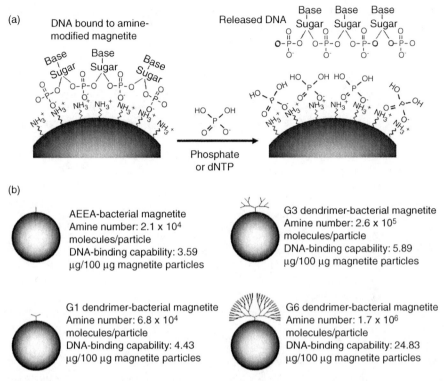

FIGURE 10.9   Scheme of DNA release from amine-modified bacterial magnetite (a). Amine numbers and DNA-binding capabilities of amine-modified bacterial magnetites (b).

Enhancement of DNA-binding capability was achieved using PAMAM-modified magnetic particles [105]. A cascading hyperbranched PAMAM dendrimer allowed enhanced extraction of DNA from fluid suspensions. Characterization of the synthesis revealed linear doubling of the surface amine charge from generations one through five starting with an amino silane initiator (Fig. 10.9b). Furthermore, TEM revealed clear dispersion of the single-domain magnetite in aqueous solution. The PAMAM dendrimer-modified magnetites have been used to carry out magnetic separation of DNA. Binding and release efficiencies increased with the number of generations and those of bacterial magnetite modified with six generation dendrimer were 7 and 11 times, respectively, as many as those of bacterial magnetite modified with AEEA.

Due to the high DNA-binding capability and good dispersion of amine-modified bacterial magnetite, this is a suitable material to use in automated DNA extraction system [99]. For example, conventional procedures for plant DNA extraction have employed cesium chloride density gradients to eliminate enzyme-inhibitory polysaccharides as well as cetyltrimethylammonium bromide (CTAB)-based protocols [116]. These methods require centrifugation, which is the most difficult process for integration into the automated system. We used the amino silane-modified particles for the

DNA extraction from maize. The yield of genomic DNA was approximately 315 ± 3 ng when 60 μg of magnetites were used. The $(A_{260}-A_{320}):(A_{280}-A_{320})$ ratio of the extracted DNA was 1.9 ± 0.1. These results suggest that our proposed DNA extraction method allowed highly accurate DNA recovery for subsequent PCR processes. This method did not require laborious pretreatment using organic solvents and permitted rapid completion of DNA extraction within 30 min, from sample application to the measurement of extracted DNA concentration and purity. Although several automated instruments for DNA extraction from plant tissues are present, these methodologies require centrifugation before sample application in order to eliminate enzyme inhibitory polysaccharides. Our newly developed instrument facilitates DNA extraction from maize powder without the centrifugation steps.

10.4.4 Bioremediation

The recovery of metals and metalloids using microorganisms has emerged as a potentially attractive and environment-friendly alternative to conventional techniques such as reclamation treatments. The uptake and crystallization of Pb(II), Ag(I), Au(III), U(VI), Se(IV), and other metals by various bacterial species have been well documented previously. Bacteria also act as efficient sorption surfaces for soluble metal ion species in the aqueous environment. However, although these bacteria have been adapted for and applied to the accumulation of various metals and metalloids, cell recovery remains a bottleneck in this approach. A more versatile method or technology is required in order to overcome this problem. Magnetosomes extracted from magnetotactic bacteria were shown to be a powerful tool for various separation and detection techniques in biotechnological applications. Magnetotactic bacterial cells containing magnetosomes are also useful to separate and collect trace substances, because the direction of movement of motile cells can be easily manipulated and they can be easily separated from the solution by attracting them to a magnet (Fig. 10.10a). To use this ability, bioremediations of metal ions were investigated using cultured magnetotactic bacteria.

As a first examination, cadmium (Cd^{2+}) recovery was investigated by using *D. magneticus* strain RS-1 with the aim of on-site heavy metal removal from the environment [117]. Strain RS-1 precipitated over 95% of cadmium in the medium when the initial Cd^{2+} concentration was 1.3 ppm in the growth medium. They were deposited on the cell surface as CdS crystals in a size range of 20–40 nm (Figs. 10.10b, 10.10c). The composition ratio of S/Cd was 0.7, which corresponds with cadmium sulfide quantum semiconductor crystallites, and can be used as material for electric component. RS-1 cells were recovered by a simple magnetic separation revealing the removal of 58% cadmium from the culture medium. The Cd^{2+}-binding ability can be enhanced by a protein modification of cell surface. Cell surface display technique in *M. magneticum* strain AMB-1 has been developed by expressing hexahistidine residues within the outer coiled loop of the membrane-specific protein (Msp1) of the strain AMB-1 [118]. The high level of expression of the Msp1 protein allows it to be a possible carrier protein or candidate in constructing an effective surface display system on AMB-1 cells. By secondary structural analysis, Msp1 was predicted

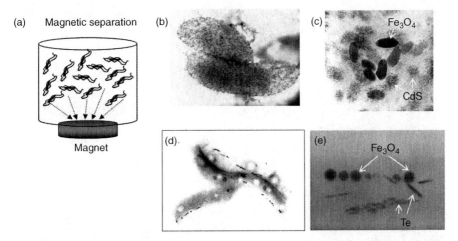

FIGURE 10.10   Scheme of magnetic separation of magnetotactic bacteria. (a). Transmission electron micrographs of *D. magneticus* strain RS-1 grown in the presence of cadmium (b, c), and *M. magneticum* strain AMB-1 grown in the presence of tellurite (d, e). Reprinted with permission from Reference 119. Copyright 2010 Applied Environmental Microbiology.

to have β-strand and coiled-loop repeats which were similar to those of OmpA. Msp1 protein also shared homology with the outer membrane porin precursor of *Rhodospirillum rubrum* and was found to be highly expressed in strain AMB-1. The optimal display site of the hexahistidine residues was identified via secondary structure prediction, immunofluorescence microscopy, and heavy metal-binding assay. The established transformant showed high immunofluorescence response, high Cd^{2+} binding, and high recovery efficiency in comparison to those of the negative control when manipulated by magnetic force.

Alternatively, bioremediation of toxic metal or metalloid ions can be achieved to use an ability that bacterial cell accumulates and mineralizes metal ions within the cells. Tellurium uptake and crystallization were investigated using strain AMB-1 [119]. After the cultivation of strain AMB-1 in a medium containing tellurite, the cells formed separately two independent crystal types, tellurium and magnetite (Fig. 10.10d). A presence of rod-shaped nanostructures (~15 nm in diameter by ~200 nm in length) was observed under TEM (Fig. 10.10e). The number of crystals within the cell increased and larger crystals were observed withincreasing initial concentration of tellurite in the medium. Elemental mapping indicated that Te in the rod-shaped crystals and Fe and O in magnetosomes were concentrated within the magnetotactic bacteria. It is worthy to note that magnetosomes and tellurite nanorods were simultaneously synthesized within the same cell, while nanorods were observed neither in the periplasmic space nor on the cell surface. In addition, by crystallizing the pollutant, approximately 70 times more molecules were bioaccumulated per cell in comparison with the cell surface adsorption approach. Moreover, there have been reports of magnetite and iron sulfide magnetosome crystals containing manganese and copper, respectively, found in environmental magnetotactic bacteria. The controlled cobalt doping of magnetite magnetosome crystals was also reported. The presence of cobalt resulted in a pronounced increase in magnetic coercive. These findings will provide an important advance in designing biologically synthesized magnetic materials with highly controlled magnetic properties.

10.5   SUMMARY AND PERSPECTIVES

Understandings of magnetosome formation mechanism have tremendously improved over the last 10 years. Nevertheless, there are many unexplained phenomena and events that have attracted much interest from various scientific fields, such as biology, chemistry, physics, geology, medical pharmaceutical sciences, biotechnology, and nanotechnology. For example, origin and evolution of magnetotactic bacteria, actual function of the magneto-aerotaxis, gene transfer mechanism of MAI region, minimal gene set for magnetosome formation, membrane invagination, iron uptake, magnetosome alignment, and crystal size and shape control mechanisms, still remain largely unknown. They will be clarified by the comprehensive and comparative studies that are now in progress by a number of researchers worldwide. Moreover, the results obtained from these fundamental studies will allow us to develop novel molecular constructs on the magnetosome surface. The constructed functional magnetosomes generated by the molecular bioengineering techniques will be used for various biotechnological applications.

ACKNOWLEDGMENTS

The authors acknowledge supports from Grant-in-Aid for Scientific Researches from the Japan Society for the Promotion of Science (JSPS), Grant-in-Aid for Scientific Research on Innovative Areas from the Ministry of Education, Culture, Sports, Science, and Technology, Japan (MEXT), and Industrial Technology Research Grant Program from the New Energy and Industrial Technology Development Organization (NEDO) of Japan.

REFERENCES

[1] Baeuerlein, E. *Biomineralization. Progress in Biology, Molecular Biology, and Application*; Wiley-VCH: Weinheim, **2004**.

[2] Frigeri, L. G.; Radabaugh, T. R.; Haynes, P. A.; Hildebrand, M. *Mol. Cell. Proteomics* **2006**, *5*, 182–193.

[3] Marsh, M. E. *Gravit. Space Biol. Bull.* **1999**, *12*, 5–14.

[4] Cha, J. N.; Shimizu, K.; Zhou, Y.; Christiansen, S. C.; Chmelka, B. F.; Stucky, G. D.; Morse, D. E. *Proc. Natl. Acad. Sci. U S A* **1999**, *96*, 361–365.

[5] Kroger, N.; Deutzmann, R.; Sumper, M. *Science* **1999**, *286*, 1129–1132.

[6] Suzuki, M.; Saruwatari, K.; Kogure, T.; Yamamoto, Y.; Nishimura, T.; Kato, T.; Nagasawa, H. *Science* **2009**, *325*, 1388–1390.

[7] Blakemore, R. P. *Science* **1975**, *190*, 377–379.

[8] Spring, S.; Schleifer, K. H. *Syst. Appl. Microbiol.* **1995**, *18*, 147–153.

[9] Kawai, N.; Ito, A.; Nakahara, Y.; Futakuchi, M.; Shirai, T.; Honda, H.; Kobayashi, T.; Kohri, K. *Prostate* **2005**, *64*, 373–381.

[10] Gleich, B.; Weizenecker, R. *Nature* **2005**, *435*, 1214–1217.

[11] Kuhara, M.; Takeyama, H.; Tanaka, T.; Matsunaga, T. *Anal. Chem.* **2004**, *76*, 6207–6213.

[12] Pardoe, H.; Clark, P. R.; St Pierre, T. G.; Moroz, P.; Jones, S. K. *Magn. Reson. Imaging* **2003**, *21*, 483–488.

[13] Plank, C.; Schillinger, U.; Scherer, F.; Bergemann, C.; Remy, J. S.; Krotz, F.; Anton, M.; Lausier, J.; Rosenecker, J. *Biol. Chem.* **2003**, *384*, 737–747.

[14] Takahashi, M.; Akiyama, Y.; Ikezumi, J.; Nagata, T.; Yoshino, T.; Iizuka, A.; Yamaguchi, K.; Matsunaga, T. *Bioconjug. Chem.* **2009**, *20*, 304–309.

[15] Thornhill, R. H.; Burgess, J. G.; Sakaguchi, T.; Matsunaga, T. *FEMS Microbiol. Lett.* **1994**, *115*, 169–176.

[16] Blakemore, R. P.; Maratea, D.; Wolfe, R. S. *J. Bacteriol.* **1979**, *140*, 720–729.

[17] Matsunaga, T.; Sakaguchi, T.; Tadokoro, F. *Appl. Microbiol. Biotechnol.* **1991**, *35*, 651–655.

[18] Schleifer, K. H.; Schuler, D.; Spring, S.; Weizenegger, M.; Amann, R.; Ludwig, W.; Kohler, M. *Syst. Appl. Microbiol.* **1991**, *14*, 379–385.

[19] Meldrum, F. C.; Mann, S.; Heywood, B. R.; Frankel, R. B.; Bazylinski, D. A. *Proc. R. Soc. Lond. B Biol. Sci.* **1993**, *251*, 231–236.

[20] Bazylinski, D. A.; Frankel, R. B.; Jannasch, H. W. *Nature* **1988**, *334*, 518–519.

[21] Lefevre, C. T.; Bernadac, A.; Yu-Zhang, K.; Pradel, N.; Wu, L. F. *Environ. Microbiol.* **2009**, *11*, 1646–1657.

[22] Sakaguchi, T.; Burgess, J. G.; Matsunaga, T. *Nature* **1993**, *365*, 47–49.

[23] Kawaguchi, R.; Burgess, J. G.; Sakaguchi, T.; Takeyama, H.; Thornhill, R. H.; Matsunaga, T. *FEMS Microbiol. Lett.* **1995**, *126*, 277–282.

[24] Sakaguchi, T.; Arakaki, A.; Matsunaga, T. *Int. J. Syst. Evol. Microbiol.* **2002**, *52*, 215–221.

[25] Lefevre, C. T.; Frankel, R. B.; Posfai, M.; Prozorov, T.; Bazylinski, D. A. *Environ. Microbiol.* **2011**, *13*, 2342–2350.

[26] Lefevre, C. T.; Menguy, N.; Abreu, F.; Lins, U.; Posfai, M.; Prozorov, T.; Pignol, D.; Frankel, R. B.; Bazylinski, D. A. *Science* **2011**, *334*, 1720–1723.

[27] Flies, C. B.; Peplies, J.; Schuler, D. *Appl. Environ. Microbiol.* **2005**, *71*, 2723–2731.

[28] Spring, S.; Amann, R.; Ludwig, W.; Schleifer, K. H.; Vangemerden, H.; Petersen, N. *Appl. Environ. Microbiol.* **1993**, *59*, 2397–2403.

[29] Farina, M.; Esquivel, D. M. S.; Debarros, H. G. P. L. *Nature* **1990**, *343*, 256–258.

[30] Mann, S.; Sparks, N. H. C.; Frankel, R. B.; Bazylinski, D. A.; Jannasch, H. W. *Nature* **1990**, *343*, 258–261.

[31] Simmons, S. L.; Bazylinski, D. A.; Edwards, K. J. *Science* **2006**, *311*, 371–374.

[32] Matsunaga, T.; Okamura, Y.; Fukuda, Y.; Wahyudi, A. T.; Murase, Y.; Takeyama, H. *DNA. Res.* **2005**, *12*, 157–166.

[33] Fukuda, Y.; Okamura, Y.; Takeyama, H.; Matsunaga, T. *FEBS Lett.* **2006**, *580*, 801–812.

[34] Schubbe, S.; Kube, M.; Scheffel, A.; Wawer, C.; Heyen, U.; Meyerdierks, A.; Madkour, M. H.; Mayer, F.; Reinhardt, R.; Schuler, D. *J. Bacteriol.* **2003**, *185*, 5779–5790.

[35] Ullrich, S.; Kube, M.; Schubbe, S.; Reinhardt, R.; Schuler, D. *J. Bacteriol.* **2005**, *187*, 7176–7184.

[36] Schubbe, S.; Williams, T. J.; Xie, G.; Kiss, H. E.; Brettin, T. S.; Martinez, D.; Ross, C. A.; Schuler, D.; Cox, B. L.; Nealson, K. H.; Bazylinski, D. A. *Appl. Environ. Microbiol.* **2009**, *75*, 4835–4852.

[37] Richter, M.; Kube, M.; Bazylinski, D. A.; Lombardot, T.; Glockner, F. O.; Reinhardt, R.; Schuler, D. *J. Bacteriol.* **2007**, *189*, 4899–4910.

[38] Jogler, C.; Kube, M.; Schubbe, S.; Ullrich, S.; Teeling, H.; Bazylinski, D. A.; Reinhardt, R.; Schuler, D. *Environ. Microbiol.* **2009**, *11*, 1267–1277.

[39] Nakamura, C.; Burgess, J. G.; Sode, K.; Matsunaga, T. *J. Biol. Chem.* **1995**, *270*, 28392–28396.

[40] Okamura, Y.; Takeyama, H.; Matsunaga, T. *J. Biol. Chem.* **2001**, *276*, 48183–48188.

[41] Nakazawa, H.; Arakaki, A.; Narita-Yamada, S.; Yashiro, I.; Jinno, K.; Aoki, N.; Tsuruyama, A.; Okamura, Y.; Tanikawa, S.; Fujita, N.; Takeyama, H.; Matsunaga, T. *Genome Res.* **2009**, *19*, 1801–1808.

[42] Arakaki, A.; Shibusawa, M.; Hosokawa, M.; Matsunaga, T. *Appl. Environ. Microbiol.* **2010**, *76*, 1480–1485.

[43] Jogler, C.; Lin, W.; Meyerdierks, A.; Kube, M.; Katzmann, E.; Flies, C.; Pan, Y.; Amann, R.; Reinhardt, R.; Schuler, D. *Appl. Environ. Microbiol.* **2009**, *75*, 3972–3979.

[44] Jogler, C.; Wanner, G.; Kolinko, S.; Niebler, M.; Amann, R.; Petersen, N.; Kube, M.; Reinhardt, R.; Schuler, D. *Proc. Natl. Acad. Sci. U S A* **2011**, *108*, 1134–1139.

[45] Suzuki, T.; Okamura, Y.; Calugay, R. J.; Takeyama, H.; Matsunaga, T. *J. Bacteriol.* **2006**, *188*, 2275–2279.

[46] Grunberg, K.; Muller, E. C.; Otto, A.; Reszka, R.; Linder, D.; Kube, M.; Reinhardt, R.; Schuler, D. *Appl. Environ. Microbiol.* **2004**, *70*, 1040–1050.

[47] Tanaka, M.; Okamura, Y.; Arakaki, A.; Tanaka, T.; Takeyama, H.; Matsunaga, T. *Proteomics* **2006**, *6*, 5234–5247.

[48] Matsunaga, T.; Nemoto, M.; Arakaki, A.; Tanaka, M. *Proteomics* **2009**, *9*, 3341–3352.

[49] Komeili, A.; Li, Z.; Newman, D. K.; Jensen, G. J. *Science* **2006**, *311*, 242–245.

[50] Schultheiss, D.; Handrick, R.; Jendrossek, D.; Hanzlik, M.; Schuler, D. *J. Bacteriol.* **2005**, *187*, 2416–2425.

[51] Zeytuni, N.; Ozyamak, E.; Ben-Harush, K.; Davidov, G.; Levin, M.; Gat, Y.; Moyal, T.; Brik, A.; Komeili, A.; Zarivach, R. *Proc. Natl. Acad. Sci. U S A* **2011**, *108*, E480–487.

[52] Komeili, A.; Vali, H.; Beveridge, T. J.; Newman, D. K. *Proc. Natl. Acad. Sci. U S A* **2004**, *101*, 3839–3844.

[53] Matsunaga, T.; Tsujimura, N.; Okamura, Y.; Takeyama, H. *Biochem. Biophys. Res. Commun.* **2000**, *268*, 932–937.

[54] Tanaka, M.; Arakaki, A.; Matsunaga, T. *Mol. Microbiol.* **2010**, *76*, 480–488.

[55] Quinlan, A.; Murat, D.; Vali, H.; Komeili, A. *Mol. Microbiol.* **2011**, *80*, 1075–1087.

[56] Uebe, R.; Junge, K.; Henn, V.; Poxleitner, G.; Katzmann, E.; Plitzko, J. M.; Zarivach, R.; Kasama, T.; Wanner, G.; Posfai, M.; Bottger, L.; Matzanke, B.; Schuler, D. *Mol. Microbiol.* **2011**, *82*, 818–835.

[57] Arakaki, A.; Webb, J.; Matsunaga, T. *J. Biol. Chem.* **2003**, *278*, 8745–8750.

[58] Wahyudi, A. T.; Takeyama, H.; Okamura, Y.; Fukuda, Y.; Matsunaga, T. *Biochem. Biophys. Res. Commun.* **2003**, *303*, 223–229.

[59] Suzuki, T.; Okamura, Y.; Arakaki, A.; Takeyama, H.; Matsunaga, T. *FEBS Lett.* **2007**, *581*, 3443–3448.

[60] Pradel, N.; Santini, C. L.; Bernadac, A.; Fukumori, Y.; Wu, L. F. *Proc. Natl. Acad. Sci. U S A* **2006**, *103*, 17485–17489.

[61] Taoka, A.; Asada, R.; Wu, L. F.; Fukumori, Y. *J. Bacteriol.* **2007**, *189*, 8737–8740.

[62] Scheffel, A.; Gruska, M.; Faivre, D.; Linaroudis, A.; Plitzko, J. M.; Schüler, D. *Nature* **2006**, *440*, 110–114.

[63] Draper, O.; Byrne, M. E.; Li, Z.; Keyhani, S.; Barrozo, J. C.; Jensen, G.; Komeili, A. *Mol. Microbiol.* **2011**, *82*, 342–354.

[64] Scheffel, A.; Schuler, D. *J. Bacteriol.* **2007**, *189*, 6437–6446.

[65] Scheffel, A.; Gardes, A.; Grunberg, K.; Wanner, G.; Schuler, D. *J. Bacteriol.* **2008**, *190*, 377–386.

[66] Lohsse, A.; Ullrich, S.; Katzmann, E.; Borg, S.; Wanner, G.; Richter, M.; Voigt, B.; Schweder, T.; Schuler, D. *PLoS One* **2011**, *6*, e25561.

[67] Murat, D.; Quinlan, A.; Vali, H.; Komeili, A. *Proc. Natl. Acad. Sci. U S A* **2010**, *107*, 5593–5598.

[68] Du, C.; Falini, G.; Fermani, S.; Abbott, C.; Moradian-Oldak, J. *Science* **2005**, *307*, 1450–1454.

[69] Lakshminarayanan, R.; Chi-Jin, E. O.; Loh, X. J.; Kini, R. M.; Valiyaveettil, S. *Biomacromolecules* **2005**, *6*, 1429–1437.

[70] Wang, W.; Bu, W.; Wang, L.; Palo, P. E.; Mallapragada, S.; Nilsen-Hamilton, M.; Vaknin, D. *Langmuir* **2012**, *28*, 4274–4282.

[71] Tanaka, M.; Mazuyama, E.; Arakaki, A.; Matsunaga, T. *J. Biol. Chem.* **2011**, *286*, 6386–6392.

[72] Matsunaga, T.; Nakamura, C.; Burgess, J. G.; Sode, K. *J. Bacteriol.* **1992**, *174*, 2748–2753.

[73] Okamura, Y.; Takeyama, H.; Sekine, T.; Sakaguchi, T.; Wahyudi, A. T.; Sato, R.; Kamiya, S.; Matsunaga, T. *Appl. Environ. Microbiol.* **2003**, *69*, 4274–4277.

[74] Nakamura, C.; Kikuchi, T.; Burgess, J. G.; Matsunaga, T. *J. Biochem.* **1995**, *118*, 23–27.

[75] Matsunaga, T.; Togo, H.; Kikuchi, T.; Tanaka, T. *Biotechnol. Bioeng.* **2000**, *70*, 704–709.

[76] Matsunaga, T.; Sato, R.; Kamiya, S.; Tanaka, T.; Takeyama, H. *J. Magn. Magn. Mater.* **1999**, *194*, 126–134.

[77] Yoshino, T.; Kato, F.; Takeyama, H.; Nakai, M.; Yakabe, Y.; Matsunaga, T. *Anal. Chim. Acta* **2005**, *532*, 105–111.

[78] Yoshino, T.; Takahashi, M.; Takeyama, H.; Okamura, Y.; Kato, F.; Matsunaga, T. *Appl. Environ. Microbiol.* **2004**, *70*, 2880–2885.

[79] Yoshino, T.; Matsunaga, T. *Appl. Environ. Microbiol.* **2006**, *72*, 465–471.

[80] Maeda, Y.; Yoshino, T.; Matsunaga, T. *Appl. Environ. Microbiol.* **2010**, *76*, 5785–5790.

[81] Maeda, Y.; Yoshino, T.; Matsunaga, T. *J. Mater. Chem.* **2009**, *19*, 6361–6366.

[82] Maeda, Y.; Yoshino, T.; Takahashi, M.; Ginya, H.; Asahina, J.; Tajima, H.; Matsunaga, T. *Appl. Environ. Microbiol.* **2008**, *74*, 5139–5145.

[83] Takahashi, M.; Yoshino, T.; Takeyama, H.; Matsunaga, T. *Biotechnol. Prog.* **2009**, *25*, 219–226.

[84] Matsunaga, T.; Maeda, Y.; Yoshino, T.; Takeyama, H.; Takahashi, M.; Ginya, H.; Asahina, J.; Tajima, H. *Anal. Chim. Acta.* **2007**, *597*, 331–339.

[85] Yoshino, T.; Matsunaga, T. *Biochem. Biophys. Res. Commun.* **2005**, *338*, 1678–1681.

[86] Yoshino, T.; Shimojo, A.; Maeda, Y.; Matsunaga, T. *Appl. Environ. Microbiol.* **2010**, *76*, 1152–1157.

[87] Jorgensen, R. A.; Berg, D. E.; Allet, B.; Reznikoff, W. S. *J. Bacteriol.* **1979**, *137*, 681–685.

[88] Matsunaga, T.; Takahashi, M.; Yoshino, T.; Kuhara, M.; Takeyama, H. *Biochem. Biophys. Res. Commun.* **2006**, *350*, 1019–1025.

[89] Yoshino, T.; Hirabe, H.; Takahashi, M.; Kuhara, M.; Takeyama, H.; Matsunaga, T. *Biotechnol. Bioeng.* **2008**, *101*, 470–477.

[90] Takahashi, M.; Yoshino, T.; Matsunaga, T. *Biomaterials* **2010**, *31*, 4952–4957.

[91] Matsunaga, T.; Tadokoro, F.; Nakamura, N. *IEEE Trans. Magnet.* **1990**, *26*, 1557–1559.

[92] Matsunaga, T.; Tsujimura, N.; Kamiya, S. *Biotechnol. Tech.* **1996**, *10*, 495–500.

[93] Yang, C.; Takeyama, H.; Tanaka, T.; Matsunaga, T. *Enzyme Microb. Technol.* **2001**, *29*, 13–19.

[94] Yang, C. D.; Takeyama, H.; Matsunaga, T. *J. Biosci. Bioeng.* **2001**, *91*, 213–216.

[95] Maruyama, K.; Takeyama, H.; Nemoto, E.; Tanaka, T.; Yoda, K.; Matsunaga, T. *Biotechnol. Bioeng.* **2004**, *87*, 687–694.

[96] Matsunaga, T.; Kawasaki, M.; Yu, X.; Tsujimura, N.; Nakamura, N. *Anal. Chem.* **1996**, *68*, 3551–3554.

[97] Tanaka, T.; Yamasaki, H.; Tsujimura, N.; Nakamura, N.; Matsunaga, T. *Mat. Sci. Eng. C-Biomim.* **1997**, *5*, 121–124.

[98] Yoshino, T.; Tanaka, T.; Takeyama, H.; Matsunaga, T. *Biosens. Bioelectron.* **2003**, *18*, 661–666.

[99] Yoza, B.; Arakaki, A.; Maruyama, K.; Takeyama, H.; Matsunaga, T. *J. Biosci. Bioeng.* **2003**, *95*, 21–26.

[100] Matsunaga, T.; Arakaki, A.; Takahoko, M. *Biotechnol. Bioeng.* **2002**, *77*, 614–618.

[101] Tanaka, T.; Takeda, H.; Kokuryu, Y.; Matsunaga, T. *Anal. Chem.* **2004**, *76*, 3764–3769.

[102] Arakaki, A.; Hideshima, S.; Nakagawa, T.; Niwa, D.; Tanaka, T.; Matsunaga, T.; Osaka, T. *Biotechnol. Bioeng.* **2004**, *88*, 543–546.

[103] Nakagawa, T.; Hashimoto, R.; Maruyama, K.; Tanaka, T.; Takeyama, H.; Matsunaga, T. *Biotechnol. Bioeng.* **2006**, *94*, 862–868.

[104] Yoza, B.; Matsumoto, M.; Matsunaga, T. *J. Biotechnol.* **2002**, *94*, 217–224.

[105] Yoza, B.; Arakaki, A.; Matsunaga, T. *J. Biotechnol.* **2003**, *101*, 219–228.

[106] Amemiya, Y.; Arakaki, A.; Staniland, S. S.; Tanaka, T.; Matsunaga, T. *Biomaterials* **2007**, *28*, 5381–5389.

[107] Prozorov, T.; Palo, P.; Wang, L.; Nilsen-Hamilton, M.; Jones, D.; Orr, D.; Mallapragada, S. K.; Narasimhan, B.; Canfield, P. C.; Prozorov, R. *ACS Nano.* **2007**, *1*, 228–233.

[108] Galloway, J. M.; Arakaki, A.; Masuda, F.; Tanaka, T.; Matsunaga, T.; Staniland, S. S. *J. Mater. Chem.* **2011**, *21*, 15244–15254.

[109] Matsunaga, T.; Kamiya, S. *Appl. Microbiol. Biotechnol.* **1987**, *26*, 328–332.

[110] Nakamura, N.; Hashimoto, K.; Matsunaga, T. *Anal. Chem.* **1991**, *63*, 268–272.

[111] Tanaka, T.; Matsunaga, T. *Anal. Chem.* **2000**, *72*, 3518–3522.

[112] Wacker, R.; Ceyhan, B.; Alhorn, P.; Schueler, D.; Lang, C.; Niemeyer, C. M. *Biochem. Biophys. Res. Commun.* **2007**, *357*, 391–396.

[113] Ronaghi, M.; Uhlen, M.; Nyren, P. *Science* **1998**, *281*, 363, 365.

[114] Yoshino, T.; Nishimura, T.; Mori, T.; Suzuki, S.; Kambara, H.; Takeyama, H.; Matsunaga, T. *Biotechnol. Bioeng.* **2009**, *103*, 130–137.

[115] Tanaka, T.; Sakai, R.; Kobayashi, R.; Hatakeyama, K.; Matsunaga, T. *Langmuir* **2009**, *25*, 2956–2961.

[116] Murray, M. G.; Thompson, W. F. *Nucleic Acids Res.* **1980**, *8*, 4321–4325.

[117] Arakaki, A.; Takeyama, H.; Tanaka, T.; Matsunaga, T. *Appl. Biochem. Biotechnol.* **2002**, *98*, 833–840.

[118] Tanaka, M.; Nakata, Y.; Mori, T.; Okamura, Y.; Miyasaka, H.; Takeyama, H.; Matsunaga, T. *Appl. Environ. Microbiol.* **2008**, *74*, 3342–3348.

[119] Tanaka, M.; Arakaki, A.; Staniland, S. S.; Matsunaga, T. *Appl. Environ. Microbiol.* **2010**, *76*, 5526–5532.

PART V

APPLICATIONS IN NANOTECHNOLOGY

11

PROTEIN CAGE NANOPARTICLES FOR HYBRID INORGANIC–ORGANIC MATERIALS

Shefah Qazi, Janice Lucon, Masaki Uchida, and Trevor Douglas

11.1 INTRODUCTION

Biominerals are often composite materials comprising both hard (inorganic) and soft (organic) components. The interaction between these two components provides the basis for controlled morphology, polymorph selection, and spatial localization in biological systems, which ultimately dictates their physical properties and biological utility. Understanding the basis for the interactions at the interface between hard and soft materials has also been significant in the design and implementation of synthetic biomimetic material systems [1–6].

Two quite different biological systems are the inspiration for the work described in this chapter. Both systems use assembled protein cage architectures to encapsulate and sequester the cargo that is either fragile or potentially toxic. These two systems are (1) the iron storage proteins in the ferritin superfamily, which biomineralize iron oxides within a protein cage as part of an iron detoxification and storage mechanism, and (2) viruses, which selectively package, transport, and protect genomic material within a protein cage architecture.

Ferritins are superficially simple systems with a protein cage that directs biomineralization of iron oxide at the interior protein interface and sequesters the resulting mineral nanoparticle. The protein forms a closed shell architecture, which incorporates all the control elements necessary for biomineralization. These include an

Coordination Chemistry in Protein Cages: Principles, Design, and Applications, First Edition.
Edited by Takafumi Ueno and Yoshihito Watanabe.
© 2013 John Wiley & Sons, Inc. Published 2013 by John Wiley & Sons, Inc.

enzymatic catalyst for molecular transformation of precursor ions, a mineral nucleation site, and an architecture that defines and constrains the overall morphology of the biomineral. In addition, the colloidal nature of the protein cage renders the final biomineral soluble and mobile, yet biochemically inert. Many of these properties and control elements are highly desired in the fabrication of synthetic materials. Some of the biochemical understanding of ferritin biomineralization, and virus packaging and assembly, has been applied toward biomimetic synthesis of cage constrained inorganic–organic hybrid materials.

Supramolecular assemblies of protein subunits which self-assemble into cage-like architectures are not unique to ferritins, and there are a large number of protein cage architectures that are all assembled from a distinct number of subunits to form precisely defined molecular containers in the 5–100 nm size range. Other examples of these cage-like architectures are chaperonins [7–9], DNA-binding proteins [10–15], and a very large class of protein cages—the viruses [16, 17]. Typically these protein cages are roughly spherical in shape and represent a range of relatively simple symmetries including tetrahedral, octahedral, and icosahedral; and are members of the Platonic solids. From a synthetic biomimetic standpoint, these cages represent novel environments by which materials can be synthesized in a size-constraining mode of encapsulation. A library of functional protein cage nanoparticles (PCNs) that serve as platforms for such purposes as biomimetic material synthesis, magnetic resonance imaging contrast agents (MRI-CAs), gene therapy, drug encapsulation, cell-specific targeting, and catalysis is being developed. As shown in Figure 11.1

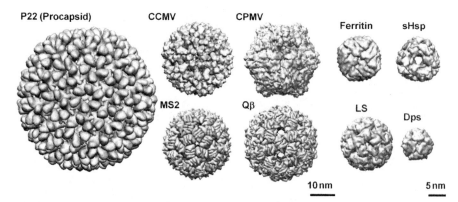

FIGURE 11.1   Space-filling images of protein cage nanoparticle including viral capsids P22 procapsid form (58 nm diameter) *PDB 3IYI*, CCMV (28 nm) *PDB 1CWP*, CPMV (30 nm) *PDB 1NYZ*, MS2 (27 nm) *PDB 2MS2*, Qβ (30 nm) *PDB 1QBE*, ferritin (12 nm) *PDB 2FHA*, small heat shock protein (sHsp, 12 nm) *PDB 1SHS*, Lumazine synthase (LS, 15 nm) *PDB 1RVV*, and DNA-binding protein from starved cells (Dps, 9 nm) *PDB 1QGH*. The images were reproduced using the UCSF Chimera package (http://www.cgl.ucsf.edu/chimera/) from the Resource for Biocomputing, Visualization, and Informatics at the University of California (supported by NIH P41 RR-01081) [18].

[18], several protein cages have been used for templated materials synthesis [19–28]. All of the PCNs represented in Figure 11.1, as well as many others, have also been probed by both chemical and genetic modification for adding nonnative functionality and exploiting the great versatility of protein cage architectures.

Conceptually, there are three different interfaces presented by all PCNs. These are the interior and exterior surfaces as well as the interfaces between subunits. A range of closed-shell protein architectures serve as size-constraining reaction vessels for nanomaterial synthesis, specifically using the interior surface of the protein cage. From the understanding of directed biomineralization in ferritin and nucleic acid packaging in virus capsids, a model has been developed for surface-induced packaging and used as a guiding principle for the synthesis of metal oxide nanoparticles as well as site-specific nucleation of metal–organic coordination polymers. The principles outlined here are not limited to the protein cage systems described here, but rather serve as a general model for protein-encapsulated biomimetic synthesis [20].

11.2   BIOMINERAL FORMATION IN PROTEIN CAGE ARCHITECTURES

11.2.1   Introduction

The ferritin superfamily of proteins are spherical protein cage architectures that are nearly ubiquitous in biology where they function to direct the biomineralization of iron as a mechanism for maintaining iron homeostasis [29, 30] and control oxidative stress. While ferritins show little homology at the primary (sequence) level, the structural homology (at the $2°$, $3°$, and $4°$ levels) is very highly conserved. Ferritin and bacterioferritins are composed of 24 structurally identical subunits that assemble into a very robust protein cage with octahedral (432) symmetry (Fig. 11.1). The external diameter of these assembled protein cages is roughly 12 nm and the architecture defines an internal cavity that is 6–8 nm in diameter. The structural motif of the ferritin subunit consists of a four-helix bundle with a fifth C-terminal helix (helix E) oriented at $60°$ to the four-helix bundle axis. In the $4°$ structure of the assembled protein cage, the fifth helix forms the fourfold axis through assembly of an intersubunit four-helix bundle [31–33].

While iron is a necessary element for life, it has a paradoxical relationship in biology due to its reactivity in forming reactive oxygen species. When iron is stored as a nanoparticle of iron oxide (ferrihydrite) inside the ferritin protein cage, it is completely sequestered and rendered inert [34]. The encapsulation and sequestration of the iron oxide nanoparticle in biological systems highlight its potential for use as a synthetic platform for materials synthesis. The cage-like property of ferritin provides an ideal size-constrained reaction environment for nanomaterial synthesis where the protein shell acts both to direct mineralization and as a passivating layer preventing unwanted particle–particle interactions.

11.2.2 Mineralization

In vivo, ferritin is responsible for sequestering and storing toxic iron as an innocuous mineral of iron oxide (ferrihydrite) through an overall protein-mediated reaction represented in Reaction 11.1 [34].

$$4Fe^{2+} + O_2 + 6H_2O \rightarrow 4FeOOH + 8H^+ \qquad (11.1)$$

The actual biological process of iron oxidation and encapsulation is considerably more complex than what Reaction 11.1 indicates, and some of the intimate steps remain unresolved. The mechanism by which iron is incorporated into ferritin *in vitro* can be described by four major events: iron entry, iron oxidation, iron oxide nucleation, and iron oxide particle growth. Iron entry into the cage-like architecture occurs via the channel (threefold symmetry) formed at the interface between subunits [35, 36]. Fe(II) oxidation is enzymatically catalyzed by reaction at the ferroxidase center resulting in the formation of Fe(III). The nucleation of an iron oxide material from this insoluble ion is facilitated at the interior protein interface, and the particle grows from this nucleus but is limited by the size constraints of the cage.

Conserved acidic residues along the threefold channel in eukaryotes have been shown to bind metals. Electrostatic calculations on the recombinant human H-chain ferritin reveal electrostatic gradients at the threefold axes that act as a guiding force directing cations through the channel toward the interior of the protein cage [35]. Specifically, the electrostatic guidance suggests a pathway between the channel and ferroxidase center. It has also been suggested that this channel may be dynamic, thus modulating the dimensions of the opening to the cage interior. This dynamic breathing of the ferritin protein cage is supported by the permeation of 7–9 Å-sized EPR spin labels into the interior of the protein cage directed predominately by charge effects of the threefold axis [37].

The biomineralization of iron oxide in ferritin occurs through several well-defined steps but the exact pathway can vary depending on the ratio of iron to ferritin protein cage. Many of these steps have been characterized and it is apparent that at low iron loading ratios (<48 Fe per cage), the mineralization is dominated by the enzymatic oxidation of Fe(II), while at higher iron loadings the process is dominated by the mineral itself [38, 39]. The multistep mineralization process is mediated at all levels by the protein cage.

The enzymatic oxidation of Fe(II) by the ferroxidase reaction of ferritin suggests that ferritin biomineralization is specific for iron *in vivo*. However, *in vitro* experiments with ferritin proteins having no ferroxidase activity have been shown to encapsulate iron oxide nanoparticles at nearly equivalent efficiencies [40]. Ferritin mutants, where the nucleation sites have been deleted but the ferroxidase sites remain intact, are also able to spatially direct mineralization within the confines of the protein cage. In mutant ferritins, lacking both the ferroxidase and the nucleation sites, there is a loss of spatial control in directing mineral formation, and bulk precipitation of iron oxides occurs [40, 41]. The synergistic effect of these two components is important. Simplistically, the ferroxidase site converts Fe(II) to Fe(III). The Fe(II) is significantly more soluble, at physiologically relevant pH, than the Fe(III). Thus, the ferroxidase center converts an undersaturated condition to a supersaturated condition inside the protein cage—this is sufficient to achieve the spatially directed mineralization observed. The high negative charge density on the interior surface of the assembled ferritin protein cage that constitutes the nucleation sites serve to aggregate ions at the protein interface. This perhaps increases the local concentration at the interface and facilitates oxidative mineralization. Each of these components is sufficient to direct mineralization, but when they are both absent all the controls are removed.

11.2.3   Model for Synthetic Nucleation-Driven Mineralization

Observing successful nucleation site-driven mineralization in ferritin suggests that the oxidation and mineralization reaction may not exhibit a high degree of specificity for Fe, and that mineralization could be expected to occur with a range of other transition metal ions, driven through purely electrostatic effects at the interior protein interface. According to this model, the very negatively charged interior surface of ferritin will accumulate counter ions (Fe(II)), at concentrations significantly higher than bulk concentration, in close proximity to this surface. The lack of specificity for Fe has been used very successfully for directing the formation of other metal oxide particles in ferritin and other protein cage architectures represented in Figure 11.1.

Using the surface-directed electrostatic model as a guiding principle, ferritin has been used as a template for the synthesis of nonnative minerals including cobalt oxides ($CoOOH$, Co_3O_4) [42–44], manganese oxides (Mn_2O_3, Mn_3O_4) [45, 46] and other materials [47–58]. The fact that these mineralization reactions are not specific to iron suggests that the electrostatic character of the interior surface of the protein cage plays an important role in mineralization.

Protein cage-directed mineralization appears to require three essential features: (1) pores in the protein shell that allows molecular access to the interior of the protein cage from the bulk solution; (2) chemically distinct interior and exterior surfaces; and (3) protein cage stability under the conditions required for the synthesis. The protein cages represented in Figure 11.1 satisfy these three basic criteria, or have been engineered to incorporate these features.

11.2.4   Mineralization in Dps: A 12-Subunit Protein Cage

To test the electrostatic model for protein-directed mineralization synthetic reactions, another member of the ferritin superfamily, Dps, with similar characteristics to ferritin was used. Dps (DNA-binding protein from nutrient-starved cells) was originally isolated from *Escherichia coli* [59], and since its discovery, structural and functional homologs have been isolated in many other bacteria [60–62] as well as archaea [13, 14, 63]. Dps produces an iron oxide core, similar to that of the typical ferritins, to protect cellular DNA from oxidative stress [64–66]. While there is some structural similarity between ferritin and Dps proteins, the Dps only assemble into 12 subunit cages with 23 symmetry [61]. The subunit structure has a four-helix bundle core, a fifth helix is in a loop connecting the B and C helices perpendicular to the four-helix bundle. The subunit dimer of a Dps from *Listeria innocua* (*Li*Dps) also contains a ferroxidase-like binding site, but it varies from ferritin as it is found at the dimer interface between subunits rather than within the four-helix bundle [10, 66, 67].

The Dps architecture has two types of threefold symmetry pores. One of these channels is lined with hydrophilic amino acids that can provide cations from bulk solution molecular access to the interior surface. The size of this pore is roughly the same size as the threefold pore of ferritin, about 0.7–0.9 nm [10, 66]. An electrostatic driving force for ion accumulation has been shown for Dps [68] that is similar to what has been seen in ferritin. It is proposed that metal cations inside the protein cage interact with the very negatively charged interior surface, which facilitates a similar oxidative mineralization reaction described for ferritin. The electrostatic surface of the interior of *Li*Dps is similar to the interior surface of ferritin, with clusters of glutamic acid residues that can be involved in mineral core nucleation. Much like ferritin, these structural characteristics of the Dps cage have been used to direct mineralization of metal oxides such as γ-Fe_2O_3 (maghemite), Co_3O_4, and $Co(O)OH$ [19, 69] as well as CdS [70], CdSe [71], and platinum [72]. The synthesis reactions of metal oxides were all performed using similar approaches where the empty Dps protein cage is incubated with M(II) ion and an oxidant (H_2O_2). The bulk reaction involving metal ion oxidation and subsequent precipitation was slow relative to the protein-catalyzed reaction and, in all cases, no bulk precipitation was observed in the protein reactions. The resulting nanoparticles are extremely small (3.75 ± 0.88 nm) and contain only a few hundred metal ions [19]. A Dps structure has been reported [73] in which the protein crystal was incubated with varying amounts of iron and small clusters of Fe ions were identifiable in the structure, and several iron ions were seen bound to specific negatively charge amino acids in the protein shell. This is the first such example of a precursor to the Dps mineral core that has been structurally identified [73]. The potential for using the Dps proteins to understand both their roles in biomineralization and biomimetic materials synthesis has been enhanced by the recent discovery and structural characterization of two Dps proteins from hyperthermophilic microorganisms, which exhibit elevated thermal stability, an ideal property for synthetic applications [13, 14].

The early stages of iron oxide and Pt nanoparticle synthesis have been investigated using noncovalent mass spectrometry (NCMS) within the *Li*Dps protein cage [72, 74]. Unlike most other analytical techniques, mass spectrometry can simultaneously detect different molecular masses present in a mixture. Thus, mass spectrometry can detect multiple transient populations and monitor their changes individually. The combination of electrospray ionization (ESI) and the time-of-flight (TOF) mass analyzer makes it possible to measure the mass of intact noncovalently associated macromolecular complexes without disturbing structures as well as the masses of individual protein components [75–79]. Accurate mass measurements have been made of metal-mineralized Dps protein cages, which allows for the quantitative examination of the effects of metal ion concentration on the final nanoparticle size (Fig. 11.2a). This was corroborated by transmission electron microscopy (TEM) and a kinetic model was used to infer the molecular details of metal ion accumulation and

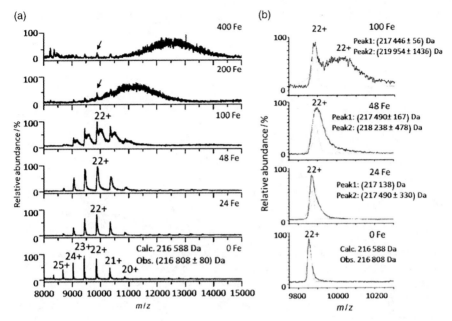

FIGURE 11.2 (a) Mass spectra of Fe-mineralized wt *Li*Dps at various nominal loading ratios of Fe(II) (0, 24, 48, 100, 200, and 400 Fe(II) per cage, bottom to top). Charged peaks are indicated in the bottom spectrum. (b) Fittings of 22+ peaks of Fe-mineralized wt *Li*Dps at lower loads either with one (0 Fe(II)) or two Gaussians. Peak 2 of 24Fe and Peak 1 of the 48Fe and 100Fe spectra correspond to a mass of *Li*Dps with 12 associated Fe. Reproduced with permission from Reference 74. Copyright Wiley-VCH Verlag GmbH & Co. KGaA.

nanoparticle formation from these final reaction products. Modeling results suggested that particle growth involves both a binding phase (with growth rate limited by ion accumulation) and a growth phase (with growth rate proportional to the particle surface area).

Based on the NCMS data, a mechanism for the process of biomimetic iron oxide formation within *Li*Dps was proposed. At low Fe(II) ion loadings, the ions are distributed evenly into *Li*Dps such that up to 12 Fe ions accumulate in each cage through the whole population of *Li*Dps (Fig. 11.2b). A subpopulation of the cages obtains a limited amount of Fe (Peak 1 of 24Fe spectrum) presumably due to depletion of available Fe(II) ions (Fig. 11.2b). With intermediate Fe(II) loading, the binding sites are completely occupied and those cages accumulating Fe beyond some critical amount begin to form small iron oxide nanoparticles (Scheme 11.1). The short-lived nucleating intermediate was not detected in the final reaction products presumably due to its rapid transition to an iron oxide particle. Sufficiently large amounts of Fe(II) ion loading allows full occupation of the binding sites followed by initiation of small iron oxide formation. Subsequently, the surface of small iron oxide may dominate iron oxide growth, through autocatalysis, resulting in larger iron oxide

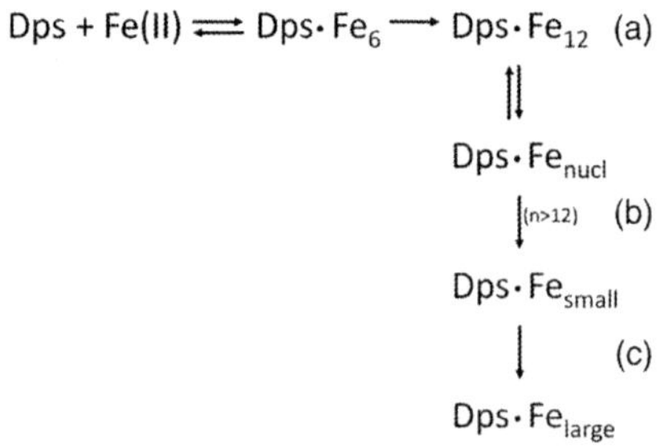

$$\text{Dps} + \text{Fe(II)} \rightleftharpoons \text{Dps} \cdot \text{Fe}_6 \longrightarrow \text{Dps} \cdot \text{Fe}_{12} \quad \text{(a)}$$

$$\Updownarrow$$

$$\text{Dps} \cdot \text{Fe}_{nucl}$$

$$\downarrow (n>12) \quad \text{(b)}$$

$$\text{Dps} \cdot \text{Fe}_{small}$$

$$\downarrow \quad \text{(c)}$$

$$\text{Dps} \cdot \text{Fe}_{large}$$

SCHEME 11.1   A proposed mechanism for the process of biomimetic iron oxide synthesis within the *Listeria innocua* Dps protein cage, based on noncovalent mass spectrometry data. Reproduced with permission from Reference 74. Copyright Wiley-VCH Verlag GmbH & Co. KGaA.

nanoparticle formation (Scheme 11.1). A small subpopulation of the cages do not reach the critical Fe accumulation point and remain with only 12 Fe ions bound. The mechanistic scheme, with a two-stage growth process, could be modeled effectively with a simple kinetic model to account for the distribution of species at the end of each reaction.

11.2.5   Icosahedral Protein Cages: Viruses

Viruses and virus-like particles, an abundant subclass of PCNs, are exquisite examples of supramolecular assembly that represents the organization of subunits into a precisely defined, stable protein cage [80]. All viruses function by using protein or protein/lipid capsids in order to transport their nucleic acid to a host cell. They are metastable structures poised between the stability required for packaging and transport and the necessary instability associated with cargo release and infection. Many viruses package their genomic nucleic acid through noncovalent electrostatic interactions with the protein cage interface or through physical packaging at extremely high pressure, which indicates the unique distinction between the interior and exterior surfaces.

11.2.6   Nucleation of Inorganic Nanoparticles Within Icosahedral Viruses

A dominant model for ferritin biomineralization invokes strong electrostatic interactions between the nascent mineral and the interior protein interface. This model has been successfully used to reengineer the interior surfaces of virus capsids (CCMV and P22) for the directed nucleation and mineralization of iron oxides. CCMV is a plant virus that assembles from 180 identical subunits into an icosahedral protein cage (Fig. 11.1), which packages its RNA genome through electrostatic interactions with the interior of the capsid. Although the surface charge on the inside of the assembled CCMV is the opposite of ferritin, the electrostatically distinct positively charged interior surface was used to direct mineralized polyoxometallates at the interior surface of the protein cage. Metal-oxo precursors ($WO_4{}^{2-}$) were incubated with the CCMV under conditions of high pH (pH 7.5, swollen conditions), where the interior surface is expected to assist in the aggregation of the negatively charged metal-oxo anions at the interface. When the pH was lowered, an acid-catalyzed polymerization of the metal-oxo ions resulted in the formation of large polyoxometallate anions, which readily crystallize at the interface to form nanoparticles, entrapped within the CCMV protein cage [21].

Mutant CCMV subunits, in which the interior surface charge was changed from net positive to net negative through replacement of nine basic residues with glutamic acid residues (CCMV-subE), still assemble to form stable icosahedral protein cages [81]. Incubation of CCMV-subE, with its altered interior surface charge, with Fe(II) ions in the presence of air results in the spatially selective mineralization of Fe_2O_3 (ferrihydrite) without bulk precipitation [20]. In contrast, the unaltered CCMV cage exhibits no tendency to direct Fe_3O_3 mineralization and only the bulk reaction is observed. Thus, the modified CCMV cage, subE, has been shown to act as a ferritin mimic for mineralization of iron oxide nanoparticles.

This mineralization work was expanded with the use of the viral capsid from the bacteriophage P22 to nucleate and form sequestered nanoparticles of Fe_2O_3. P22 assembles from 420 identical copies of a coat protein with the assistance of a variable number of a 303 residue scaffold protein (SP) (100–300 copies are encapsulated per capsid). The *C*-terminal region (65 residues) of the scaffold protein is required for the templated assembly of the capsid and the *N*-terminal region of the scaffold protein can be truncated without loss of assembly [82–86]. In addition, *N*-terminal fusions to the SP can be engineered to incorporate a polyanionic peptide appending repeats of a five-amino acid sequence, $(ELEAE)_n$, to the SP, which is localized to the interior of the viral cavity. This peptide repeat was chosen in order to mimic the electrostatic gradient between the interior and exterior environments found in the native ferritin system, which facilitates the formation of Fe_2O_3 within the cage interior. These modifications to the SP had no effect on the assembly of the P22 capsid. Incubation of these modified capsids with Fe(II) in the presence of air resulted in the formation of Fe_2O_3 particles encapsulated within the P22 capsid [87]. The outer capsid layer of the P22 assembly physically constrains the growth of the Fe_2O_3 particle limiting the particle size. Removal of the scaffold protein destroyed the ability of the capsid to direct nucleation and growth (Fig. 11.3).

11.3   POLYMER FORMATION INSIDE PROTEIN CAGE NANOPARTICLES

11.3.1   Introduction

The use of protein–polymer composite materials for medical and materials applications is a growing field, which aims to take advantage of the exquisite monodispersity and bioactivity of biomolecules while imparting new material properties via polymer conjugation. Through careful selection of the protein and polymer, new thermo,

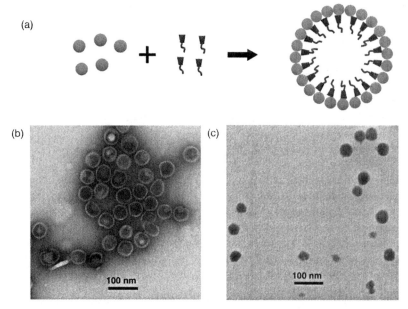

FIGURE 11.3 (a) Schematic of glutamic acid-rich peptide fused to scaffold protein in P22. (b) Uranyl acetate-stained TEM image of P22 with truncated SP–EA peptide fusion. (c) TEM images of P22 with truncated SP–EA peptide fusion mineralized with 20,000Fe per cage. Reproduced with permission from Reference 87. Copyright The Royal Society of Chemistry.

light, and pH sensitive macromolecular materials can be produced to more specifically control the activity and phase solubility of the biomolecule [88, 89]. When the polymer is exposed rather than buried inside a protein, it can impart improved retention, lowered immunogenicity, and increased bioavailability to the modified biomolecule [90, 91]. To attain the desired final material properties, not only is the spatial relationship between the polymer and the overall protein architecture important but also the careful selection of both the protein and polymer components is essential.

Viruses, which in their infectious state comprise at least both protein and nucleic acid polymers, provide a biological example of an internally confined protein–polymer construct, where the protein protects and constrains the hidden polymer (DNA or RNA). Depending on the particular virus, the protein–polymer composite may be formed and stabilized via electrostatic interactions of the genomic nucleic acid with the inner protein cage surface or in other viruses, physical packaging of the DNA or RNA occurs resulting in extremely high pressures on the protein cage. In both methods, the nucleic acid polymer is generated before being sequestered inside the protein or protein/lipid shell.

Using this biological process as an inspiration, a variety of nonnatural polymeric cargos have been entrapped within viruses taking advantage of the available interior space. In the most biologically analogous methods, electrostatics are used to engulf different nucleic acid strands or other negatively charged polymers [21]. In cases where the virus coat protein monomers can be disassembled and subsequently reassembled into the cage architecture, the complete capsid can be assembled around polymers or polymer–nanoparticle composites [92–94]. Unlike the natural system, if instead the polymer is grown within the protein cage, a wider range of synthetic polymers can be attained [95–97]. Preferably, in the final case, an engineered approach for filling protein cages with a synthetic polymer allows for the incorporation of functional groups that can be used or modified to introduce new functionalities to the protein–polymer composite.

11.3.2   Azide–Alkyne Click Chemistry in sHsp and P22

The first method used to form an anchored polymer network inside protein cages was azide–alkyne click chemistry (AACC), as it provides a means to produce covalently linked branched oligomers using aqueous-phase chemistry compatible with modification of biopolymers [96, 98, 99]. Considerable work has been done to improve the compatibility and efficiency of click chemistry with biomolecules employed for the addition of peptides, fluorophores, and glycopolymers on the surface of protein cages [100–102]. Under biomolecule amenable conditions, the alkyne and azide functional groups are coupled through hetero [3+2] cycloaddition reactions mediated by a Cu(I) catalyst in the presence of a Cu-binding ligand, THPTA [103]. Using this chemistry, two complementary monomers have been constructed where free amines have been designed into the azide component allowing for modification of the network after polymerization, while a branched alkyne was utilized to allow a cross-linked oligomer to form.

A small heat shock protein (sHsp) from *Methanococcus jannaschii* was modified with these two monomeric pieces to form a branched oligomer with sites for internal functionalization of the protein cage [96]. Heat shock is heterologously expressed in *E. coli* and self-assembles into a porous, octahedral complex from 24 subunits with an exterior diameter and interior diameter of 12 nm and 6.5 nm, respectively. The sHsp cage has eight 3 nm pores and six 1.7 nm pores, which allow for the exchange of small molecules between the cage interior and exterior [9]. A mutant of this protein has been recombinantly expressed in which a glycine (residue 41) on the interior surface of the cage has been substituted for a cysteine, sHsp(G41C) [22]. Since sHsp lacks endogenous cysteines, this mutation provides a unique site for chemical conjugation of molecules to the cage interior.

To initiate the synthesis of the AACC oligomer, sHsp(G41C) is first labeled with *N*-propargyl bromoacetamide, termed generation zero (G0.0). From this site, oligomer growth is directed to the interior in a step-wise fashion alternatively using monomers 2-azido-1-azidomethyl ethylamine and tripropargyl amine. The synthesis of G0.5 is initiated by addition of monomer 2-azido-1-azidomethyl ethylamine. Next, G1.0 is produced by reacting tripropargyl amine with 2-azido-1-azidomethyl ethylamine to create a branched oligomer that promotes cross-linking between sHsp subunits. Higher generations are achieved through further cycling of the two monomers. The

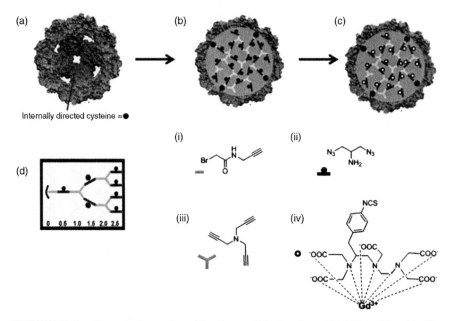

FIGURE 11.4 General scheme for azide–alkyne click chemistry (AACC) in sHsp. (a) sHsp (G41C) with internally mutated cysteines. (b) Oligomer growth across the entire cage, G2.5. (c) Addition of Gd-DTPA-SCN. (d) Schematic of G0.0–G0.5. (i–iv) AACC monomers. (i) *N*-propargyl bromoacetamide, (ii) 2-azido-1-azidomethyl ethylamine, (iii) tripropargyl amine, (iv) 2-(4-isothiocyanatobenzyl)-diethylene-triamine pentaacetic acid coordinated to gadolinium.

interior oligomer growth sufficiently occupies the interior volume of sHsp reaching completion by G2.5 (Fig. 11.4).

The designed objective of the branched oligomer was to label primary amines on the diazido amine monomer units with imaging or drug modalities. It was found to be heat stable up to 120°C at G2.5 due to internal covalent cross-linking of the AACC oligomer [96]. To evaluate how many amine groups on the dendritic structures were addressable, the amines were labeled with fluorescein isothiocyanate (FITC) which is similar in size to many anticancer drugs and can be detected using UV-vis spectroscopy. Results indicate that 142 FITC molecules were covalently bound to the polymer at G2.5. Since the polymer was designed to introduce 168 amines into the cage, the FITC labeling results show that roughly 85% of the newly installed amines in the cavity were addressable.

More recently, the AACC oligomer synthesis has been extended to a significantly larger PCN, the P22 virus-like particle. Recombinant expression of the *Salmonella typhimurium bacteriophage P22* requires co-expression of the coat protein (46.6 kD) and scaffold protein (33.6 kD) for self-assembly [104, 105]. The P22 capsid is capable of transformation into a series of distinct morphologies including procapsid (PC, 58 nm) which contains scaffold protein, empty shell (ES, 58 nm) where the

(a) PC and ES   (b) EX   (c) WB

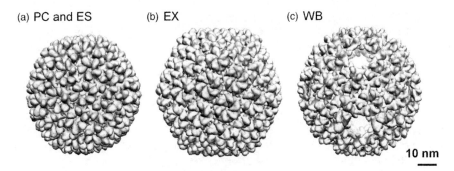

10 nm

FIGURE 11.5 Three different morphologies of P22 capsid, Procapsid (PC), empty shell (ES), expanded shell (EX), and wiffleball (WB). PC and ES forms have identical surface morphology with or without scaffold proteins. (a) PDB 3IYI, (b) PDB 2XYZ, (c) PDB 3IYH [106, 108].

scaffold protein has been removed, expanded shell (EX, 64 nm), and wiffleball (WB, 64 nm) where all 12 pentamers have been removed (Fig. 11.5) [106, 107]. The scaffold protein is removed using guanidine–HCl extractions followed by heating at 65°C to generate the capsid in its expanded morphology [108–111]. The EX form most closely mimics the state found in the DNA-packaged infectious virion. Extended heating at 75°C for 20 min induces the selective release of the 60 subunits at the 12 fivefold vertices to produce the WB morphology. The large 10 nm diameter pores of WB at the 12 vertices ensure free molecular exchange between the interior and exterior environments of the capsid.

The P22 is 200 times larger in volume than sHsp and thus has the potential to be loaded with much higher number density of functionalities. The P22 coat protein has been genetically modified at residue 118 from lysine to cysteine [112] to provide an internal initiation site for AACC [96, 100]. In the case of the P22 PCN, the cage has been partially filled by cycling the AACC reaction through G4.5. By TEM and dynamic light scattering (DLS), the size of the P22 remained unchanged indicating that the polymerization is directed to the interior. Evidence of mass increase at each generation was provided by denaturing gel electrophoresis showing a shift in the apparent mass of the subunit bands. At higher generations, covalent cross-linking of subunits as oligomer growth extends throughout the cage results in a further contribution to the observed mass increase. The stepwise polymer growth was additionally characterized by multi-angle laser light scattering combined with high-performance liquid chromatography (MALLS-HPLC). These data suggest that particles are monodisperse and polymer growth occurs exclusively on the interior of the P22 PCN [113].

11.3.3 Atom Transfer Radical Polymerization in P22

While the "click" chemistry method is an effective means to build molecular scaffolds within protein cages, the generational nature of the polymerization makes this process demanding, so a different route has been developed. To make addressable polymer networks for directed cargo loading in a few simple steps, within the confines of a PCN macro-initiator, atom transfer radical polymerization (ATRP) has been employed. Of the several continuous biomolecule-anchored polymerization methods, the use of ATRP is particularly suited for improved formation of polymers inside protein cages. This method is not only rapid, monomer promiscuous, and one that results in products with relatively low polydispersity but also the simplicity of the ATRP initiator means that it can be readily modified for attaching to biomolecules. By combining ATRP with a container-like protein, the formation of a polymer scaffold constrained to the interior of a PCN architecture can be afforded in a single short reaction.

As a functional example, ATRP has been used to make addressable cross-linked strands within the confines of the bacteriophage P22-based PCN (Fig. 11.6) [97]. This method and the selection of the primary amine rich 2-aminoethyl methacrylate (AEMA) monomer results in PCN-confined strands that act as a scaffold for the attachment of primary amine-reactive molecules of interest, such as a dye, drug, or contrast agent introducing new functionality to the construct and resulting in very high density loading of the capsid. Specifically, an internal cysteine containing viral coat protein, P22(S39C), modified with a thiol-reactive ATRP initiator [114, 115] is used as macroinitiator and a size-constrained reaction vessel for ATRP-driven growth of poly(AEMA) strands cross-linked with *bis*-acrylamide inside the virus-like particle. For this application, the scaffold protein, which is co-purified from *E. coli*, is removed using guanidine extractions followed by heating at 65°C to make the coat protein EX state. The single point mutation in the wild-type P22 coat protein, S39C, introduced a new addressable thiol, specifically designed such that the thiol is exposed exclusively to the interior, for attachment of a maleimide-modified ATRP initiator.

The internal P22–AEMA polymer construct can be made from the P22-macroinitiator using ATRP biomolecule conditions in standard use and the resulting polymer is confined to the interior of the cage. This synthesis utilizes a Cu(I)/bpy catalyst and the reaction was halted after 3 h with good polymer growth. Purification of the protein–polymer conjugate (P22–AEMA) away from remaining AEMA monomer and copper catalyst was accomplished by pelleting the protein in an ultracentrifuge, which easily separates large macromolecular complexes from small molecular species. The resulting construct exhibited increased subunit mass by denaturing gel analysis, as indicated by a shift to higher molecular mass. When this construct is analyzed by TEM and DLS, the particle morphology and diameter remain unchanged before and after the polymerization reaction indicating that the polymer is internally constrained.

To make a functional material and to demonstrate that the introduced amines are addressable, FITC was used as a labeling agent. Both the P22–AEMA and a P22-initiator control were incubated with an excess of FITC per subunit with unreacted FITC removed via ultracentrifuge-driven protein pelleting to give P22–AEMA–FITC and P22–FITC, respectively. The P22–AEMA–FITC contained more than 8000 FITC/PCN and P22–FITC, where only the endogenous lysines are modified, contained nearly 400 FITC/PCN. This increase of over 20-fold in the degree of labeling is significant enough to be readily discernable even by eye. This minimal

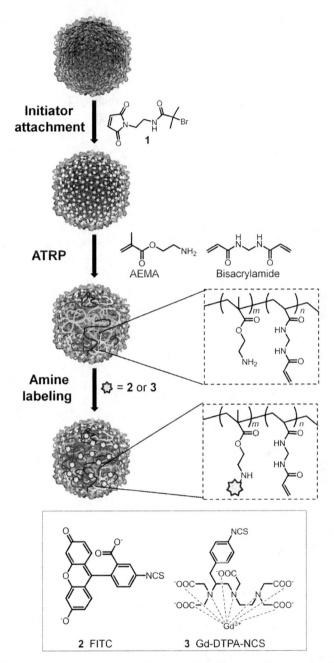

FIGURE 11.6   P22 containing an internal thiol labeled with a cysteine-reactive ATRP initiator (**1**) can be used as a macroinitiator and size-constrained reaction vessel for the ATRP-driven formation of poly(AEMA) strands within the virus-like particle. This polymer scaffold can subsequently be modified with primary amine reactive agents (**2**, **3**) introducing new functionality to the construct.

reactivity of FITC with endogenous lysines is advantageous because it means that in the P22–AEMA sample the vast majority (>95%) of the addressable sites are on the polymer, so any cargo molecule added is largely hidden on the interior of the construct. Furthermore, the resulting crowded environment inside the cage results in a dramatic decrease in the observed fluorescence intensity of the construct, highlighting the potential use of this construct for confined environment studies.

The application of ATRP methods for site-directed polymer formation inside a PCN results in an anchored network that is unparalleled for labeling purposes and utilizes the previously largely untapped interior volume of the P22 PCN. The P22(S39C)-based macroinitiator effectively directs polymer growth to the PCN interior resulting in confined growth as the protein shell acts as a barrier, limiting polymer size and leaving only the protein exposed to the bulk solution. By selecting an appropriate macroinitiator and monomer, the new multimeric protein–polymer composite acts as a scaffold for the attachment of molecules of interest such as the fluorophore FITC, or the MRI contrast agent Gd-DTPA-NCS. This introduced scaffold results in a significantly increased number of labels per cage compared to other PCN-based systems. The improvement in labeling is beneficial for the delivery of materials on a per-particle basis allowing for high concentration delivery of contrast agent or other molecules of interest. Due to the simplicity, modular nature, and loading level of the ATRP-based approach taken to make these P22-polymer internal composites, this same method can be employed to make a range of novel PCN-polymer composites with biomedical and catalytic applications.

11.3.4   Application as Magnetic Resonance Imaging Contrast Agents

The effectiveness of MRI-CAs for T1-enhanced images can be improved by attaching a small molecule contrast agent to a supramolecular platform. Nanoparticles such as dendrimers, liposomes, and supramolecular protein complexes are desirable platforms for next generation contrast agent development as they provide a large contrast payload and are readily modified for addition of targeting functionalities [98, 116]. The potential for large magnetic material payloads and the ideal rotational properties of protein cages, along with advantages of structural rigidity and water solubility over other nanoparticle platforms make them extremely well suited for MRI contrast enhancement [28, 117].

Several groups working with a variety of PCNs have made important contributions in the development of these supramolecular structures as MRI-CAs. To date, the range of these PCN-based CAs can be divided into four major categories: (1) endogenous metal-binding sites [118], (2) genetic insertion of a metal-binding peptide [117], (3) chemical attachment of small molecule chelates [27, 28, 119], and (4) PCN-branched polymer hybrid particles [96]. The majority of these complexes have been synthesized by chemically attaching derivatives of clinically employed CAs, such as Gd-DOTA (Dotarem), Gd-DTPA (Magnevist), and Gd-HOPO [120], directly to functional groups on these PCNs. Some variation in ionic relaxivity (relaxivity per Gd ion) between PCN-based CAs can be attributed to the differences in three important parameters (Chart 11.1) that control the performance of the CA: the rotational correlation time of the Gd^{3+} ion (τ_R), the exchange lifetime of the Gd^{3+} bound

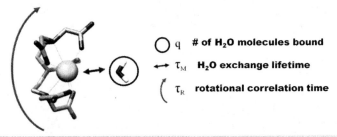

	r_1 ionic (sec^{-1} mM^{-1})	r_1 particle (sec^{-1} mM^{-1})	Gd^{3+}/particle
DTPA-Gd, 43 MHz [140]	3.4	3.4	1.0
sHsp G2.5 AACC, 31MHz [96]	25.4	4000	160
P22 G2.5 AACC, 28MHz [113]	21.7	41,300	1900
P22 ATRP, 60MHz [97]	22.0	200,000	9100

CHART 11.1 Comparison of PCN-CAs discussed in this section as macromolecular contrast agents.

water (τ_M), and number of waters bound to the Gd^{3+} ion (q) [27, 28, 96, 121]. Longer τ_R values are better for efficient relaxation of water proton spins, resulting in a more sensitive contrast agents. The τ_M and q also need to be optimized to attain maximum relaxivity. While larger q leads to higher relaxivity, this generally correlates with lower stability of a Gd-ligand complex which becomes an issue for *in vivo* applications as free Gd^{3+} ion has been found to be significantly toxic.

The large surface area and interior volume of protein cages can accommodate numerous Gd^{3+} ions. Both AACC and ATRP methods of polymerization provide a multitude of primary amine groups which can be functionalized with Gd-chelates. For PCN-based MRI-CAs, the molecule most commonly appended to the supramolecular platform is small molecule Gd-DTPA-NCS (Fig. 11.4 iv). By anchoring Gd-chelates to the protein cage, the τ_R or mobility of the Gd^{3+} ions is decreased, which directly increases the ionic and particle relaxivity (relaxivity rate per PCN) [27, 28, 117, 119, 122]. Targeting groups can be presented on the exterior of the PCN to direct localization of the protein cages to receptors of interest resulting in novel MR probes to improve the visualization of tissues of interest [25, 123–137].

Modifying the protein–polymer constructs described above with Gd-DTPA-NCS results in constructs with high per-particle relaxivities, as shown in Chart 11.1. When the AACC-based systems are labeled with Gd-DTPA-NCS, the sHsp AACC construct shows a per-particle relaxivity of 4000 mM^{-1} sec^{-1}, while the P22-AACC shows a per-particle relaxivity value of greater than 40,000 mM^{-1} sec^{-1}, an order of magnitude greater than its smaller counterpart. The ATRP-based system for P22-AEMA has a Gd-DTPA loading of approximately 9000 Gd per cage. This is significantly more Gd^{3+} both on a per-cage and per-subunit basis than any previous report using PCNs, with labeling greater than five times more than other reported samples by any Gd^{3+} incorporation method [27, 28, 98, 117–119, 122, 138, 139]. The relaxivity of these highly loaded particles was measured at 60 MHz and found to have a per-particle relaxivity of 200,000 mM^{-1} s^{-1}, which exceeds the previous reports of PCN-based MRI-CAs by an order of magnitude [27, 28, 98, 117–119, 122, 138, 139]. The ionic relaxivity is 22 mM^{-1} s^{-1}, which is consistent with enhancement over free Gd-DTPA [140] by tethering the chelate to a large supramolecular particle with a large rotational correlation time. Due to the potential for enhancement of the loading capacity of protein architectures by building polymer scaffolds inside the empty inner sphere of protein cages, these protein–polymer composites hold promise as metal chelate carriers for next generation MRI-CAs.

11.4   COORDINATION POLYMERS IN PROTEIN CAGES

11.4.1   Introduction

Coordination polymers and metal–organic framework materials have become a burgeoning field in materials science over the last decade because of their unique properties and wide variety of potential applications [141–144]. For some applications, including selective gas absorption and storage, bulk material with infinite structure would be desirable [145]. For some other applications, coordination polymers with well-controlled size and morphological distribution in the nanometer range would be preferable [146]. In particular, it is critical for biomedical applications to maintain control over the size and shape of a material because *in vivo* distribution and pharmacokinetics are heavily affected by these parameters. Utilization of a template material to control polymer growth is a promising strategy to synthesize these nanometer-scaled coordination polymers [147, 148]. PCNs are expected to be an ideal material for these templates because they have homogenous size, shape, and structure. Already, the use of engineered specific chemical modification sites on the cage architecture allows for improved control over the number and location of attached dyes, organometallics, tethered targeting moieties, and initiation sites of polymers, while also acting to discriminate and protect their interior cargos from the external environment. By taking advantage of the features of protein cages as synthetic containers, new routes toward the coordination polymer synthesis with uniform particle size and multifunctionality can be developed.

11.4.2   Metal–Organic Branched Polymer Synthesis by Preforming Complexes

One early attempt to create an encapsulated coordination polymer within a protein cage utilized metal coordination complexes as building blocks for polymerization through a Cu(I)-catalyzed "click" reaction via pendant azide and alkyne moieties resulting in the formation of a 1,4-disubstituted 1,2,3-triazole-linked oligomer [99]. In particular, this ligand-based method allows for specific stepwise additions of different metal coordination centers that may not otherwise polymerize under mild aqueous reaction conditions. This coupling reaction was selected due to its demonstrated compatibility with proteins and high reaction specificity having been used extensively for the attachment of a variety of ligands to proteins and the generation of related branched organic polymers within PCNs [96, 100].

This ligand–ligand coupling approach, as summarized in Figure 11.7, was applied to create an example, branched coordination polymer, which was grown in a stepwise manner with discrete generations, within the PCN sHsp(G41C). The genetically introduced interior cysteine (G41C) was the oligomer initiation site, established by the reaction with *N*-propargyl bromoacetamide to generate a protein-bound alkyne functionality, identified as generation 0.5 (G0.5). The first Cu(I)-catalyzed click reaction was between this cage-bound alkyne and a modified tris-phenanthroline-Fe(II) coordination complex, bearing pendant azide groups, to form generation 1.0 (G1.0).

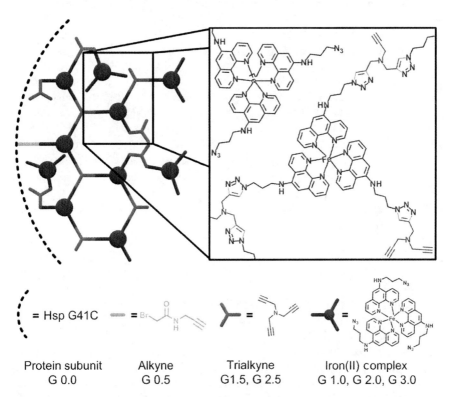

Protein subunit	Alkyne	Trialkyne	Iron(II) complex
G 0.0	G 0.5	G1.5, G 2.5	G 1.0, G 2.0, G 3.0

FIGURE 11.7 Model scheme for the stepwise formation of a ligand-based coordination polymer inside sHsp(G41C) starting with the introduction of an alkyne (G0.5) at the internal cysteine of each protein subunit. The oligomer grows from the initial alkyne generationally through the attachment of either an alkyne or azide functional group containing compound using "click" chemistry. Reproduced with permission from Reference 99. Copyright The Royal Society of Chemistry.

The exposed branching azide functional groups on the coordination complex were subsequently reacted with tripropargylamine to form generation 1.5 (G1.5), with further branching of the oligomer. The azide coordination complex and alkyne linkers were added and coupled alternatively through G3.0 at which point no further increase in polymerization was observed.

Thus, each generation of the oligomer (G1.0–G3.0) was constructed by click-based coupling to the previous generation of either Fe(azide-phen)$_3^{2+}$ (G1.0, G2.0, G3.0) or tripropargylamine (G1.5, G2.5) under biomolecule AACC conditions [103]. No precipitation occurred with the protein in solution, suggesting that the polymerization reaction is spatially localized to within the cage-like structure of the sHsp. However, when the reaction components were mixed at an equivalent concentration, in the absence of the protein cage, a dark red precipitate formed over the same reaction period, due to an unconstrained bulk polymerization.

To further confirm that the protein acts as a barrier for the oligomer, the size distributions of the protein-coordination polymer samples were measured and found to be nearly identical to the native sHsp(G41C)-G0.0 when investigated by SEC, DLS, and TEM. Collectively, these data confirm that the cage structure remains intact during the synthesis with the oligomer constrained to the interior of the cage. Furthermore, size exclusion chromatography elution profiles of each generation exhibited nearly identical elution times to that of the starting material sHsp(G41C). Also, the absorbance of the Fe(azide-phen)$_3^{2+}$ complex, which has a maxima at 530 nm, co-eluted with the protein, indicating that the complex is tightly associated with the protein cage. Due to the small (6.5 nm) internal diameter of this cage, only a few generations were required to reach completion of the polymerization. The strong absorbance of the Fe(azide-phen)$_3^{2+}$ moieties was used to track the addition of the growing branched oligomer with each integer generation. By G3.0, the oligomer formation was complete as only a small change in the Fe to protein ratio was observed compared to G2.0. Using the complex extinction coefficient at 530 nm (8.8 mM^{-1}) to determine the concentration of Fe(azide-phen)$_3^{2+}$ in conjunction with protein quantification, the amount of Fe in each generation was calculated to be 8, 9, 20, 23, and 24 Fe per cage for G1.0, G1.5, G2.0, G2.5, and G3.0, respectively. No further increase in the number of irons incorporated into generation G3.0 is an evidence that all the accessible alkynes have been reacted. The volume occupied by the coordination polymer, predicted by molecular modeling of the monomers, and based on the number of irons per cage, is about 20% of the interior volume of the cage. This filled volume compares well with the "click"-based organic polymer inside sHsp(G41C) with a nearly equivalent (18%) portion of the interior volume occupied at the final generation [96].

Similar to the "click"-based organic oligomers, the formation of a coordination network inside the cage acts to stabilize the cage structure. When acid–urea denaturing conditions are used to dissociate the native protein cage into subunits, the monomer which is dominant in G0.0 and G0.5 gradually decreases with a parallel increase in intensity at the top of the gel at later generations. This shift to higher molecular weights indicates that there is a steady increase in the mass of the sHsp subunits with increasing generation. In addition, the coordination polymer is covalently linked to the cage, as pale pink color could be seen at the top bands in lanes G1.5–G3.0

indicating the presence of the $Fe(az\text{-}phen)_3{}^{2+}$ oligomer complex. Treating G3.0 with the strong iron chelator, deferoxamine, in a ratio of 10:1 of deferoxamine to iron, removed the iron from the coordination complex eliminating the pink sample color.

In summary, a test case example of a branched coordination polymer, which extends across the interior cavity of the sHsp(G41C) PCN, has been demonstrated. This polymer was initiated at the genetically introduced cysteine residue on each subunit and preceded by first forming the coordination complex and then synthesizing the coordination polymer using a ligand–ligand AACC-based reaction allowing the cage architecture to be retained while cross-linking the subunits and filling the interior cavity.

11.4.3   Coordination Polymer Formation from Ditopic Ligands and Metal Ions

An alternative approach for forming a coordination polymer within a PCN has been demonstrated recently by using ditopic ligands to connect metal ions forming a three-dimensional network structure within the P22 PCN. In this study, a cysteine-reactive modified 1,10-phenanthroline (phen) ligand [149] was conjugated to unique cysteine residues on the interior of a P22(S39C) protein cage, making P22-phen, which serves as an initiation site for subsequent coordination polymer formation. A phenanthroline based ditopic ligand, 1,3-di-1,10-phenanthrolin-5-ylthiourea (di-phen), was synthesized to act as a bridge between metal ions [150], while Ni^{2+} was chosen for the metal ion vertices of the polymer, because it is known to coordinate with phenanthroline stronger than most other transition metals, including Fe^{2+} and Co^{3+}.

To attain the coordination polymer formation, Ni^{2+} and the di-phen ligand were added to P22-phen alternately, as shown in Figure 11.8(a) [151]. The Ni-phenanthroline complex exhibits UV absorbance around 270 nm, so if the complex is formed inside the P22-phen, the UV absorbance of the cage is expected to increase. Indeed, the UV absorbance of the P22-phen around 270 nm increased until G4.0, then slightly decreased at G4.5 (Fig. 11.8b). It should be noted that the incremental change in absorbance from G0.5 to G1.5, G1.5 to G2.5, and G2.5 to G3.5 gradually increased suggesting that the polymer is growing in the expected branching manner. Further analysis of the samples by DLS, MALLS-HPLC, and analytical ultra centrifugation revealed that the molecular weight of the P22 PCN increased until coordination polymer G4.0 without significant aggregation of the protein. Retention of the initial PCN diameter indicates that the coordination polymer is constrained inside the P22 capsid until G4.0. Beyond this generation, some aggregation is observed potentially due to some branches of the polymer reaching the exterior of the capsid and causing the formation of inter-capsid aggregates. In addition, according to a structure model based on a pair distribution function analysis and *ab initio* quantum mechanical calculations [150], the interior surface of the cage is fully covered with the polymer at G4.0. Therefore, the supply of new di-phen molecules to the inside of the P22 PCN could be hindered by the polymer layer and as a consequence further interior polymer growth might be suppressed.

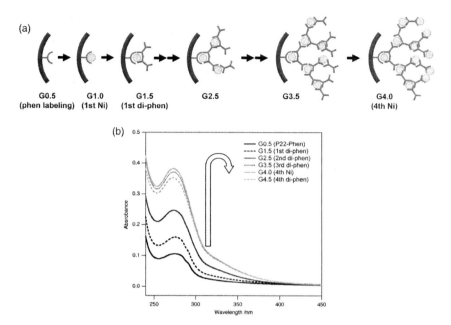

FIGURE 11.8   (a) Schematic illustration of a step-by-step coordination polymer formation. (b) Monitoring of coordination polymer growth through the process. UV-Vis absorbance around 270 nm was increasing up to G4.0. This is due to binding of di-phen to the cage construct and indicates growth of a Ni/di-phen coordination polymer. Absorbance was decreased at the G4.5 (4th di-phen) because of some protein precipitation likely due to inter-cage binding via the metal–ligand coordination.

Although the results demonstrated here is a proof of concept experiment and the obtained material exhibits no practical functionality, similar PCN-coordination polymer constructs could likely be obtained from a combination of other metals and ditopic ligands. Using this metal–ligand binding process, desired functionalities, such as metals for MR imaging capability [147, 148, 152], could be constructed by choosing appropriate metal and ditopic ligands.

11.4.4   Altering Protein Dynamics by Coordination: Hsp-Phen-Fe

Proteins can be considered as versatile ligands in nature because they possess a diverse capability of metal binding [153–155]. This ability to associate metals with proteins is one key feature of biology, imparting metal-based enzyme functionalities and allowing proteins to construct higher order structures. Tremendous interest has recently been focused on exploiting metal–protein coordination to impart new functions and/or to guide protein assembly in design [156–158]. These studies are discussed in more detail in other chapters of this book.

   Coordination chemistry has recently been used to alter the stability and dynamics of a PCN [159]. In this study, a modified 1,10-phenanthroline (phen) ligand was introduced to genetically engineered cysteine residues (E102C) of sHsp cage, which is

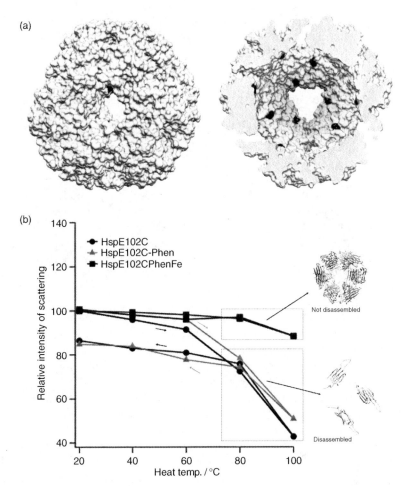

FIGURE 11.9   (a) Space-filling representation of sHsp exterior surface (left) and half cutaway view along the threefold axis (right). Amino acid position 102, which is replaced with cysteine, is colored black. (b) SLS intensity of Hsp(E102C), Hsp(E102C)-phen, and Hsp(E102C)-phen-Fe during a heating–cooling process. Intensity of Hsp(E102C) and Hsp(E102C)-phen significantly dropped at high temperature suggesting the cage-like structure of these samples were disassembled into subunits at high temperatures. Hsp(E102C)-phen-Fe retained its structure even at high temperatures due to cross-linking of subunits around the threefold axis via coordination of phen and Fe.

located around the threefold axis of the cage (Fig. 11.9a). Titration of the Hsp(E102C)-phen construct with Fe(II) resulted in 0.3 Fe(II)/subunit with an associated increase in absorbance at 517 nm, suggesting the formation of a Fe(phen)$_3$$^{2+}$ complex with three adjacent subunits contributing the phenanthrolines.

A remarkable, though sometimes undesirable, feature of the wild-type sHsp is that it exchanges its subunits at temperature above 60°C [160]. Static light scattering measurement of Hsp(E102C) and the Hsp(E102C)-phen indicates that these mutants also exhibit the same subunit exchange phenomena. When the scattered light intensity of these cages was measured on samples treated at increasing temperatures, the intensity significantly dropped with temperatures above 60°C, then recovered to nearly the initial level once the temperature was reduced back to ambient temperature (20°C), suggesting the cages were dissociated to subunit monomers or dimers at higher temperatures and reassembled again when they were cooled down, as charted in Figure 11.9(b). Interestingly, the Fe-treated Hsp(E102C)-phen changed little in scattering intensity even at 100°C indicating that the $Fe(phen)_3^{2+}$ coordination prevents heat-induced dissociation of the cage to subunits. Furthermore, digestion of these Hsp cages using trypsin revealed that Hsp(E102C)-phen-Fe was significantly less digested compared to the Hsp(E102C) and Hsp(E102C)-phen. In particular, digestion at trypsin recognition sites located near the inside of the threefold axis seemed to be especially suppressed, which may be due to the hindrance of trypsin access to these sites from cross-linking the subunit with the $Fe(phen)_3^{2+}$ coordination bond. Together, these results demonstrate that coordination chemistry can be exploited to enhance physical and biochemical stability of PCNs. Stabilization of PCNs in this or similar manners could expand the use of PCNs as a platform of material syntheses and biomedical applications.

11.5   SUMMARY AND PERSPECTIVES

By working to exploit the interior volume of a range of PCN architectures, several new means to incorporate a wide variety of functionalities to PCNs have been developed. Improving the understanding of mineralization process allows for the expansion of electrostatically driven mineralization to PCNs beyond the ferritin superfamily and the incorporation of an expanded range of mineral compositions. Building organic oligomers inside PCNs provides not only a means to stabilize the architecture but also a means to create a high density of attachment sites for active molecules such as Gd-chelates. Forming coordination polymers through interior-directed polymerization can also be used to modulate the PCN properties and act as a means to introduce new metal sites in the PCN. Future PCN materials will likely sample the methods from each of these broad approaches to mineralization, organic polymerization, and coordination polymerization to make active nanomaterial composites.

ACKNOWLEDGMENTS

This work was supported by a grant from the National Institutes of Health (NIBIB R01 EB012027).

REFERENCES

[1] Lee, S. W.; Mao, C. B.; Flynn, C. E.; Belcher, A. M. *Science* **2002**, *296*, 892–895.
[2] Nam, K. T.; Kim, D. W.; Yoo, P. J.; Chiang, C. Y.; Meethong, N.; Hammond, P. T.; Chiang, Y. M.; Belcher, A. M. *Science* **2006**, *312*, 885–888.

[3] Sun, J.; DuFort, C.; Daniel, M. C.; Murali, A.; Chen, C.; Gopinath, K.; Stein, B.; De, M.; Rotello, V. M.; Holzenburg, A.; Kao, C. C.; Dragnea, B. *P. Natl. Acad. Sci. USA* **2007**, *104*, 1354–1359.

[4] Aizenberg, J.; Fratzl, P. *Adv. Mater.* **2009**, *21*, 387–388.

[5] Aizenberg, J.; Muller, D. A.; Grazul, J. L.; Hamann, D. R. *Science* **2003**, *299*, 1205–1208.

[6] Sarikaya, M.; Tamerler, C.; Jen, A. K.; Schulten, K.; Baneyx, F. *Nat. Mater.* **2003**, *2*, 577–585.

[7] Koeck, P. J. B.; Kagawa, H. K.; Ellis, M. J.; Hebert, H.; Trent, J. D. *Biochim. Biophys. Acta* **1998**, *1429*, 40–44.

[8] Trent, J. D. *FEMS Microbiol. Rev.* **1996**, *18*, 249–258.

[9] Kim, K. K.; Kim, R.; Kim, S. H. *Nature* **1998**, *394*, 595–599.

[10] Bozzi, M.; Mignogna, G.; Stefanini, S.; Barra, D.; Longhi, C.; Valenti, P.; Chiancone, E. *J. Biol. Chem.* **1997**, *272*, 3259–3265.

[11] Grant, R. A.; Filman, D. J.; Finkel, S. E.; Kolter, R.; Hogle, J. M. *Nat. Struct. Biol.* **1998**, *5*, 294–303.

[12] Grove, A.; Wilkinson, S. P. *J. Mol. Biol.* **2005**, *347*, 495–508.

[13] Ramsay, R.; Wiedenheft, B.; Allen, M.; Gauss, G. H.; Lawrence, C. M.; Young, M.; Douglas, T. *J. Inorg. Biochem.* **2006**, *100*, 1061–1068.

[14] Wiedenheft, B.; Mosolf, J.; Willits, D.; Yeager, M.; Dryden, K. A.; Young, M.; Douglas, T. *P. Natl. Acad. Sci. USA* **2005**, *102*, 10551–10556.

[15] Zhao, G.; Ceci, P.; Ilari, A.; Giangiacomo, L.; Laue, T. M.; Chiancone, E.; Chasteen, N. D. *J. Biol. Chem.* **2002**, *277*, 27689–27696.

[16] Bancroft, J. B.; Hiebert, E.; Bracker, C. E. *Virology* **1969**, *39*, 924–930.

[17] Crick, F. H. C.; Watson, J. D. *Nature* **1956**, *177*, 473–475.

[18] Pettersen, E. F.; Goddard, T. D.; Huang, C. C.; Couch, G. S.; Greenblatt, D. M.; Meng, E. C.; Ferrin, T. E. *J. Comput. Chem.* **2004**, *25*, 1605–1612.

[19] Allen, M.; Willits, D.; Mosolf, J.; Young, M.; Douglas, T. *Adv. Mater.* **2002**, *14*, 1562–1565.

[20] Douglas, T.; Strable, E.; Willits, D.; Aitouchen, A.; Libera, M.; Young, M. *Adv. Mater.* **2002**, *14*, 415–418.

[21] Douglas, T.; Young, M. *Nature* **1998**, *393*, 152–155.

[22] Flenniken, M. L.; Willits, D. A.; Brumfield, S.; Young, M.; Douglas, T. *Nano Lett.* **2003**, *3*, 1573–1576.

[23] Shenton, W.; Mann, S.; Cölfen, H.; Bacher, A.; Fischer, M. *Angew. Chem. Int. Edit.* **2001**, *40*, 442–445.

[24] Laufer, B.; Steinmetz, N. F.; Hong, V.; Manchester, M.; Kessler, H.; Finn, M. G. *Biopolymers* **2009**, *92*, 323–323.

[25] Raja, K. S.; Wang, Q.; Finn, M. G. *ChemBioChem* **2003**, *4*, 1348–1351.

[26] Wang, Q.; Lin, T. W.; Tang, L.; Johnson, J. E.; Finn, M. G. *Angew. Chem. Int. Edit.* **2002**, *41*, 459–462.

[27] Datta, A.; Hooker, J. M.; Botta, M.; Francis, M. B.; Aime, S.; Raymond, K. N. *J. Am. Chem. Soc.* **2008**, *130*, 2546–2552.

[28] Hooker, J. M.; Datta, A.; Botta, M.; Raymond, K. N.; Francis, M. B. *Nano Lett.* **2007**, *7*, 2207–2210.

[29] Harrison, P. M.; Arosio, P. *Biochim. Biophys. Acta* **1996**, *1275*, 161–203.

[30] Matias, P. M.; Tatur, J.; Carrondo, M. A.; Hagen, W. R. *Acta Crystallogr. Sect. F. Struct. Biol. Cryst. Commun.* **2005**, *61*, 503–506.

[31] Hempstead, P. D.; Yewdall, S. J.; Fernie, A. R.; Lawson, D. M.; Artymiuk, P. J.; Rice, D. W.; Ford, G. C.; Harrison, P. M. *J. Mol. Biol.* **1997**, *268*, 424–448.

[32] Kurtz, D. M., Jr. *J. Biol. Inorg. Chem.* **1997**, *2*, 159–167.

[33] Lawson, D. M.; Artymiuk, P. J.; Yewdall, S. J.; Smith, J. M. A.; Livingstone, J. C.; Treffry, A.; Luzzago, A.; Levi, S.; Arosio, P.; Cesareni, G.; Thomas, C. D.; Shaw, W. V.; Harrison, P. M. *Nature* **1991**, *349*, 541–544.

[34] Mayer, D. E.; Rohrer, J. S.; Schoeller, D. A.; Harris, D. C. *Biochemistry* **1983**, *22*, 876–880.

[35] Douglas, T.; Ripoll, D. R. *Protein Sci.* **1998**, *7*, 1083–1091.

[36] Theil, E. C.; Takagi, H.; Small, G. W.; He, L.; Tipton, A. R.; Danger, D. *Inorg. Chim. Acta* **2000**, *297*, 242–251.

[37] Yang, X.; Arosio, P.; Chasteen, N. D. *Biophys. J.* **2000**, *78*, 2049–2059.

[38] Bauminger, E. R.; Treffry, A.; Quail, M. A.; Zhao, Z.; Nowik, I.; Harrison, P. M. *Biochemistry* **1999**, *38*, 7791–7802.

[39] Lindsay, S.; Brosnahan, D.; Lowery, T. J., Jr.; Crawford, K.; Watt, G. D. *Biochim. Biophys. Acta* **2003**, *1621*, 57–66.

[40] Juan, S. H.; Aust, S. D. *Arch. Biochem. Biophys.* **1998**, *350*, 259–265.

[41] Santambrogio, P.; Levi, S.; Cozzi, A.; Corsi, B.; Arosio, P. *Biochem. J.* **1996**, *314*, 139–144.

[42] Douglas, T.; Stark, V. T. *Inorg. Chem.* **2000**, *39*, 1828–1830.

[43] Kim, J. W.; Choi, S. H.; Lillehei, P. T.; Chu, S. H.; King, G. C.; Watt, G. D. *Chem. Commun.* **2005**, *32*, 4101–4103.

[44] Tsukamoto, R.; Iwahor, K.; Muraoka, M.; Yamashita, I. *B. Chem. Soc. Jpn.* **2005**, *78*, 2075–2081.

[45] Mann, S.; Meldrum, F. C. *Adv. Mater.* **1991**, *3*, 316–318.

[46] Meldrum, F. C.; Douglas, T.; Levi, S.; Arosio, P.; Mann, S. *J. Inorg. Biochem.* **1995**, *58*, 59–68.

[47] Douglas, T.; Dickson, D. P. E.; Betteridge, S.; Charnock, J.; Garner, C. D.; Mann, S. *Science* **1995**, *269*, 54–57.

[48] Ensign, D.; Young, M.; Douglas, T. *Inorg. Chem.* **2004**, *43*, 3441–3446.

[49] Kramer, R. M.; Li, C.; Carter, D. C.; Stone, M. O.; Naik, R. R. *J. Am. Chem. Soc.* **2004**, *126*, 13282–13286.

[50] Meldrum, F. C.; Heywood, B. R.; Mann, S. *Science* **1992**, *257*, 522–523.

[51] Sano, K.; Sasaki, H.; Shiba, K. *J. Am. Chem. Soc.* **2006**, *128*, 1717–1722.

[52] Wong, K. K. W.; Mann, S. *Adv. Mater.* **1996**, *8*, 928–932.

[53] Ueno, T.; Suzuki, M.; Goto, T.; Matsumoto, T.; Nagayama, K.; Watanabe, Y. *Angew. Chem. Int. Ed.* **2004**, *43*, 2527–2530.

[54] Okuda, M.; Iwahori, K.; Yamashita, I.; Yoshimura, H. *Biotechnol. Bioeng.* **2003**, *84*, 187–194.

[55] Iwahori, K.; Yoshizawa, K.; Muraoka, M.; Yamashita, I. *Inorg. Chem.* **2005**, *44*, 6393–6400.

[56] Klem, M. T.; Resnick, D. A.; Gilmore, K.; Young, M.; Idzerda, Y. U.; Douglas, T. *J. Am. Chem. Soc.* **2007**, *129*, 197–201.

[57] Galvez, N.; Sanchez, P.; Dominguez-Vera, J. M.; Soriano-Portillo, A.; Clemente-Leon, M.; Coronado, E. *J. Mater. Chem.* **2006**, *16*, 2757–2761.

[58] Coronado, E.; Clemente-Leon, M.; Soriano-Portillo, A.; Galvez, N.; Dominguez-Vera, J. M. *J. Mater. Chem.* **2007**, *17*, 49–51.

[59] Almiron, M.; Link, A. J.; Furlong, D.; Kolter, R. *Gene. Dev.* **1992**, *6*, 2646–2654.

[60] Ceci, P.; Ilari, A.; Falvo, E.; Chiancone, E. *J. Biol. Chem.* **2003**, *278*, 20319–20326.

[61] Ilari, A.; Savino, C.; Stefanini, S.; Chiancone, E.; Tsernoglou, D. *Acta. Crystallogr. D.* **1999**, *55*, 552–553.

[62] Marjorette, M.; Pena, O.; Bullerjahn, G. S. *J. Biol. Chem.* **1995**, *270*, 22478–22482.

[63] Reindel, S.; Schmidt, C. L.; Anemuller, S.; Matzanke, B. F. *Biochem. Soc. Trans.* **2002**, *30*, 713–715.

[64] Chiancone, E.; Ceci, P.; Ilari, A.; Ribacchi, F.; Stefanini, S. *Biometals* **2004**, *17*, 197–202.

[65] Ilari, A.; Latella, M. C.; Ceci, P.; Ribacchi, F.; Su, M. H.; Giangiacomo, L.; Stefanini, S.; Chasteen, N. D.; Chiancone, E. *Biochemistry* **2005**, *44*, 5579–5587.

[66] Ilari, A.; Stefanini, S.; Chiancone, E.; Tsernoglou, D. *Nat. Struct. Biol.* **2000**, *7*, 38–43.

[67] Stefanini, S.; Cavallo, S.; Montagnini, B.; Chiancone, E. *Biochem. J.* **1999**, *338*, 71–75.

[68] Jolley, C. C.; Douglas, T. *Biophys. J.* **2010**, *99*, 3385–3393.

[69] Allen, M.; Willits, D.; Young, M.; Douglas, T. *Inorg. Chem.* **2003**, *42*, 6300–6305.

[70] Iwahori, K.; Enomoto, T.; Furusho, H.; Miura, A.; Nishio, K.; Mishima, Y.; Yamashita, I. *Chem. Mater.* **2007**, *19*, 3105–3111.

[71] Okuda, M.; Suzumoto, Y.; Iwahori, K.; Kang, S.; Uchida, M.; Douglas, T.; Yamashita, I. *Chem. Commun.* **2010**, *46*, 8797–8799.

[72] Kang, S.; Lucon, J.; Varpness, Z.; Liepold, L.; Uchida, M.; Willits, D.; Young, M.; Douglas, T. *Angew. Chem. Int. Ed.* **2008**, *47*, 7845–7848.

[73] Zeth, K.; Offermann, S.; Essen, L. O.; Oesterhelt, D. *P. Natl. Acad. Sci. USA* **2004**, *101*, 13780–13785.

[74] Kang, S.; Jolley, C. C.; Liepold, L. O.; Young, M.; Douglas, T. *Angew. Chem. Int. Ed.* **2009**, *48*, 4772–4776.

[75] Benesch, J. L.; Robinson, C. V. *Curr. Opin. Struc. Biol.* **2006**, *16*, 245–251.

[76] Esteban, O.; Bernal, R. A.; Donohoe, M.; Videler, H.; Sharon, M.; Robinson, C. V.; Stock, D. *J. Biol. Chem.* **2008**, *283*, 2595–2603.

[77] Fandrich, M.; Tito, M. A.; Leroux, M. R.; Rostom, A. A.; Hartl, F. U.; Dobson, C. M.; Robinson, C. V. *P. Natl. Acad. Sci. USA* **2000**, *97*, 14151–14155.

[78] Heck, A. J. R.; van den Heuvel, R. H. H. *Mass Spectrom. Rev.* **2004**, *23*, 368–389.

[79] Kitagawa, N.; Mazon, H.; Heck, A. J. R.; Wilkens, S. *J. Biol. Chem.* **2008**, *283*, 3329–3337.

[80] Douglas, T.; Young, M. *Science* **2006**, *312*, 873–875.

[81] Brumfield, S.; Willits, D.; Tang, L.; Johnson, J. E.; Douglas, T.; Young, M. *J. Gen. Virol.* **2004**, *85*, 1049–1053.

[82] Parker, M. H.; Casjens, S.; Prevelige, P. E. *J. Mol. Biol.* **1998**, *281*, 69–79.

[83] Parker, M. H.; Prevelige, P. E. *Virology* **1998**, *250*, 337–349.

[84] Tuma, R.; Parker, M. H.; Prevelige, P. E.; Thomas, G. J. *Biophys. J.* **1998**, *74*, A72-A72.

[85] Thuman-Commike, P. A.; Greene, B.; Malinski, J. A.; Burbea, M.; McGough, A.; Chiu, W.; Prevelige, P. E.; *Biophys. J.* **1999**, *76*, 3267–3277.

[86] Weigele, P. R.; Sampson, L.; Winn-Stapley, D.; Casjens, S. R. *J. Mol. Biol.* **2005**, *348*, 831–844.

[87] Reichhardt, C. R.; O'Neil, A.; Uchida, M.; Prevelige, P. E.; Douglas, T. *Chem. Commun.* **2011**, *47*, 6326–6328.

[88] Grover, G. N.; Maynard, H. D. *Curr. Opin. Chem. Biol.* **2010**, *14*, 818–827.

[89] Krishna, O. D.; Kiick, K. L. *Biopolymers* **2010**, *94*, 32–48.

[90] Thordarson, P.; Le Droumaguet, B.; Velonia, K. *Appl. Microbiol. Biot.* **2006**, *73*, 243–254.

[91] Klok, H. A. *Macromolecules* **2009**, *42*, 7990–8000.

[92] Comellas-Aragones, M.; de la Escosura, A.; Dirks, A. J.; van der Ham, A.; Fuste-Cune, A.; Cornelissen, J.; Nolte, R. J. M. *Biomacromolecules* **2009**, *10*, 3141–3147.

[93] Dixit, S. K.; Goicochea, N. L.; Daniel, M. C.; Murali, A.; Bronstein, L.; De, M.; Stein, B.; Rotello, V. M.; Kao, C. C.; Dragnea, B. *Nano Lett.* **2006**, *6*, 1993–1999.

[94] Hu, Y. F.; Zandi, R.; Anavitarte, A.; Knobler, C. M.; Gelbart, W. M. *Biophys. J.* **2008**, *94*, 1428–1436.

[95] Abe, S.; Hirata, K.; Ueno, T.; Morino, K.; Shimizu, N.; Yamamoto, M.; Takata, M.; Yashima, E.; Watanabe, Y. *J. Am. Chem. Soc.* **2009**, *131*, 6958–6960.

[96] Abedin, M. J.; Liepold, L.; Suci, P.; Young, M.; Douglas, T. *J. Am. Chem. Soc.* **2009**, *131*, 4346–4354.

[97] Lucon, J.; Qazi, S.; Uchida, M.; Bedwell, G.; LaFrance, B.; Prevelige, P. E.; Douglas, T. *Nat. Chem.* **2012**, *4*, 781–788.

[98] Liepold, L. O.; Abedin, M. J.; Buckhouse, E. D.; Frank, J. A.; Young, M. J.; Douglas, T. *Nano Lett.* **2009**, *9*, 4520–4526.

[99] Lucon, J.; Abedin, M. J.; Uchida, M.; Liepold, L.; Jolley, C. C.; Young, M.; Douglas, T. *Chem. Commun.* **2010**, *46*, 264–266.

[100] Gupta, S. S.; Kuzelka, J.; Singh, P.; Lewis, W. G.; Manchester, M.; Finn, M. G. *Bioconjug. Chem.* **2005**, *16*, 1572–1579.

[101] Kolb, H. C.; Finn, M. G.; Sharpless, K. B. *Angew. Chem. Int. Edit.* **2001**, *40*, 2004–2021.

[102] Wang, Q.; Chan, T. R.; Hilgraf, R.; Fokin, V. V.; Sharpless, K. B.; Finn, M. G. *J. Am. Chem. Soc.* **2003**, *125*, 3192–3193.

[103] Hong, V.; Presolski, S. I.; Ma, C.; Finn, M. G. *Angew. Chem. Int. Ed.* **2009**, *48*, 9879–9883.

[104] King, J.; Lenk, E. V.; Botstein, D. *J. Mol. Biol.* **1973**, *80*, 697–731.

[105] Prevelige, P. E., Jr.; Thomas, D.; King, J. *J. Mol. Biol.* **1988**, *202*, 743–757.

[106] Parent, K. N.; Khayat, R.; Tu, L. H.; Suhanovsky, M. M.; Cortines, J. R.; Teschke, C. M.; Johnson, J. E.; Baker, T. S. *Structure* **2010**, *18*, 390–401.

[107] Teschke, C. M.; McGough, A.; Thuman-Commike, P. A. *Biophys. J.* **2003**, *84*, 2585–2592.

[108] Chen, D. H.; Baker, M. L.; Hryc, C. F.; DiMaio, F.; Jakana, J.; Wu, W. M.; Dougherty, M.; Haase-Pettingell, C.; Schmid, M. F.; Jiang, W.; Baker, D.; King, J. A.; Chiu, W. *P. Natl. Acad. Sci. USA* **2011**, *108*, 1355–1360.

[109] Earnshaw, W.; Casjens, S.; Harrison, S. C. *J. Mol. Biol.* **1976**, *104*, 387–410.

[110] Jiang, W.; Li, Z.; Zhang, Z.; Baker, M. L.; Prevelige, P. E., Jr.; Chiu, W. *Nat. Struct. Biol.* **2003**, *10*, 131–135.

[111] Tuma, R.; Prevelige, P. E.; Thomas, G. J. *P. Natl. Acad. Sci. USA* **1998**, *95*, 9885–9890.

[112] Kang, S.; Uchida, M.; O'Neil, A.; Li, R.; Prevelige, P. E.; Douglas, T. *Biomacromolecules* **2010**, *11*, 2804–2809.

[113] Qazi, S.; Liepold, L.; Abedin, J.; Johnson, B.; Prevelige, P.; Frank, J.; Douglas, T. *Mol. Pharm.* **2012**

[114] Heredia, K. L.; Bontempo, D.; Ly, T.; Byers, J. T.; Halstenberg, S.; Maynard, H. D. *J. Am. Chem. Soc.* **2005**, *127*, 16955–16960.

[115] Mantovani, G.; Lecolley, F.; Tao, L.; Haddleton, D. M.; Clerx, J.; Cornelissen, J.; Velonia, K. *J. Am. Chem. Soc.* **2005**, *127*, 2966–2973.

[116] Mulder, W. J. M.; Strijkers, G. J.; van Tilborg, G. A. F.; Griffioen, A. W.; Nicolay, K. *NMR Biomed.* **2006**, *19*, 142–164.

[117] Liepold, L.; Anderson, S.; Willits, D.; Oltrogge, L.; Frank, J. A.; Douglas, T.; Young, M. *Magn. Reson. Med.* **2007**, *58*, 871–879.

[118] Allen, M.; Bulte, J. W. M.; Liepold, L.; Basu, G.; Zywicke, H. A.; Frank, J. A.; Young, M.; Douglas, T. *Magn. Reson. Med.* **2005**, *54*, 807–812.

[119] Anderson, E. A.; Isaacman, S.; Peabody, D. S.; Wang, E. Y.; Canary, J. W.; Kirshenbaum, K. *Nano Lett.* **2006**, *6*, 1160–1164.

[120] Datta, A.; Raymond, K. N. *Acc. Chem. Res.* **2009**, *42*, 938–947.

[121] Pierre, V. C.; Botta, M.; Aime, S.; Raymond, K. N. *Inorg. Chem.* **2006**, *45*, 8355–8364.

[122] Prasuhn, D. E.; Yeh, R. M.; Obenaus, A.; Manchester, M.; Finn, M. G. *Chem. Commun.* **2007**, *12*, 1269–1271.

[123] Destito, G.; Yeh, R.; Rae, C. S.; Finn, M. G.; Manchester, M. *Chem. Biol.* **2007**, *14*, 1152–1162.

[124] Flenniken, M. L.; Willits, D. A.; Harmsen, A. L.; Liepold, L. O.; Harmsen, A. G.; Young, M. J.; Douglas, T. *Chem. Biol.* **2006**, *13*, 161–170.

[125] Kaiser, C. R.; Flenniken, M. L.; Gillitzer, E.; Harmsen, A. L.; Harmsen, A. G.; Jutila, M. A.; Douglas, T.; Young, M. J. *Int. J. Nanomed.* **2007**, *2*, 715–733.

[126] Kaltgrad, E.; O'Reilly, M. K.; Liao, L.; Han, S.; Paulson, J. C.; Finn, M. G. *J. Am. Chem. Soc.* **2008**, *130*, 4578–4579.

[127] Koudelka, K. J.; Rae, C. S.; Gonzalez, M. J.; Manchester, M. *J. Virol.* **2007**, *81*, 1632–1640.

[128] Koudelka, K. J.; Destito, G.; Plummer, E. M.; Trauger, S. A.; Siuzdak, G.; Manchester, M. *PLoS Pathog.* **2009**, *5*, e1000417.

[129] Kovacs, E. W.; Hooker, J. M.; Romanini, D. W.; Holder, P. G.; Berry, K. E.; Francis, M. B. *Bioconjug. Chem.* **2007**, *18*, 1140–1147.

[130] Prasuhn, D. E., Jr.; Singh, P.; Strable, E.; Brown, S.; Manchester, M.; Finn, M. G. *J. Am. Chem. Soc.* **2008**, *130*, 1328–1334.

[131] Rae, C. S.; Khor, I. W.; Wang, Q.; Destito, G.; Gonzalez, M. J.; Singh, P.; Thomas, D. M.; Estrada, M. N.; Powell, E.; Finn, M. G.; Manchester, M. *Virology* **2005**, *343*, 224–235.

[132] Raja, K. S.; Wang, Q.; Gonzalez, M. J.; Manchester, M.; Johnson, J. E.; Finn, M. G. *Biomacromolecules* **2003**, *4*, 472–476.

[133] Shriver, L. P.; Koudelka, K. J.; Manchester, M. *J. Neuroimmunol.* **2009**, *211*, 66–72.

[134] Singh, P.; Prasuhn, D.; Yeh, R. M.; Destito, G.; Rae, C. S.; Osborn, K.; Finn, M. G.; Manchester, M. *J. Control Release* **2007**, *120*, 41–50.

[135] Steinmetz, N. F.; Manchester, M. *Biomacromolecules* **2009**, *10*, 784–792.

[136] Uchida, M.; Flenniken, M. L.; Allen, M.; Willits, D. A.; Crowley, B. E.; Brumfield, S.; Willis, A. F.; Jackiw, L.; Jutila, M.; Young, M. J.; Douglas, T. *J. Am. Chem. Soc.* **2006**, *128*, 16626–16633.

[137] Uchida, M.; Kosuge, H.; Terashima, M.; Willits, D. A.; Liepold, L. O.; Young, M. J.; Mcconnell, M. V.; Douglas, T. *ACS Nano.* **2011**, *5*, 2493–2502.

[138] Pokorski, J. K.; Breitenkamp, K.; Liepold, L. O.; Qazi, S.; Finn, M. G. *J. Am. Chem. Soc.* **2011**, *133*, 9242–9245.

[139] Dunand, F. A.; Borel, A.; Helm, L. *Inorg. Chem. Commun.* **2002**, *5*, 811–815.

[140] Powell, D. H.; Dhubhghaill, O. M. N.; Pubanz, D.; Helm, L.; Lebedev, Y. S.; Schlaepfer, W.; Merbach, A. E. *J. Am. Chem. Soc.* **1996**, *118*, 9333–9346.

[141] Yaghi, O. M.; O'Keeffe, M.; Ockwig, N. W.; Chae, H. K.; Eddaoudi, M.; Kim, J. *Nature* **2003**, *423*, 705–714.

[142] Kitagawa, S.; Matsuda, R. *Coordin. Chem. Rev.* **2007**, *251*, 2490–2509.

[143] Robin, A. Y.; Fromm, K. M. *Coordin. Chem. Rev.* **2006**, *250*, 2127–2157.

[144] Bureekaew, S.; Shimomura, S.; Kitagawa, S. *Sci. Technol. Adv. Mat.* **2008**, *9*, 014108.

[145] Kuppler, R. J.; Timmons, D. J.; Fang, Q.-R.; Li, J.-R.; Makal, T. A.; Young, M. D.; Yuan, D.; Zhao, D.; Zhuang, W.; Zhou, H.-C. *Coordin. Chem. Rev.* **2009**, *253*, 3042–3066.

[146] Spokoyny, A. M.; Kim, D.; Sumrein, A.; Mirkin, C. A. *Chem. Soc. Rev.* **2009**, *38*, 1218–1227.

[147] Taylor, K. M.; Rieter, W. J.; Lin, W. *J. Am. Chem. Soc.* **2008**, *130*, 14358–14359.

[148] Rieter, W. J.; Taylor, K. M. L.; An, H.; Lin, W.; Lin, W. *J. Am. Chem. Soc.* **2006**, *128*, 9024–9025.

[149] Chen, C.; Sigman, D. *P. Natl. Acad. Sci. USA* **1986**, *83*, 7147–7151.

[150] Jolley, C. C.; Lucon, J.; Uchida, M.; Reichhardt, C.; Vaughn, M.; LaFrance, B.; Douglas, T. *J. Coord. Chem.* **2011**, *64*, 4301–4317.

[151] Uchida, M.; Morris, D. S.; Kang, S.; Jolley, C. C.; Lucon, J.; Liepold, L.; Prevelige, P. E.; Douglas, T. *Langmuir* **2011**, *28*, 1998–2006.

[152] Keskin, S.; Kizilel, S. *Ind. Eng. Chem. Res.* **2011**, *50*, 1799–1812.

[153] Ghosh, D.; Pecoraro, V. L. *Curr. Opin. Chem. Biol.* **2005**, *9*, 97–103.

[154] Radford, R. J.; Brodin, J. D.; Salgado, E. N.; Tezcan, F. A. *Coordin. Chem. Rev.* **2011**, *255*, 790–803.

[155] Lu, Y.; Yeung, N.; Sieracki, N.; Marshall, N. M. *Nature* **2009**, *460*, 855–862.

[156] Lin, Y.-W.; Yeung, N.; Gao, Y.-G.; Miner, K. D.; Lei, L.; Robinson, H.; Lu, Y. *J. Am. Chem. Soc.* **2010**, *132*, 9970–9972.

[157] Radford, R.; Tezcan, F. *J. Am. Chem. Soc.* **2009**, *131*, 9136–9137.

[158] Shiga, D.; Nakane, D.; Inomata, T.; Funahashi, Y.; Masuda, H.; Kikuchi, A.; Oda, M.; Noda, M.; Uchiyama, S.; Fukui, K.; Kanaori, K.; Tajima, K.; Takano, Y.; Nakamura, H.; Tanaka, T. *J. Am. Chem. Soc.* **2010**, *132*, 18191–18198.

[159] Uchida, M.; Kang, S.; Liepold, L.; Lucon, J.; Douglas, T. In preparation.

[160] Bova, M. P.; Huang, Q.; Ding, L.; Horwitz, J. *J. Biol. Chem.* **2002**, *277*, 38468–38475.

12

NANOPARTICLES SYNTHESIZED AND DELIVERED BY PROTEIN IN THE FIELD OF NANOTECHNOLOGY APPLICATIONS

Ichiro Yamashita, Kenji Iwahori, Bin Zheng, and Shinya Kumagai

12.1 NANOPARTICLE SYNTHESIS IN A BIO-TEMPLATE

Protein cages are used as templates to synthesize nanoparticles (NPs) for nanotechnology, especially nanoelectronic devices. The biomineralization capability has been extended and NPs of materials needed for the nanotechnology and nanoelectronic devices, such as metal, metal oxide compound, semiconductor, oxide semiconductor, magnetic material, were artificially synthesized in the cavity and used in the devices. In this section, NP synthesis, NP placement, and electronic device fabrications are described.

12.1.1 NP Synthesis by Cage-Shaped Proteins for Nanoelectronic Devices and Other Applications

NPs are one of the most fundamental materials in the nanotechnology and are also essential for the nanoelectronic devices because NPs have an extremely large surface-to-volume ratio and have unprecedented and attractive properties which cannot be realized by bulk materials. There is a wide range of NP applications, such as magnetic recording materials, catalytic materials, fluorescent markers, drug delivery systems,

Coordination Chemistry in Protein Cages: Principles, Design, and Applications, First Edition.
Edited by Takafumi Ueno and Yoshihito Watanabe.

sensors, ink, and quantum electronics. Lots of methods have been developed for making NPs—physical, chemical, and biological methods. Researchers selected the best synthesis method which satisfies their requirements. Recently, the biological method, in which NPs are synthesized using biomolecules as templates (bio-template), has been attracting researchers' attention. Bio-template method had been considered to be inferior to the other methods from an engineering point of view, but in reality, it has essential merits for synthesizing NPs.

The bio-templates used for NPs synthesis so far include mammalian protein cages [1], small protein cages from bacteria [2], and viral protein cages [3]; new forms of non-spherical cages such as rings are also being proposed [4]. Their sizes range from a few nanometers to hundreds of nanometers. Viral protein cages, the structures of which have been extensively studied, have a wide variety of sizes and therefore the potential to produce NPs of various sizes. The most intensively studied bio-template for NPs synthesis is apoferritin. Apoferritin is a ubiquitous protein from bacteria to human being that plays a critical role for the homeostasis of iron ion concentration [1]. There are many kinds of apoferritins, and, among them, a horse spleen apoferritin (HsAFr) has been used most often. The outer and inner diameters of the protein shell are 12 and 7 nm, respectively. There are eight threefold channels connecting outside and inside, through which ions enter the cavity [1] (Fig. 12.1). Another protein cage, a Dps protein (DNA-binding protein from starved cell) from a bacterium, *Listeria innocua* (LiDps) has also been used. LiDps has a protein shell with an outer diameter of 9 nm and an inner diameter of 4.5 nm [2] (Fig. 12.1).

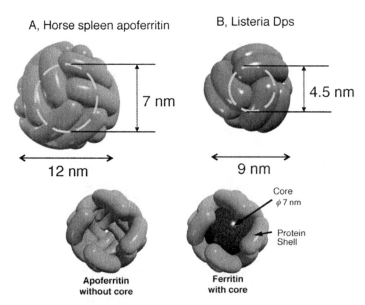

FIGURE 12.1 The schematic drawing of the ferritin and Listeria ferritin (LiDps). Apoferritin without core and ferritin with core.

NPs are now used in many applications from catalysis to fluorescence markers. Among them, one of the most promising applications is the nanoelectronic devises. From the nanoelectronic device point of view, the synthesis of NPs with diameter less than 10 nm is desired because they would have quantum effects. Moreover, the diameter of the NPs synthesized in the protein cages is determined by the cavity size and NPs should be homogeneous. These facts make the NPs produced by using the HsAFr and LiDps as bio-templates very attractive.

12.1.2   Metal Oxide or Hydro-Oxide NP Synthesis in the Apoferritin Cavity

The synthesis of NP in apoferritin cavity dates back to the 1980s [5]. Historically, several kinds of metal complexes, metal oxides and hydroxides, were sequestered in the native apoferritin cavity *in vivo* and *in vitro*. S. Mann, the pioneer in this field, made a variety of NPs ranging from manganese oxides to iron oxides [6]. Since then, there have been many reports of NP synthesis in the cavity of apoferritin, and many types of metal, metal oxide, magnetic, and semiconductor NPs accommodated by apoferritin are now available.

Cobalt oxide or hydro-oxide NPs have been synthesized. T. Douglas et al. reported the biomineralization of (Co(O)OH) or Co_3O_4 cores NPs in the apoferritin cavity for the first time. Core NPs were formed in HsAFr by oxidizing Co(II) ions with H_2O_2 while the reaction solution was dynamically titrated at pH 8.5 using NaOH [7]. They also reported the synthesis of smaller cobalt NPs using an apoferritin from *LiDps* [8]. Two types of homogeneous small cobalt NPs were synthesized at two different reaction temperatures (65°C and 23°C) in the cavity of *LiDps*. Electron diffraction measurement revealed that crystalline Co_3O_4 phase cores were synthesized at 65°C and amorphous Co(O)OH phase cores at 23°C. It is notable that the crystallinity of Co NPs in the restrictive nanomold could be controlled simply by changing the reaction temperature. R. Tsukamoto et al. reported the one-pot synthesis of Co_3O_4 NP cores in the HsAFr and recombinant ferritin, Fer-8 [9], which adopt pH control by buffer agent and is very suitable for the mass production of homogeneous Co_3O_4 NPs for industrial use. The essentially important point is the reaction solution conditions, that is, mixing 3 mM Co(II) ion with 0.5 mg/mL apoferritin in 100 mM HEPES pH 8.3 buffer solution followed by the addition of hydrogen peroxide (H_2O_2) at 50°C. X-ray photoemission spectroscopy (XPS), electron energy-loss spectroscopy (EELS) analyses, X-ray powder diffraction (XRD) structure analysis, and high-resolution transmission electron microscopy (TEM) confirmed Co_3O_4 core formation. R. Tsukamoto et al. also reported that the PEGylated apoferritin is very stable for Co_3O_4 synthesis [10]. Co_3O_4 is a wide-gap semiconductor and can be used for electronic components. In addition, Co_3O_4 NPs can be easily reduced to metallic cobalt by several methods, such as heat treatment under reductive gas or heat treatment after embedment in the SiO_2 layer. These techniques have been used to fabricate a memory device using synthesized Co oxide NPs [11]; see the next section.

M. Okuda et al. fabricated nickel hydroxide NPs in HsAFr cavities by incubating ammonium nickel sulfate solution with HsAFr [12]. Energy dispersive spectrometry (EDS) verified the existence of nickel in the core. The optimized conditions for

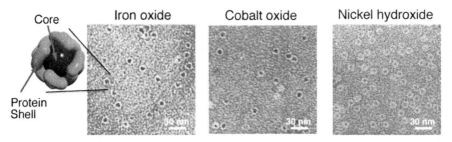

FIGURE 12.2   Typical TEM images of iron, cobalt, and nickel NPs synthesized in the HsAFr cavity. NPs were observed by TEM with 1% aurothioglucose staining. The negatively stained TEM image shows many dots (black parts) surrounded by ferritin protein shells (white parts). As aurothioglucose is too big to penetrate through narrow channels, the dots are attributed to the individual metal cores.

nickel core formation were 0.3 mg/mL HsAFr and 5 mM ammonium nickel sulfate in water containing dissolved carbon dioxide. The pH was maintained at 8.6 using a combination of two buffer solutions (HEPES-CAPSO) with 20 mM ammonia. Interestingly, the nickel oxide core formation in the apoferritin cavity needs carbonate ions, which leads to a core formation ratio (CFR) of nearly 100% (Fig. 12.2) [12]. The synthesized Ni NPs (7 nm diameter) in the apoferritin were used as catalysts to fabricate a polycrystalline silicon (Si) thin film with low Ni contamination level from amorphous Si thin film [13]. It was also found that carbonate ions were needed in the synthesis of chromium oxide or hydroxide NPs in HsAFr. The optimized conditions were 0.1 mg/mL apoferritin, 1 mM ammonium chromium sulfate, 100 mM HEPES (pH7.5) containing carbonate ions [12]. The CFR for the chromium core was about 80%. The carbonate ions may coordinate nickel ions and stabilize them in the bulk solution and accelerate the hydroxylation in the apoferritin cavity. Their exact role in the chemical reaction for core formation is still not clear. They also succeeded to synthesize indium oxide NPs in the apoferritin cavity [14].

12.1.3   Compound Semiconductor NP Synthesis in the Apoferritin Cavity

Semiconductor NPs synthesis in the apoferritin cavity is of great interest because the homogeneity in shape and size less than 10 nm makes electron energy levels of a semiconductor NP split at the same level. This quantum effect is an ideal character-istic of the nanoelectronic device components, such as quantum dots or fluorescent markers. It had been very hard, however, to synthesize compound semiconductors in apoferritin cavities because two component source ions have to be introduced into the cavity simultaneously and stoichiometrically. The first semiconductor NP synthesis was reported by K. K. W. Wong et al. [15]. They synthesized small cad-mium sulfide (CdS) cores in the apoferritin cavity by incrementally adding Cd ions followed by S ions. The development of CdS NPs was controlled by the sequential addition of 55 Cd atoms per apoferritin molecule. The diameters of the synthesized CdS NPs were 2.5 and 4.0 nm when 110 and 275 Cd (II) ions per apoferritin molecule were added to the reaction solution, respectively. This approach for controlling CdS diameter attracted much interest. The synthesized CdS NPs were analyzed by EDS, TEM, and electron diffraction, which confirmed the obtained core was a zinc blende structure CdS.

I. Yamashita et al. first reported the one-pod synthesis of semiconductor NP which fully developed in apoferritin cavity [16]. They synthesized NPs of cadmium selenide (CdSe) in the apoferritin cavity by applying a newly designed chemical reaction system (SCRY). Since the chemical reaction of Cd^{2+} and Se^{2-} in aqueous solution proceeds too fast for those ions to approach and go through the threefold channel, entering the apoferritin cavity, they stabilized Cd^{2+} by excess ammonia to form positively charged tetraaminecadmium ions $(Cd(NH_3)_4^{2+})$. Se^{2-} ions were supplied from selenourea which is slightly unstable in aqueous solution and slowly degrades. This slow chemical reaction (SCRY) suppressed the bulk precipitation in the reaction solution. Inner surface of the cavity has a lot of negatively charged amino acid residues and there is electrostatic difference, inside negative. Cd^{2+} ions generated from $Cd(NH_3)_4^{2+}$ in the vicinity of the threefold channel entrance were sucked into the cavity by the electrostatic potential difference between outside and inside. Cd^{2+} ions were highly concentrated inside and Se^{2-} ions followed after electrostatic compensation finished. This made CdSe nuclei formation inside cavity much faster than outside and a high CFR for the CdSe NP core was obtained. The size of CdSe was homogeneous and the standard deviation of the diameter was less than 10% of the average diameter. XPS measurement showed two main peaks at 413.5 eV for Cd 3d and 55.4 eV for Se 3d. XRD measurement showed that they are mixture of the cubic phase and hexagonal phase (wurtzite). After 500°C heat treatment under nitrogen gas for 1 h, cores became a single cubic phase and crystal core.

Adopting the slow chemical reaction system, K. Iwahori et al. succeeded in synthesizing homogeneous zinc selenide (ZnSe) NPs in the apoferritin cavity [17]. ZnSe is a promising material for n-type semiconductors. ZnSe NPs could be used as a quantum label in the biological field because fluorescent light is not easily quenched. They modified the slow chemical reaction synthesis process and determined the optimum reaction conditions to be 0.3 mg/mL HsAFr, 1 mM zinc acetate, 40 mM ammonium acetate, 7.5 mM ammonia water, and 10 mM selenourea. The ammonium ions stabilized Zn ions, forming tetraaminezinc ions (Figs. 12.3 and 12.4). The CFR of ZnSe NPs was as high as 90%. The characterization of the synthesized ZnSe NPs by XRD and EDS revealed that the synthesized NPs are a collection of cubic ZnSe polycrystals. It was shown that the 500°C heat treatment for 1 h under nitrogen gas transformed the polycrystalline ZnSe core into the single crystal and free of protein shell.

K. Iwahori et al. further investigated the mechanism of ZnSe NP synthesis in the apoferritin cavity using three recombinant apoferritins, Fer-8, Fer-8A, and Fer-8AK. Fer-8A is derived from Fer-8. Threefold channels of the Fer-8A are replaced by neutral alanine residues. Fer-8AK is derived from Fer-8A. Glutamic acid residues on the inner surface of Fer-8AK were replaced with positively charged lysine residue [17].

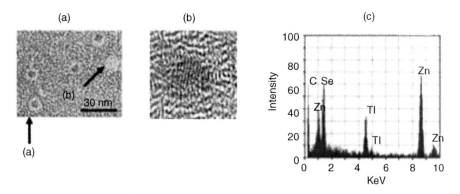

FIGURE 12.3   TEM image and EDS spectra of ZnSe NPs synthesized in the apoferritin cavity. (a) The images are stained by 1% aulothioglucose and the cores are surrounded by white rings which are negatively stained protein shells. The arrows indicate (a) an apoferritin with a ZnSe core and (b) an apoferritin without a CdSe core. (b) High-resolution TEM images of ZnSe NPs without stain. The lattice image of ZnSe NPs shows the poly ZnSe crystal. (c) EDS spectra of synthesized ZnSe NPs. The C and Ti peaks are attributed to the TEM grid.

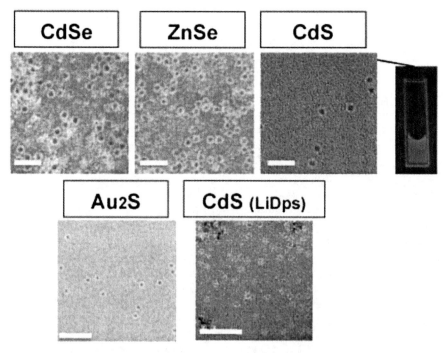

FIGURE 12.4   The TEM images of apoferritin with CdSe, ZnSe, CdS, Au_2S NPs and LiDps with CdS NPs. CdS-ferritin shows the red photoluminescence. The bars are 50 nm.

Three factors were found to be important:

(a) the threefold channel that selectively introduces Zn ions into the apoferritin cavity;

(b) the apoferritin internal potential that favors Zn ion accumulation in the cavity;

(c) the nucleation site that nucleates ZnSe inside the cavity.

It was also demonstrated that the addition sequence of Zn^{2+} and Se^{2-} ions is crucial. The addition of Se^{2-} ions after Zn^{2+} ions produced a ZnSe NP core, but the reverse sequence lowered the CFR significantly. This confirmed that Zn^{2+} ions should be concentrated in the apoferritin cavity before the synthesis reaction takes place.

The SCRY also succeeded in the synthesis of many kinds of semiconductor NPs, for example, CdS [18], ZnS, Au_2S [19], CuS [20], in the apoferritin cavity. CdS NPs can also be synthesized in the small LiDps (Fig. 12.4) using the SCRY [21]. By these works, it can be claimed that the SCRY is universal use for the synthesis of compound semiconductor NPs in the apoferritin and LiDps cavity.

These synthesized semiconductor NPs have attractive advantages. Since they have a hydrophilic protein shell, they can disperse easily in aqueous solution. They can behave as supramolecules in a living body like an enzyme or nucleic acid, and, therefore, can be used naturally in a living body. The small-size dispersion and the long fluorescence lifetime of synthesized semiconductor NPs are important for many kinds of applications. In addition, recently, it was demonstrated that the CdS NPs synthesized in the apoferritin show the circular polarized luminescence (CPL). This is the first report where CPL was observed in semiconductor NPs [22]. This CPL could not be observed in chemically synthesized CdS NPs without ferritin protein shell. We think this reason why the configuration of amino acids on the nucleation sites at the inner surface occurs to the CPL. The shell of ferritin protein plays an exciting role, not only as a size control of NPs but also in CPL.

12.1.4 NP Synthesis in the Apoferritin with the Metal-Binding Peptides

A. Belcher et al. demonstrated that the phage display method could select peptides with specific affinity against semiconductor surfaces and many such target-specific peptides are available now as a result of researchers' efforts. The target materials range from semiconductors to metal oxides. It is notable that a peptide with specific affinity against a certain inorganic material has the ability to enhance the nucleation of the material in solution.

Using this property, R. M. Kramer et al. reported the synthesis of metal NPs in apoferritin by displaying the Ag-specific peptides on the inner surface of the cavity [23]. They used genetically engineered human L-subunit ferritin (LCF) with a silver-binding peptide, AG4. They selected AG4, a dodecapeptide (Asn-Pro-Ser-Ser-Lue-Phe-Arg-Tyr-Leu-Pro-Ser-Asp), from a phage peptide display library, and showed that it can reduce the silver ions to metallic silver at neutral pH and synthesize Ag NPs. They constructed a chimera protein, that is, the AG4 peptide sequence fused to the C-terminal of L-chain ferritin sequences. This chimera protein makes 24-mer cage-shaped protein with the AG4 peptide displayed on the inner cavity surface. They incubated purified empty LCF-AG4 protein cages in the presence of 0.4 mM AgNO$_3$ for several hours at 37°C and found that Ag NP grew in the cavity. EDS and HR-TEM confirmed the single crystalline structure of Ag NPs with a diameter of 5 nm.

K. Sano et al. also reported that the Ti-binding peptide on the apoferritin could bind to the Ti-board and it also has mineralization ability of Ti, Ag, and Si as oxidized materials [24]. Many types of metal-binding peptides are innovative materials for the synthesis of NPs in the bio-templates.

Recently, M. Uchida et al. fabricated a ferritin targeting cancer cells which had a ferromagnetic NP inside. They genetically incorporated a cell-specific targeting peptide, RGD-4C peptide, which binds $\alpha_V b_3$ integrins upregulated on tumor vasculature, on the exterior surface of human H-chain ferritin [25]. This bioconjugate has multifunctions, such as cell targeting, imaging, and therapeutic agents, and this work extends the applications of protein cages with NP core into the field of nanomedicine.

These isolated peptides can be applied to other protein bio-templates, for example, rod-shaped bio-template such as TMV or virus cages. In many researches, gold-binding peptides are studied intensively. Q. Wang bound the gold-binding peptide to the CCMV virus cage and they fabricated nanoelectrode for electronic devices [26]. H. Kirimura et al. fabricated a conjugation of ferritin and Ti-binding peptides and they succeed in the selective deposition of one ferritin molecule on the Ti-board [27]. This Ti-binding peptide conjugated ferritin was used for construction of the nanoelectronic devices (see the next chapter). Since many kinds of peptides recognizing metal or other materials can be isolated from peptide libraries, these peptides will open up the possibility of synthesizing a variety of homogeneous NPs and making bioconjugates, ferritins with synthesized NPs, targeting various materials and cells. Namely, if we select a suitable peptide and make fusion-protein cages with the peptide displayed on its inner surface or outer surface, we will be able to guide the bioconjugates which have versatile NPs to the desired applications.

12.2 SITE-DIRECTED PLACEMENT OF NPs

NPs synthesized themselves have specific and attractive characteristics. However, if they were delivered to the desired position or arrayed on a substrate as designed, they will do their work much more efficiently or will do the functions which are never realized by NP in the solution. In such a sense, the delivery and fixation of NPs as designed by protein is not merely transferring NPs. It is an important part of the nanodevice fabrications.

12.2.1 Nanopositioning of Cage-Shaped Proteins

Positioning of NPs to nanoscaled sites is an important issue during nanofabrication process. The conventional approach is known by chemically modifying NPs followed by letting them adsorb to chemically pre-modified substrate. Recently, biomolecule-based NPs delivery has attracted increasing interests. For instance, DNA origami technique has been reported for its precise programmed positioning of DNA-functionalized NPs on the DNA-embedded substrate by DNA base pairing [28–35]. Protein-directed NPs delivery is another fascinating approach in which inorganic material recognizing peptides are usually genetically fused to the protein to recognize the targets and adsorb to them [27, 36, 37]. To date, a number of functional peptides have been found by phage display or other methods, which makes it possible for us to select a proper one according the targets [38]. Since a variety of inorganic NPs can be fabricated in ferritin's interior cavity as described above, further efforts have been paid to immobilize these NPs to a desired substrate to fabricate nanoscaled functional block using the protein frame. The immobilization of NP-filled ferritin can be achieved by several distinct ways such as controlling the surface potentials of protein and substrate [39] or adding the protein to a Ti substrate recognizing peptide (TBP, RKLPDA) [27]. The TBP shows strong affinity to Ti, Si, and Ag but not to Au, Cr, Pt, Sn, Zn, Cu, and Fe; and TBP genetically attached to the outer surface of L-ferritin endows the protein shell with the adsorption property to Si and Ti. In addition, a surfactant Tween-20 significantly reduces the affinity to Si and enhances the selective adsorption to Ti surface on Si substrate. To understand the mechanism of TBP's selective binding to Ti surface, K. Sano et al. created a series of alanine-substituted mutants of TBP. As a result, the initial Arg (R1), Pro (P5), and Asp (D5) were found to be essential to the adsorption. Therefore, they explained this adsorption phenomenon focusing on the electronic interactions between hydroxyl groups in the TBP and Ti surface, which consists of $-Ti-OH_2^+$ and $-Ti-O^-$ islands alternately. The positive-charged initial Arg (R1) adsorbs to the negative-charged Ti islands; and negative-charged Asp (D5) adsorbs to the positive-charged Ti islands. Also, the cis-peptide bond of Pro (P4) plays an important role to direct R1 and D5 to the same surface which makes it convenient for TBP to interact with Ti surface [40]. T. Hayashi et al. further investigated the role of surfactant Tween-20 and demonstrated that the hydrophilicity or hydrophobicity is an important factor to govern the adhesion force between TBP-fused ferritin and the substrate. Against a hydrophobic surface, the strength of the adhesion exceeds the strength of the specific binding between TBP-fused ferritin and Ti, indicating that minT1-LF might not distinguish the target when the target is mixed with hydrophobic objects. Adding surfactant such as Tween-20 to the solution can avoid such a situation [41].

12.2.2   Nanopositioning of Au NPs by Porter Proteins

The robust spherical conformation of ferritin is a superior property which plays an important role to control the internalized NP's size and shape. However, it may also limit the encapsulated NPs only to be spherical and less than 7 nm in diameter. To get free from the limit, a ferritin-based encapsulation-delivery system was established, which can accommodate and deliver NPs of various sizes and shapes. Ferritin's pH-dependent dissociation-assembly property enables us to realize the method.

It is known that ferritin dissociates into subunit dimers in an acidic environment and the subunit dimers can reassemble into the 24-mer ferritins when pH is increased

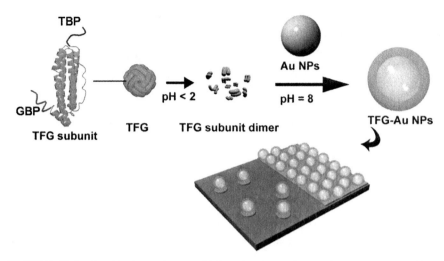

FIGURE 12.5 Graphical description of bifunctional protein-based encapsulation/delivery system for selective nanoscale positioning of targeted NPs on a certain substrate surface.

to 7 [42]. Several groups have succeeded in loading NPs into the cavity of ferritin during its reassembly process [43–46]. Based on this knowledge, a versatile bifunctional ferritin-based encapsulation-delivery system was proposed where Au NPs were selected as encapsulated NPs due to their fascinating physical properties such as surface-enhanced Raman spectroscopy (SERS) and colorimetric shift for detection application, and Ti was selected as the immobilizing site [47]. As shown in Figure 12.5, a bifunctional horse L-ferritin mutant, TFG, which is derived from horse L-chain ferritin mutant Fer-8, is engineered to contain a gold-binding peptide (GBP, MHGKTQATSGTIQS) at the C-terminus, and a TBP at the N-terminus. It is anticipated that GBP motif may bind to Au NPs for the formation of the Au-protein hybrids during the pH-dependent dissociation-reassembly process, and TBP can anchor these hybrid NPs onto TiO$_2$-patterend substrates subsequently. Therefore, TFG is also called as porter protein because it can fetch targeted NPs to the desired Ti sites.

Size exclusion chromatography (SEC) showed that TFG ferritin dissociates into subunit dimers when lowering solution pH to 2.0, and the subunit dimers can be reassembled into the 24-mer ferritins when the pH is increased to 8 again. By simply mixing such a dissociated TFG with 5-nm or 15-nm diameter Au NPs and allowing the encapsulation carry out overnight, Au-protein hybrids can be obtained after removing the excessive protein. TEM observation showed that not only 5-nm diameter Au NPs but also 15-nm diameter Au NPs are completely encapsulated by TFG protein which have a GBP but not by Fer-8 which have no GBP (Fig. 12.6). Au NPs encapsulation was also confirmed by using FG, which has only GBP but not TBP. The gold-binding peptide endows ferritin with this binding function. It is anticipated that the encapsulation takes place in a continuous mood: First GBP recognizes Au NPs and adsorbs to their surfaces. Simultaneously, self-reassembly of subunit dimers

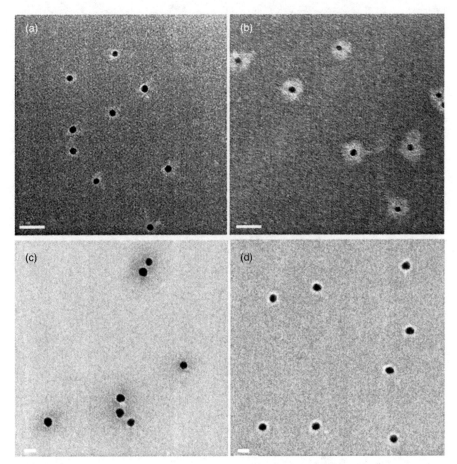

FIGURE 12.6 TEM images of Au-protein hybrids: (a) 5 nm diameter Au-Fer-8 hybrids; (b) 5 nm diameter Au-TFG hybrid; (c) 15 nm diameter Au-Fer-8 hybrids; (d) 15 nm diameter Au-TFG hybrid. All samples were stained by 3% PTA. The bars are 20 nm.

takes place, which enables porter proteins to efficiently cover Au NPs' surface to form a protein layer. For this reason, TFG can encapsulate not only 5-nm diameter Au NPs but also 15-nm diameter Au NPs. Moreover, the encapsulation of several shape-different Au NPs, such as triangular, cubic NPs, was also confirmed.

In addition to the encapsulation, selective positioning of the 15-nm diameter Au-TFG hybrid on a Ti-patterned silicon substrate was investigated because noble-metal NPs at least 10 nm in diameter are required to obtain superior plasmonic properties. Since TFG subunit dimers bind to Au NPs via GBP at C-terminus, TBP at N-terminus which is located at opposite side of the subunit dimer is expected to be displayed on the outer surface of the obtained TFG-Au-NPs conjugate, which should contribute to the adsorption property of Au-TFG hybrid to Ti or Si. Before the delivery of

FIGURE 12.7 SEM images of controlled placement of Au NPs on Ti-patterned Si substrates for large-scale devices that utilize a great number of NPs (a) and for point-contact devices that utilize a small number of NPs (b).

Au-TFG hybrid, a substrate with patterned 2-nm-thick Ti layer deposited on a thermally oxidized Si substrate was prepared by photolithography and the lift-off process. The substrate was cleaned by UV/ozone treatment prior to use and the top Ti surface was oxidized to TiOx. A mixture solution containing TFG-Au NPs, 50 mM Tris-HCl, and 150 mM NaCl was dropped to the cleaned substrate, followed by washing by MilliQ water. Field-emission scanning electron microscopy (FE-SEM) observation showed that TFG-Au NPs adsorbed to both Si and Ti surfaces, whereas the protein-free Au NPs did not adsorb to either the Si or Ti surface; this suggests that the TFG subunit endows Au NPs with a high affinity to the Si and Ti surfaces. The adsorption selectivity can be drastically improved and TFG-Au NPs adsorbed only to the Ti surface by adding 0.5 % of Tween-20 to the TFG-Au-NPs-containing mixture solution (Fig. 12.7a). Despite the high adsorption density, TFG-Au NPs remain isolated to each other on the substrate, with little aggregation. Moreover, some TFG-Au NPs immobilized on the Si surface along the Ti line edge were observed, which bound to Ti surface only via small contact. Therefore, the adsorption force of TBP is large enough to anchor the 20-nm diameter TFG-Au NPs. This selective adsorption property was also applicable to realize highly selective positioning of the internalized NPs on Ti substrates with nanometer-level accuracy. Immobilization of a single TFG-Au NP onto 20-nm-diameter Ti islands at 400-nm interval was confirmed by simply dropping the solution to substrate (Fig. 12.7b).

Because the Au NP monolayer can directly reflect the intrinsic shape-dependent plasmonic properties, the fabrication of a single layer of a discrete array of Au NPs onto solid substrates is required for LSPR-related applications, such as biosensors, biochips, colloidal substrates for SERS, and surface plasmon-enhanced fluorescence in organic light-emitting diode (OLED) [48]. The TFG-Au NPs' superior property of precisely controlled positioning as well as densely disturbing assembly, as mentioned above, is desired for the large-scale fabrication of parallel LSPR device arrays—an array consisting of a single layer of dense TFG-Au NPs which was immobilized on an ultrathin Ti layer was fabricated. The ultrathin Ti layer was thought to negligibly

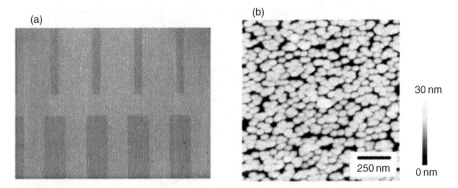

FIGURE 12.8   (a) Photograph of Ti micro-patterns (L&S: 160 μm & 720 μm and 440 μm & 440 μm) with a thickness of approximately 1 nm on a quartz substrate after TFG-Au NPs had been site-selectively absorbed. (b) AFM image of the same sample containing single layer of TFG-Au NPs array.

affect the surface plasmonic properties. The target substrate was prepared with 1-nm-thick Ti patterns which have a line-to-space distance of 160 μm/720 μm and 440 μm/440 μm on a quartz substrate, respectively. After dropping TFG-Au NPs solution followed by washing treatment, only the Ti surface area showed Au NP's plasmonic color, which clearly shows the selective formation of TFG-Au NPs layer (Fig. 12.8a). The array of as-deposited TFG-Au NPs was characterized by atomic force microscopy (AFM) and Figure 12.8b shows a typical AFM image of the TFG-Au NP array. The TFG-Au NPs were assembled into densely packed arrays containing small area with random orientation. The TFG-Au NPs density on the substrate surface reached as high as ca. 10^{11} particles/cm^2, which was evaluated by a combination of AFM observation and image analysis.

To further explore the isolation of the TFG-Au NPs in two dimensions, the plasmonic properties of the as-deposited array was characterized by UV-vis spectroscopy. The extinction spectra taken from the TFG-Au NP monolayer exhibited a single intense sharp surface plasmon band at 525 nm, which fitted well with the extinction spectrum of the original colloidal solution used for assembly. These results strongly suggest that each TFG-Au NP in the closely packed arrangement should be plasmonically insulated by the outer shell.

12.3   FABRICATION OF NANODEVICES BY THE NP AND PROTEIN CONJUGATES

Artificial synthesis and delivery of NPs by proteins, which is described above, is a highly potential method to fabricate a lot of NP-based nano functional structures. The method can be employed in a wide range of applications. One of the most promising applications is fabrication of electronic devices. Size-controlled NPs are suitable for

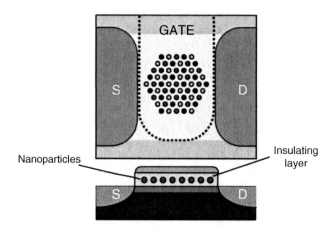

FIGURE 12.9   Schematic drawing of FNGM. NPs embedded in the insulating layer function as charge storage nodes.

charge storage nodes. This characteristic was used to fabricate floating gate memory which has a memory NPs array synthesized and produced by cage-shaped proteins. The NPs are also small enough to show ideal quantum effects and can be used to fabricate a single-electron transistor (SET), which has an ideally positioned NP as a quantum well. In this section, floating nanodot gate memory (FNGM) and SETs are described in details.

12.3.1   Fabrication of Floating Nanodot Gate Memory

FNGM attracts considerable attention as one of the emerging devices [49–52]. Structure of FNGM is schematically shown in Figure 12.9. In the conventional plate-type floating gate memory, once the insulating gate layer is damaged, the stored electrons leak to Si substrate through the damaged area. On the other hand, the distributed nanodots in the FNGM structure store the charges independently; the local charge leakage does not affect the memory function substantially. Even though some of the nanodots lose the charges, the rest of the nanodots retain the memory function. Therefore, high-density array of homogeneous nanodots or NPs is required. Ferritin molecules are suited for obtaining high-density NP array. Using the chemical flexibility of the protein shell of ferritin, high-density two-dimensional inorganic NP array is obtained [53–55]. The protein shell prevents the inner NPs from aggregating with each other. The protein shell can be removed by heat treatment so that the array of independent NPs are obtained [56]. Moreover, various kinds of NPs are synthesized so far [57], and it is possible to design the FNGM characteristics in terms of NP materials, or work function.

A. Miura et al. used ferritin molecules with Co_3O_4 cores to fabricate of MOS capacitor and MOS FET. Over the channel region, ferritin molecules with Co_3O_4 cores are adsorbed. To achieve the high-density NP adsorption, we should take into

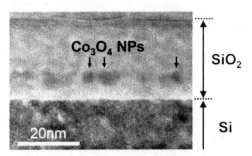

FIGURE 12.10   Cross-sectional TEM image of Co_3O_4 NPs embedded in SiO_2 layer.

account the interaction between ferritin protein and solid surface. Around neutral pH, the ferritin molecule has net negative charge, and the surface of Si substrate displays negative charges through a dissociation reaction of the surface groups. Therefore, repulsive interaction between the negative charges prevented the ferritin molecules from being adsorbed on the surface of Si substrate. To change the situation, the surface was modified to display positive charges. Surface modification of the Si substrate by a polyethyleneimine (PEI) [11] or 3-aminopropyltriethoxysilane (APTES) [58] was useful approach to introduce positive charges on the Si substrate. An Si substrate with 3 nm thermally oxidized layer was modified to display the positive charges. The solution of ferritin molecules with Co_3O_4 cores was cast on the Si substrate. Due to the attractive interaction, ferritin molecules with Co_3O_4 cores were adsorbed on the Si substrate with high density. Adsorption density of the NPs was more than $6.0 \times 10^{11} cm^{-2}$, which corresponded to 80% of the theoretically estimated maximum density [11, 59]. The use of smaller cage-shaped protein, *LiDps*, increased adsorption density of NPs. K. Yamada et al. used the LiDps molecules with Fe core to fabricate MOS capacitor and achieved the higher NP adsorption density of $1.5 \times 10^{12} cm^{-2}$ [58]. After the adsorption, UV/ozone treatment or heat treatment at 500°C under O_2 gas for protein shell elimination was carried out to suppress unwanted charge trapping at the protein shell. Complete elimination of the protein shell was confirmed by AFM, FTIR, and XPS measurements [56, 60, 61]. Control SiO_2 layer of 17 nm was deposited on the sample with Co_3O_4 NPs. Typical cross-sectional TEM image of the NP array embedded in SiO_2 layer is shown in Figure 12.10. NPs independently existed in the SiO_2 layer. Control gate electrode of Al or Ti is formed by evaporation and lift-off method. Then, the sample was annealed at 450°C for 1 h under H_2 gas diluted to 10% by N_2 gas addition. This annealing process improved Omic contact of the electrode and partially reduced the Co_3O_4 NPs embedded in the SiO_2 layer.

To evaluate the charging and discharging properties of the Co_3O_4 NPs, capacitance-voltage (C-V) measurement of the Co_3O_4 NP embedded MOS capacitor was carried out. Voltage sweeping frequency was 1 MHz. As shown in Figure 12.11, in C-V curve with small-gate voltage scanning between −3 V and 1 V, no hysteresis was observed. On the contrary, wide-gated voltage scanning between −10 V and 6 V generated a clear hysteresis. It was explained by the differences of the applied electric fields. The total electric field is determined by the total thickness of SiO_2

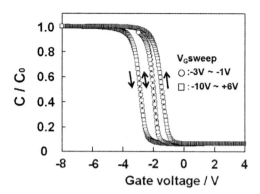

FIGURE 12.11   C-V curve of Co_3O_4 NP embedded MOS capacitor.

layer (20 mm) and effective applied voltage from flat band voltage (in this case, -2.0 V). Strengths of the electric fields were calculated to be 0.5 MV/cm and 4.0 MV/cm for narrow scanning (-3 V/-1 V) and wide scanning (-10 V/6 V), respectively. The high electric field induced tunneling of the charges to the embedded NPs and the stored charges caused the hysteresis in the C-V curve. The counterclockwise hysteresis behavior was an evidence of charging and discharging in the embedded NPs. From the C-V curve, memory window was evaluated to be 1.5 V. This memory window was maintained whenever sweeping frequency was varied. Therefore, it was indicated that the charges were not trapped at the surface levels. Annealing under the H_2 reduced the Co_3O_4 NPs. Conductive and semiconductive states were existed in the Co_3O_4 NPs. XPS analysis showed metal cobalt peak in the spectrum [59].

Current-voltage (I-V) characteristics of the Co_3O_4 NPs embedded MOS FET was investigated as shown in Figure 12.12 [11, 59]. A narrow-gate voltage sweep (-1 V/4 V) produced typical I-V curves of n-channel MOS FET. When wide-gate voltage sweep (-10 V/10 V) was applied, I-V curve showed clear hysteresis.

Endurance of the fabricated FNGM was investigated [11, 59]. Stressing was performed by applying ±8 V pulse bias with 100 ms duration. Negligible degradation

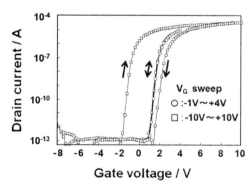

FIGURE 12.12   I-V curve of Co_3O_4 NP embedded MOS FET.

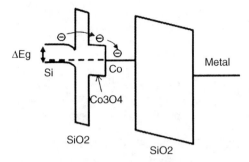

FIGURE 12.13   Schematic drawing of electron injection process to Co_3O_4 NP. Biasing metal electrode induces direct tunneling of electrons to the Co_3O_4 NP.

was confirmed up to 10^5 cycles of program and erase operations. Charge retention was investigated. Trapped charges were maintained up to 10,000 s. It is enough time for the practical use.

The Co_3O_4 NPs embedded MOS FET functions as FNGM device. The Co_3O_4 NPs trapped the charges to exhibit memory property. Charging process of electron injection to the Co_3O_4 is shown in Figure 12.13, with energy band diagram for Si substrate, tunneling SiO_2, Co_3O_4, metal-Co, control SiO_2, and gate metal [59, 62]. When the gate electrode is biased, electrons are injected to the embedded Co_3O_4 NP by direct tunneling. The injected electrons immediately fall into the potential well of metal-Co, and back-tunneling of the stored electrons is inhibited by the energy gap ΔE_{gap} between metal-Co and Si substrate. This charge confinement successfully functions for long charge retention.

12.3.2   Fabrication of Single-Electron Transistor Using Ferritin

Among the NP-based devices, SETs have been attracting much attention due to their potential for achieving extremely small size and low power consumption [63]. The operation of the SETs is based on a Coulomb blockade phenomenon in electron transportation. Technical challenges in the fabrication of SET structure are in making a Coulomb island and sandwiching it between the two tunnel junctions. So far, there are a number of reports on making SETs. Some important methods include NP deposition by chemical vapor deposition or colloidal method; oblique evaporation; pattern-dependent oxidation in Si nanostructure. However, these methods include delicate process tuning [63–69]. Sometimes the methods require successfully arranged structure. Therefore, producibility is one of the key issues in the fabrication of SET structure.

The use of apoferritin molecules gives us solutions for the above challenges [70]. Using apoferritin molecules enables us to homogeneously synthesize NPs that can work as Coulomb islands, because the size of NPs is regulated by the size of apoferritin cavity. Since various kinds of NPs have been synthesized, wide variety of materials are available for the Coulomb islands [57]. Remaining task for SET fabrication is

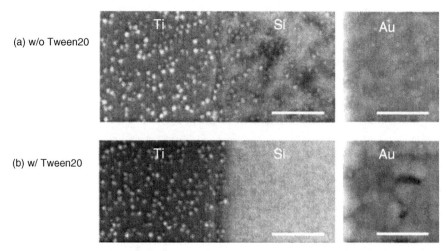

(a) w/o Tween20

(b) w/ Tween20

FIGURE 12.14 Adsorption of ferritin molecules (a) without and (b) with Tween-20 surfactant in solution. NP core of the ferritin is Co_3O_4. Scale bars are 100 nm.

to sandwich the ferritin NP by two-tunnel junction, that is, to place the ferritin NP between a pair of electrodes with keeping nanogaps. S. Kumagai and S. Yoshii et al. solved this geometric challenge by using the target-specific affinity of apoferritin molecules [70]. Details are described as follows.

The affinity of the recombinant apoferritin molecule, Fer-8, was investigated. Patterns of Ti and Au were prepared on an Si substrate. Apoferritin molecules with NP cores were adsorbed on the Si substrate (Fig. 12.14a). NPs were found on the three surfaces of Ti, Si, and Au. When surfactant of Tween-20 was added to the ferritin solution, the adsorption characteristics significantly altered (Fig. 12.14b). The ferritin molecules were found on the Ti surfaces whereas the ferritin molecules were not found on the Si surface. On the Au surface, a few ferritin molecules were not found. Because the surfactant reduces nonspecific interaction, for example, hydrophobic/hydrophilic interactions, between ferritin and solid surface, selective adsorption onto the Ti surface was realized.

Based on the above experimental results, SET structure was designed as shown in Figure 12.15a [70]. A pair of electrodes separated by a nanogap that was as large as ferritin molecule was prepared on an Si substrate. Each electrode consisted of Ti and Au layers and Au layer was stacked on the Ti layer. The Au layer played a role of conducting electrons. The Ti layer adsorbed or caught the ferritin molecules when the ferritin molecules approached the Au/Ti electrodes. Under the condition of surfactant in solution, the ferritin molecules adhered to the Ti layer in lateral direction whereas few ferritin molecules adsorbed on the Au surface. When a ferritin molecule reached to the nanogap, this molecule adhered to both the electrodes by target-specific affinity, as shown in Figure 12.15b. The process is self-aligning. The ferritin brought the inner NP core at the middle of the nanogap and prevented the NP from directly attaching to the electrode. The NP was separated by the protein shell whose thickness is exactly

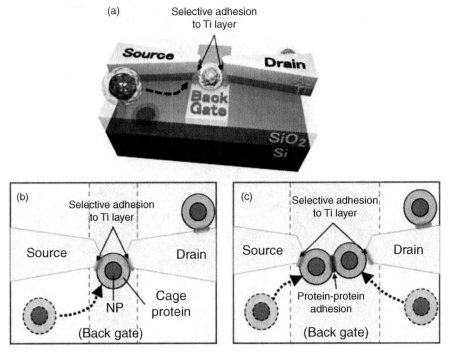

FIGURE 12.15 Schematic illustration of SET device using selective adhesion of ferritin molecules to the surface of Ti layer. (a) Bird's eye view. (b) Nanogap size is less than ferritin molecule. One ferritin molecule is trapped. (c) Nanogap size is larger than one ferritin molecule but less than two. Two ferritin molecules are trapped in the nanogap.

defined. The separation distance functioned as tunnel gaps. Changing the nanogap width allowed multiple ferritin adsorptions. A pair of electrodes with the gap size larger than one ferritin but less than two was prepared. The nanogap trapped two ferritin molecules, as shown in Figure 12.15c.

Electrodes of SET were fabricated as follows. Doped poly-Si layer was deposited on a thermally oxidized Si substrate. Gate electrode was patterned in the poly-Si layer. Surface of the gate electrode was thermally oxidized. A pair of the Au/Ti electrode was prepared on the gate electrode substrate through electron beam lithography, electron beam evaporation of Au(20 nm)/Ti(2 nm) layers, and lift-off process.

Since the target-specific affinity is short-range interaction, the ferritin molecules must approach the surface of the target materials. In the solution, the long-range interaction is electrostatic interaction. The characteristic interaction length (Debye length, λ) is given as a function of ionic strength in solution ($\lambda = \sqrt{\varepsilon\varepsilon_0 k T /2e^2 I}$, ε, relative dielectric constant; ε_0, dielectric constant in vacuum; k, Boltzmann constant; T, temperature; e, elementary charge; I, ionic strength). Debye length in pure water at neutral pH was calculated to be 1 μm. When ionic strength was adjusted to 100 mM,

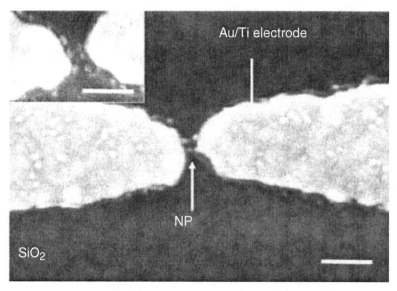

FIGURE 12.16 Ferritin (In oxide core) adsorption into the nanogap. Inset shows multiple ferritin (Co_3O_4 core) adsorption. Scale bars are 50 nm.

the Debye length was reduced to 1 nm. As a result, ferritin molecules easily reached the surface to exhibit their target-specific affinity. During a number of approach trials, one ferritin molecule was trapped between the Au/Ti electrodes. The protein shell prevented the inner NP from directly contacting with the electrodes and kept gaps. After the NP adsorption, the protein shell was eliminated by UV/Ozone treatments. The two nanogaps that were formed between the electrode and NP functioned as tunnel gaps in electron transport.

The fabricated SET structures are shown in Figure 12.16. NPs were clearly observed between the electrodes. One ferritin molecule was adsorbed at the narrowest gap. On the other hand, the nanogap as shown in the inset was larger but less than two ferritin molecules, and multiple ferritin NP adsorptions were realized.

Current-voltage (*I-V*) characteristics of the fabricated SET devices were investigated. Temperature was set to 4.2 K. Voltage was swept from −600 mV to 600 mV. Back-gate voltage was set to 0 V. One of the I-V curves is shown in Figure 12.17. With increasing bias voltage, current increased stepwise. This stepwise increase corresponds to Coulomb staircase. With increasing bias voltage between the source-drain electrodes, electron tunneled to the NP and was stored there. The stored electron inhibited the second electron incoming to the NP of the Coulomb island. With further increase in the bias voltage, the second electron overcame the repulsive potential from the first stored electron and could be transferred to the NPs. Step positions in the I-V curve was determined by the peak positions in the dI/dV curve. Intervals of the Coulomb staircase are 260 mV. From the orthodox theory ($\Delta V = e/C$, ΔV, interval of the staircase; e, elementary charge; C, capacitance)

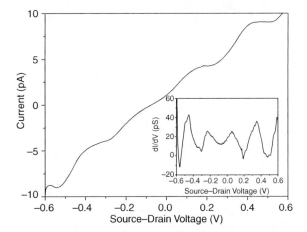

FIGURE 12.17   Current–voltage curve. Gate voltage is set to 0 V. SET device with In oxide NP. Inset is dI/dV curve.

with the measured intervals of Coulomb staircase, capacitance of the SET devices was calculated to be on the order of 10^{-19} F. Evaluating the capacitance of $\phi 7$ nm NP from $C = 4\pi\varepsilon\varepsilon_0 r$ (r, radius of NP), it was a reasonable value.

Drain current was measured as a function of gate voltage, as shown in Figure 12.18. Source-drain voltage was set 0.4 V. With increasing gate voltage, Coulomb oscillation was observed. Drain current increases and decreases periodically. The gate voltage can vary the chemical potential in the NP. The electrons are transferred to NP when the chemical potential in the NP was matched with the chemical potential of either electrode.

Nanoelectronics device of SET was demonstrated. The biological method successfully fabricated the SET structure and the SET worked perfectly. It was experimentally

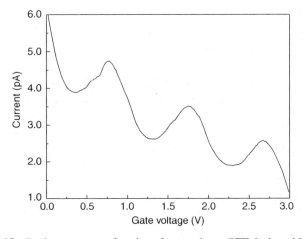

FIGURE 12.18   Drain current as a function of gate voltage. SET devise with In oxide NP.

proved that the combination of the conventional semiconductor microfabrication process and synthesizing and assembling nano components by the biological method can produce nanoelectronics devices, which was thought to be impossible before. The groundless belief that biological molecules cannot make nanostructures was broken and this biological path is now a promising way to make nanoelectronics devices.

REFERENCES

[1] Harrison, M. P.; Arosio, P. *Biochem. Biophys. Acta.* **1996**, *1275*, 161–203.

[2] Bozzi, M.; Mignogna, G.; Stefanini, S.; Barra, D.; Longhi, C.; Valenti, P.; Chiancone, E. *J. Biol. Chem.* **1997**, *272*, 3259.

[3] Douglas, T.; Young, M. *Nature* **1998**, *393*, 152–155.

[4] Heddle, J. G.; Yokoyama, T.; Yamashita, I.; Park, S. Y.; Tame, J. R. *Structure* **2006**, *14*, 925–933.

[5] Mann, S.; Williams, J. W.; Treffry, A.; Harrison, P. M. *J. Mol. Biol.* **1987**, *198*, 405–416.

[6] Meldrum, F. C.; Wade, V. J.; Nimmo, D. L.; Heywood, B. R.; Mann, S. *Nature* **1991**, *349*, 684–687.

[7] Douglas, T.; Stark, V. T. *Inorg. Chem.* **2000**, *39*, 1828–1830.

[8] Allen, M.; Willits, D.; Young, M.; Douglas, T. *Inorg. Chem.* **2003**, *42*, 6300–6305.

[9] Tsukamoto, R.; Iwahori, K.; Muraoka, M.; Yamashita, M. I. *Bull. Chem. Soc. Jpn.* **2005**, *78* (Suppl. 11), 2075–2081.

[10] Tsukamoto, R.; Muraoka, M.; Fukushige, Y.; Nakagawa, H.; Kawaguchi, T.; Nakatsuji, Y.; Yamashita, I. *Bull. Chem. Soc. Jpn.* **2008**, *81* (Suppl. 12), 1669–1674.

[11] Miura, A.; Hikono, T.; Matsukawa, T.; Yano, H.; Hatayama, T. T.; Uraoka, Y.; Fuyuki, T.; Yoshii, S.; Yamashita, I. *Jpn. J. Appl. Phys.* **2006**, *45*, L1–L3.

[12] Okuda, M.; Iwahori, K.; Yamashita, I.; Yoshimura, H. *Biotech. Bioeng* **2003**, *84*, 187–194.

[13] Kirimura, H.; Uraoka, Y.; Fuyuki, T.; Okuda, M.; Yamashita, I. *Appl. Phys. Lett.* **2005**, *86*, 262106.

[14] Okuda, M.; Kobayashi, Y.; Suzuki, K.; Sonoda, K.; Kondoh, T.; Wagawa, A.; Kondo, A.; Yoshimura, H. *Nano Lett.* **2005**, *5*, 991–993.

[15] Wong, K. K. W.; Mann, S. *Adv. Mater.* **1996**, *8*, 928–932.

[16] Yamashita, I.; Hayashi, J.; Hara, H. *Chem. Lett.* **2004**, *33*, 1158–1159.

[17] Iwahori, K.; Yoshizawa, K.; Muraoka, M.; Yamashita, I. *Inorg. Chem.* **2005**, *44*, 6393–6400.

[18] Iwahori, K.; Yamashita, I. *Nanotechnology* **2008**, *19*, 495601.

[19] Yoshizawa, K.; Iwahori, K.; Sugimoto, K.; Yamashita, I. *Chemi. Lett.* **2006**, *35* (Suppl. 10), 1192–1193.

[20] Iwahori, K.; Takagi, R.; Kishimoto, N.; Yamahsita, I. *Mater. Lett.* **2011**, *65*, 21–22.

[21] Iwahori, K.; Enomoto, T.; Furusho, H.; Miura, A.; Nishio, K.; Mishima, Y.; Yamashita, I. *Chem. Mater.* **2007**, *19*, 3105–3111.

[22] Naito, M.; Iwahori, K.; Miura, A.; Yamane, M.; Yamashita, I. *Angew. Chem. Int. Edt.* **2010**, *49*, 7006–7009.

[23] Kramer, R. M.; Li, C.; Carter, D. C.; Stone, M. O.; Naik, R. R. *J. Am. Chem. Soc.* **2004**, *126*, 13282–13286.

[24] Sano, K.; Ajima, K.; Iwahori, K.; Yudasaka, M.; Iijima, S.; Yamashita, I.; Shiba, K. *Small* **2005**, *1* (Suppl. 8–9), 826–832.

[25] Uchida, M.; Flenniken, M. L.; Allen, M.; Willits, D. A.; Croley, B. E.; Brumfield, S.; Willis, A. F.; Jackiw, L.; Jutila, M.; Young, M. J.; Douglas, T. *J. Am. Chem. Soc.* **2006**, *128*, 16626–16633.

[26] Wang, Q.; Lin, T. W.; Tang, L.; Johnson, J. E.; Finn, M. G. *Angew. Chem. Int. Edt.* **2002**, *41*, 459–462.

[27] Yamashita, I.; Kirimura, H.; Okuda, M.; Nishio, K.; Sano, K.; Shiba, K.; Hayashi, T.; Hara, M.; Mishima, Y. *Small* **2006**, *2* (Suppl. 10), 1148–1152.

[28] Zheng, J.; Constantinou, P. E.; Micheel, C.; Alivisatos, A. P.; Kiehl, R. A.; Seeman, N. C. *Nano Lett.* **2006**, *6*, 1502–1504.

[29] Sharma, J.; Chhabra, R.; Cheng, A.; Brownell, J.; Liu, Y.; Yan, H. *Science* **2009**, *323*, 112–116.

[30] Pal, S.; Deng, Z.; Ding, B.; Yan, H.; Liu, Y. *Angew. Chem. Int. Ed.* **2010**, *49*, 2700–2704.

[31] Zhao, Z.; Jacovetty, E. L.; Liu, Y.; Yan, H. *Angew. Chem. Int. Ed.* **2011**, *50*, 2041–2044.

[32] Rothemund, P. W. *Nature* **2006**, *440*, 297–302.

[33] Mirkin, C. A.; Letsinger, R. L.; Mucic, R. C.; Storhoff, J. J. *Nature* **1996**, *382*, 607–609.

[34] Nykypanchuk, D.; Maye, M. M.; van der Lelie, D.; Gang, O. *Nature* **2008**, *451*, 549–552.

[35] Park, S. Y.; Lytton-Jean, A. K.; Lee, B.; Weigand, S.; Schatz, G. C.; Mirkin, C. A. *Nature* **2008**, *451*, 553–556.

[36] Sano, K.; Sasaki, H.; Shiba, K. *J. Am. Chem. Soc.* **2006**, *128*, 1717–1722.

[37] Ishikawa, K.; Yamada, K.; Kumagai, S.; Sano, K.; Shiba, K.; Yamashita, I.; Kobayashi, M. *Appl. Phys. Express* **2008**, *1*, 034006.

[38] Shiba, K. *Curr. Opin. Biotechnol.* **2010**, *21*, 412–425.

[39] Yoshii, S.; Kumagai, S.; Nishio, K.; Kadotani, A.; Yamashita, I. *Appl. Phys. Lett.* **2009**, *95*, 133702.

[40] Sano, K.; Shiba, K. *J. Am. Chem. Soc.* **2003**, *125*, 14234–14235.

[41] Hayashi, T.; Sano, K.; Shiba, K.; Kumashiro, Y.; Iwahori, K.; Yamashita, I.; Hara, M. *Nano. Lett.* **2006**, *6*, 515–519.

[42] Webb, B.; Frame, J.; Zhao, Z.; Lee, M. L.; Watt, G. D. *Arch. Biochem. Biophys.* **1994**, *309*, 178–183.

[43] Aime, S.; Frullano, L.; Geninatti, C. S. *Angew. Chem. Int. Ed. Engl.* **2002**, *41*, 1017–1019.

[44] Simsek, E.; Kilic, M. A. *J. Magn. Magn. Mater.* **2005**, *293*, 509–513.

[45] Hennequin, B.; Turyanska, L.; Ben, T.; Beltran, A. M.; Molina, S. I.; Li, M.; Mann, S.; Patane, A.; Thomas, N. R. *Adv. Mater.* **2008**, *20*, 3592–3596.

[46] Zheng, B.; Yamashita, I.; Uenuma, M.; Iwahori, K.; Kobayashi, M.; Uraoka, Y. *Nanotechnology* **2010**, *21*, 045305.

[47] Zheng, B.; Zettsu, N.; Fukuta, M.; Uenuma, M.; Hashimoto, T.; Gamo, K.; Uraoka, Y.; Yamashita, I.; Watanabe, H. *Chem. Phys. Lett.* **2011**, *506*, 76–80.

[48] Fujiki, A.; Uemura, T.; Zettsu, N.; Akai-Kasaya, M.; Saito, A.; Kuwahara, Y. *Appl. Phys. Lett.* **2010**, *96*, 043367.

[49] Tiwari, S.; Rana, F.; Hanafi, H.; Hartstein, A.; Crabbe, E. F.; Chan, K. *Appl. Phys. Lett.* **1996**, *68*, 1377.

[50] Kohno, A.; Murakami, H.; Ikeda, M.; Miyazaki, S.; Hirose, M. *Jpn. J. Appl. Phys.* **2001**, *40*, L721–L723.

[51] Salonidou, A.; Nassiopoulou, A. G.; Giannakopoulos, K.; Travlos, A.; Ioannou-Sougleridis, V.; Tsoi, E. *Nanotechnology* **2004**, *15*, 1233–1239.

[52] Song, Y. H.; Bea, J. C.; Tanaka, T.; Koyanagi, M. *Jpn. J. Appl. Phys.* **2010**, *49*, 074201-1–074201-5.

[53] Matsui, T.; Matsukawa, N.; Iwahori, K.; Sano, K.-I.; Shiba, K.; Yamashita, I. *Jpn. J. App. Phys.* **2007**, *46*, L713–L715.

[54] Matsui, T.; Matsukawa, N.; Iwahori, K.; Sano, K.-I.; Shiba, K.; Yamashita, I. *Langmuir* **2007**, *23*, 1615–1618.

[55] Ikezoe, Y.; Kumashiro, Y.; Tamada, K.; Matsui, T.; Yamashita, I.; Shiba, K.; Hara, M. *Langmuir* **2008**, *22*, 12836–12641.

[56] Yamashita, I. *Thin Solid Films* **2001**, *393*, 12–18.

[57] Yamashita, I.; Iwahori, K.; Kumagai, S. *Biochim. Biohys. Acta* **2010**, *1800*, 846–857.

[58] Yamada, K.; Yoshii, S.; Kumagai, S.; Fujiwara, I.; Nishio, K.; Okuda, M.; Matsukawa, N.; Yamashita, I. *Jpn. J. Appl. Phys.* **2006**, *45*, 4259–4264.

[59] Miura, A.; Tsukamoto, R.; Yoshii, S.; Yamashita, I.; Uraoka, Y.; Fuyuki, T. *Nanotechnology* **2008**, *19*, 255201-1–255201-6.

[60] Hikono, T.; Uraoka, Y.; Fuyuki, T.; Yamashita, I. *Jpn. J. Appl. Phys.* **2003**, *42*, L398–L399.

[61] Yoshii, S.; Yamada, K.; Matsukawa, N.; Yamashita, I. *Jpn. J. Appl. Phys.* **2005**, *44*, 1518–1523.

[62] Miura, A.; Tanaka, R.; Uraoka, Y.; Matsukawa, N.; Yamashita, I.; Fuyuki, T. *Nanotechnology* **2009**, *20*, 125702-1–125702-9.

[63] Ono, Y.; Fujiwara, A.; Nishiguchi, K.; Inokawa, H.; Takahashi, Y. *J. Appl. Phys.* **2005**, *97*, 031101-1–031101-19.

[64] Dutta, A.; Lee, S. P.; Hayafune, Y.; Hatatani, S.; Oda, S. *Jpn. J. Appl. Phys.* **2000**, *39*, 264–267.

[65] Sato, T.; Ahmed, H.; Brown, D.; Johnson, F. G. *J. Appl. Phys.* **1997**, *82*, 696–701.

[66] Hu, S. F.; Yeh, R. L.; Liu, R. S. *J. Vac. Sci. Technol. B* **2004**, *22*, 60–64.

[67] Lee, J.-H.; Cheon, J.; Lee, S. B.; Chang, Y.-W.; Kim, S.-I.; Yoo, K.-H. *J. Appl. Phys.* **2005**, *98*, 084315-1–084315-3.

[68] Chung, S.-W.; Ginger, D. S.; Morales, M. W.; Zhang, Z.; Chandrasekhar, V.; Ratner, M. A.; Mirkin, C. A. *Small* **2005**, *1*, 64–69.

[69] Kubatkin, S. E.; Danilov, A. V.; Bogdanov, A. L.; Claeson, T. *Appl. Phys. Lett.* **1998**, *73*, 3604–3606.

[70] Kumagai, S.; Yoshii, S.; Matsukawa, N.; Nishio, K.; Tsukamoto, R.; Yamashita, I. *Appl. Phys. Lett.* **2009**, *94*, 083103.

13

ENGINEERED "CAGES" FOR DESIGN OF NANOSTRUCTURED INORGANIC MATERIALS

PATRICK B. DENNIS, JOSEPH M. SLOCIK, AND RAJESH R. NAIK

13.1 INTRODUCTION

Biology provides a wealth of biomolecular templates with diverse architectures, sizes, structures, and physiochemical properties. In nature, these are routinely used as building blocks for the synthesis of exquisitely structured inorganic materials as exemplified by the protein-directed silica frustule of diatoms and the calcium carbonate shells of mollusks [1]. Additionally, biology provides a source of many more biomolecular templates which are not associated with the synthesis of inorganic materials such as with viruses, chaperonin proteins, and high molecular weight polymeric proteins [2]. Instead, these possess an interior cavity or cage, are offered in an assortment of sizes and shapes (cylindrical vs. spherical), and are composed of defined amino acid sequences. Consequently, this diversity is particularly useful in material science for controlling different aspects of nanoparticle synthesis, imparting molecular recognition into abiotic materials, and in the assembly of ordered crystalline structures [3]. More importantly, these non-inorganic templates in nature can be genetically modified to accommodate an inorganic material or to participate in nanoparticle synthesis. In these proteins, sequences can be manipulated by a single amino acid modification or can be engineered to include a small peptide fusion at the N- or C-terminus to confer nanoparticle binding or synthesis capability. In particular, proteins with cage-like architectures are well suited for genetic modifications and the synthesis of inorganic materials given their precisely defined nanoscale dimensions.

Coordination Chemistry in Protein Cages: Principles, Design, and Applications, First Edition.
Edited by Takafumi Ueno and Yoshihito Watanabe.
© 2013 John Wiley & Sons, Inc. Published 2013 by John Wiley & Sons, Inc.

The design and synthesis of inorganic nanomaterials has greatly benefited from advances in the selection and screening of inorganic-binding peptides, the recombinant expression of protein cage systems, and the abundance of different templates. As a result, the addition of metal-binding peptides to proteins has expanded the functionality of protein cages to encompass the synthesis of atypical inorganic nanoparticle substrates, patterning of protein cages on 2D surfaces, and hybrid assembly. In the following chapter, we will describe how the incorporation of specific metal-binding peptides into discrete protein cage scaffolds and loosely defined polymeric protein cages can lead to controlled deposition of nanoparticles, control over NP sizes and crystallinity, and the fabrication of hybrid composites (Fig. 13.1). We will highlight the modification of ferritin and heat-shock proteins with metal-binding peptide fusions as model protein systems for nanoparticle synthesis and assembly. Additionally, we will describe how large polymeric proteins such as silkworm silk have been engineered to entrap and disperse a variety of inorganic nanomaterials in complex protein matrices.

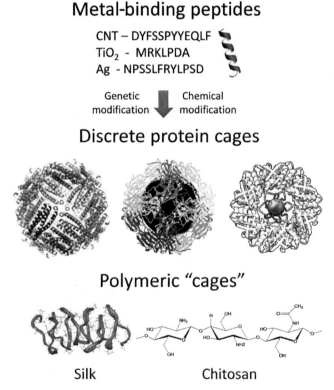

Metal-binding peptides

CNT – DYFSSPYYEQLF
TiO_2 - MRKLPDA
Ag - NPSSLFRYLPSD

Genetic modification   Chemical modification

Discrete protein cages

Polymeric "cages"

Silk   Chitosan

FIGURE 13.1   Addition of metal-binding peptides to discrete and polymeric protein cages. Discrete protein cages are represented by the ribbon structures of ferritin (right) and heat-shock proteins (middle and left). Polymeric "cages" are represented by the ribbon structure of spider silk and molecular structure of chitosan.

13.2   METAL-BINDING PEPTIDES

The interaction of peptides with inorganic materials has been used to create many different types of materials ranging from single peptide coated nanoparticles to heterogeneous multicomponent nanomaterial assemblies. Also, the properties of these materials are readily modified by peptide binding to produce an enhanced or newly emergent property. For example, the peptide functionalization of gold nanoparticles resulted in a new plasmon-induced circular dichroism peak [4] whereas peptide-coated palladium nanoparticles have been successful at modulating catalytic activity for Stille coupling [5]. In both cases, the peptides were selected to exhibit high binding affinities for gold and palladium by means of combinatorial peptide libraries. Given the importance of highly specific peptide-nanoparticle interactions, we will briefly discuss the identification of these peptide components for binding inorganic materials by using combinatorial peptide libraries, different screening procedures, and provide a few examples of well-characterized metal-binding peptides. Notably, these nanoparticle-peptide interactions have higher binding affinities than non-biological or polymer-based ligands.

   To date, the number of nanoparticle-binding peptide sequences is staggering and continues to increase with new high-affinity sequences at an expanding rate [6, 7]. This is primarily due to advancements in combinatorial peptide screening approaches and availability of high-throughput commercial peptide libraries. Generally, peptide libraries are constructed of up to 10^9 short random peptide sequences (7-mer or 12-mer peptides) displayed on the protein capsid of genetically engineered bacteriophages (phage-displayed peptide libraries) or on the surface of cells (cell-surface-displayed peptide libraries). For the identification of metal-binding peptides, the library is mixed with the target material of interest and is subjected to multiple rounds of "panning" and increasingly stringent conditions. These successive rounds of panning ensure that only peptides with the highest binding affinities remain bound to the target. In total, the peptides have been identified that bind to metals [8], carbon nanomaterials [9], metal oxides [10], metal sulfides, high-temperature ceramics, and mixed metal alloys [7].

   For gold, there have been several peptide sequences reported to exhibit high binding affinities. By comparison, each possesses inherently different amino acid sequences (i.e., AYSSGAPPMPPF [8] derived from phage display and QAT-SIGVEKLAGMAESKPTKT [11] identified from cell surface display). Compared to most gold-binding ligands or peptides which contain a thiol group and interact with gold through a covalent Au-S bond, these combinatorially selected peptides lack a cysteine residue. For these materials, the peptide-gold binding interactions have been well characterized and applied to the synthesis of peptide-coated gold nanoparticles [12]. Alternatively, Brown et al. have used a repeating polypeptide library displayed on the surface of *E. coli* bacterium to search for peptides capable of dissolving highly insoluble HoPO$_4$ and Ho$_2$O$_3$ (K_{sp} of $< 10^{-24}$) through multiple enrichment steps [10]. Brown et al. started by finding peptides that bind specifically to Ho$_2$O$_3$, and these peptides were then further enriched for their ability to release phosphate from HoPO$_4$ and at the same time transfer Ho^{3+} from Ho$_2$O$_3$ to a nitrilotriacetic acid chelator. Notably, the best Ho$_2$O$_3$ peptide dissolvers contained the recurring sequence of basic acids (Thr-Arg-Arg) which suggested that holmium ions were released upon substitution with the positively charged peptide. Ultimately, the substrate specificity of metal-binding peptides can then be programmed into a protein cage or as a fusion with polymeric proteins and can be used for synthesis, functionalization, and/or assembly with inorganic materials.

13.3 DISCRETE PROTEIN CAGES

In nature, the biological function of protein cages is to preserve and protect genetic information (i.e., RNA or DNA containing viruses) to regulate intracellular trafficking, to stabilize and refold denatured proteins (heat-shock proteins), and to serve as a storage reservoir for iron oxide particles as in the case of the ferritin family of proteins. For these reasons, protein cages exist in a variety of geometrical shapes and sizes. However, other than the iron oxide containing ferritin proteins, protein cages generally lack metal-binding specificity or intrinsic mineralization activity. In spite of this, protein cages are highly valued as platforms and containers for nanoparticle synthesis, as described in previous chapters and numerous comprehensive reviews [7]. This is because they are highly monodisperse, abundant in nature and easily attainable, possess well-defined cages and cavities, exhibit structures with addressable molecular components, and are amenable to genetic engineering. Many different proteins can therefore be genetically or chemically modified with specific metal-binding peptides at distinct locations on the protein, while still exploiting the physical size properties of the cage. As a result, peptide-modified protein cages have been used for the size-constrained synthesis of monodisperse nanoparticles, decontamination of heavy metal ions, as nanocatalysts for photoreduction, magnetic separation, and in drug delivery. For example, Naik et al. has utilized the silver-binding and mineralization activity of a phage-derived peptide (Ag4) to create a peptide ferritin fusion [13]. Here the silver-binding peptide was inserted as a C-terminal fusion in the light-chain subunit of ferritin and displayed on the inside of the self-assembled protein cage as 24 separate peptides lining the interior cavity (Fig. 13.2). Bacteria expressing the Ag4 peptide-modified ferritin demonstrated a 20% increase in viability compared to control bacteria upon exposure to high concentrations of Ag^+ ions. In an alternative approach, a carbon-affinity peptide was fused to the N-terminus of the Dps protein from *Listeria innocua* [14]. The Dps protein cage, modified with the carbon-affinity peptide, was used to decorate single-walled carbon nanotubes (SWNT), whereas the hollow unmodified cavity of the Dps cage was filled with a cobalt oxide nanoparticle core via the addition of Co^{2+} salts. Upon assembly, this created a hybrid material composed of a SWNT coated with Co-oxide-filled protein cages that showed good current-voltage properties characteristic of a memory device. Ferritin cages were modified with an amino terminal 12-mer titanium-binding peptide that was used to mineralize titania nanoparticle layers on and around ferritin, resulting in multilayer structures of protein and titania [15].

To further increase the functionality and association of proteins with different materials, the light-chain subunit of ferritin was genetically modified at both the

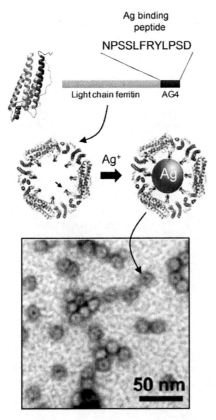

FIGURE 13.2 Constrained synthesis of Ag NPs using peptide-modified ferritin proteins. Light-chain ferritin subunit (top, left) with AG4 silver-binding peptide added to C-terminus. AG4-modified light-chain ferritin subunits self-assembled as protein cage with AG4 peptides displayed in interior cavity (arrow). Ag NPs are formed within ferritin cage by diffusion of Ag^+ ions into interior cavity and binding/nucleation at AG4 peptides. Low-voltage TEM micrograph (bottom) showing ferritin cages loaded with Ag NPs (electron-dense dark cores) and empty cages (light cores). Reprinted with permission from Reference 13. Copyright 2004 American Chemical Society.

amino- and carboxy-termini to produce a protein construct with one type of peptide displayed on the interior (gold-binding peptide) and the other (titanium-binding peptide) expressed on the exterior of the protein subunit [16]. Consequently, this bifunctional protein subunit was used to encapsulate and coat the surface of a preformed 15 nm gold particle using the affinity of the gold-binding peptide to direct reassembly of the subunits whereas the titanium peptide displayed on the exterior of the subunit was used to fabricate an array of gold nanoparticles on a Ti-patterned SiO_2 substrate. Notably, the large size of the gold NP was used to expand the interior volume of the reassembled cage from ∼6 nm of natively assembled ferritin to ∼15 nm after reassembly to accommodate the larger NP.

In addition to the metal-binding peptides displayed on the exterior surface, ferritin has also been chemically modified with DNA for the sterically controlled assembly of complementary DNA-functionalized gold NPs [17]. For DNA conjugation to the surface, Yamashita et al. mutated Ser86 to Cys86 in the light chain of ferritin whereby, upon self-assembly of the subunits, 24 thiol groups were introduced and presented around the ferritin cage for attachment to maleamide-derivatized DNA molecules consisting of 17 base pairs. The DNA-modified ferritin was subsequently decorated with 3–10 gold NPs after DNA hybridization.

Whereas metal-binding peptides and DNA impart specificity to protein cages either on the interior or exterior surface such as with ferritin, there are many more instances where unmodified apoferritin (lacking the iron core) or ferritin-like proteins have been used in place of peptide fusions to produce a variety of nanomaterials. In these systems, the interior cavity of the unmodified protein cage is lined with multiple negatively charged amino acid residues (aspartic acid, glutamic acid) which are able to bind and accumulate metal cation precursors within the cage much like Fe^{2+} ions *in vivo*. These precursors resulted in protein-encapsulated nanomaterials composed of platinum [18, 19], gold [20], CdS quantum dots [21], cobalt nickel nanoparticles [22], and Au/Pd bimetallic nanoparticles [23]. For many of these protein-encapsulated nanomaterials, enhanced properties were obtained due to protein confinement. For example, CdS quantum dots synthesized within the cavity of apoferritin exhibited circularly polarized luminescence from the formation of a chiral crystal structure, whereas 1–2 nm platinum nanoparticles encapsulated by ferritin showed HRP-like peroxidase and catalase activity. In another example, the uniformity of protein apo-ferritin cages was exploited for the formation and synthesis of monodisperse cobalt nanoparticles as catalysts for carbon nanotube growth [24]. In this case, the monodispersity of cobalt oxide catalyst and template was critical in determining the quality and size of carbon nanotubes produced. Consequently, Naik et al. used iron oxide encapsulated by a Dps protein (DNA-binding protein similar to ferritin) to catalytically produce 10 nm single-walled carbon nanotubes [25].

S-layer proteins share similar physical dimensions with ferritin, but possess alternate protein architecture and pore structure. S-layer proteins compose the cellular envelope of bacteria and archaea and function mainly in a protective capacity (bacteriophages, pH resistance, mechanical stabilization), but differ from traditional protein cages by containing a central pore instead of a fully enclosed cavity that is accessible the top and bottom sides. Also, they possess an ability to form a monomolecular 2D protein array vs. 3D template which is useful for patterning over large areas. Consequently, S-layer templates have been used to pattern silica structures over micrometer length scales [26] and to fabricate silicon nanopillar arrays by serving as an etching mask [27].

13.4   HEAT-SHOCK PROTEINS

The cost and technical limitations of nanolithography approaches aimed at producing organized arrays of functional nanoparticles at scale lengths below 100 nm is a serious impedance to these "top down" methods. This has led to the study of bacterial chaperonins and their orthologs which have shown the ability to self-organize at length scales below 100 nm, producing highly ordered arrays of nanocomposites. Toward the refinement of this "bottom up" approach, two thermoresistant protein complexes have been studied extensively for their ability to nucleate inorganic nanoparticles and to facilitate the arrangement of these nanoparticles into highly ordered arrays. One of these thermoresistant proteins, the 60 kDa beta-subunit of HSP60 (or TF55β), was isolated from the geothermal hot spring archaeon, *Sulfolobus shibatae*, whose optimal growth conditions are 83°C at a pH of 3 [28, 29]. In the presence of ATP and Mg^{2+} ions, TF55β will self-organize to form an octadecameric ring structure made up of two, stacked nonameric rings. The stacked-ring structure is 17 nm in diameter and encloses a 3 nm pore at its center [30]. Interestingly, under certain conditions, the octadecameric rings of TF55β can further assemble into nanowire filaments, suggesting a potential role for this protein as a cytoskeletal scaffold [41]. The second thermostable protein widely studied, termed Stable Protein 1 (SP1), is a 12.4 kDa protein originally isolated from the Aspen tree (*Populus tremula*) which demonstrates increased expression as a result of a number of environmental stresses, including elevated temperatures, desiccation, and high salinity [31]. SP1 has a quaternary structure that is very similar to that of TF55β in which 12 SP1 monomers self-assemble into a dodecameric multimer that comprises two hexameric rings, resulting in a stacked structure with an outer diameter of 11 nm and a central pore of 2–3 nm [32]. Thus, both the TF55β octadecamer and SP1 dodecamer can be looked at as open-ended protein cages. The TF55β and SP1 complexes have a number of properties that make them promising starting points for the creation of novel nanomaterials. First, both proteins demonstrate resistance to high temperatures, with SP1 also demonstrating resistance to the ionic detergents and the presence of organic solvents, thus allowing their use in the creation of nanomaterials synthesized under the conditions that would denature other proteins [33]. Second is their ability to spontaneously self-assemble into higher-order structures after denaturation, which can be exploited to create heterogeneous materials with complex functions [33, 34]. Third, the geometry of the TF55β octadecamer and SP1 dodecamer allows close packing that can lead to highly ordered, 2D arrangements of protein-based nanoparticles with regularly spaced pores. Last, the TF55β and SP1 monomers contain loops and termini that project into the central pore region defined by the stacked rings, allowing for facile modification of the central space to direct localization of various inorganic nanoparticle species. In fact, modifications of the monomeric subunits throughout the coding region have been employed for a number of different applications involving both thermotolerant proteins, which is discussed below.

A first example of highly ordered nanocomposite array synthesis based on the self-organization of a thermoresistant protein complex was demonstrated with the TF55β octadodecamer. TF55β monomers contain multiple sites, an apical loop as well as the amino- and carboxy-termini, which allow modification of the pore region in the assembled octadodecamer. In this study, a cysteine was inserted into the apical loop oriented toward the 3 nm pore of the stacked-ring structure, resulting in the addition of 18 thiol-containing amino acid side chains that facilitated interaction of a 5 nm gold nanoparticle with the pore region [30]. If the apical pore-facing loop was removed, leaving the inserted cysteine, the pore size was increased to 9 nm and a larger gold quantum dot of 10 nm could be bound [30]. As a proof of versatility, the 3 nm gold-binding pore of TF55β was also shown to bind 4.5 nm sized semi-conducting CdSe-ZnS core-shell quantum dots through the interaction of the pore cysteines with the ZnS shell [30]. This strategy was later expanded using a form of TF55β where the apical loop was deleted and a pore-oriented decahistidine tag was added to the amino-terminus [34]. CdSe-ZnS quantum dots were capped with dihydrolipoic acid to provide enhanced interaction capability to the histidine tag, resulting in a functionalized pore with the ability to bind larger quantum dots with high affinity thereby generating a stable, fluorescent nanocomposite. If a cysteine was added to the water-accessible apical region of the TF55β nanocomposite and functionalized with a biotin group, specific binding of the nanocomposite to streptavidin beads was achieved, demonstrating a proof-of-concept approach for future targeting applications [34].

A similar approach for the generation of ordered nanocomposites was used with SP1 in which the SP1 amino-terminus was modified by the addition of a hexahistidine repeat (6His-SP1). Due to its metal-binding properties, this repeat can be used in both the rapid purification of recombinant forms of SP1 and as a nucleation point for the synthesis or placement of nanoparticles specifically in the pore region [35]. The hexahistidine peptide at the SP1 amino-terminus was first used for the binding of preformed, 1.8 nm gold nanoparticles coupled to Ni-NTA (nitrilotriacetic acid) ligands, resulting in gold nanoparticle incorporation into the SP1 pore. Interestingly, like the TF55β octadecamers, the Au-SP1 nanoparticle complexes self-assembled into nanowire filaments which demonstrated 4 nm of separation between the pore-bound gold nanoparticles. Truncation of five amino acids after the hexahistidine tag resulted in a form of SP1 that bound the same-sized gold nanoparticles, but reduced gold nanoparticle separation in the nanowires by 0.2–0.5 nm [35], exemplifying how the primary structure of a protein cage scaffold can be altered to subtly change the dimensions of the resulting composite nanostructure. Recent work on the Au-SP1 nanocomposites has also suggested an interesting biomedical application. 6His-SP1 dodecamers were shown to insert into the lipid bilayer of neuronal cells and decrease membrane polarization by acting as passive ion channels via the pore region of SP1. This depolarization activity could be effectively blocked by the addition of the Ni-NTA-coupled gold nanoparticles to the pore, suggesting that Au-SP1 nanocomposites may be used to provide ohmic coupling of the neuronal interior to a sensing electrode [36].

TF55β and SP1 have also been employed as scaffolds for magnetically and catalytically active nanostructures. An example of this was demonstrated with the *de novo* synthesis of palladium-SP1 nanoparticles utilizing the amino-terminal hexahistidine tag as a metal-binding site for Pd^{2+} ions added by incubating the purified SP1 with Na_2PdCl_4. Once bound, the Pd^{2+} ions were reduced using dimethylamine borane, generating clusters of Pd^0 nanoparticles specifically localized to the pore region of SP1. As was observed with Au-SP1 nanocomplexes, electron microscopy of the resulting Pd-SP1 nanoparticles showed the formation of nanowires made up

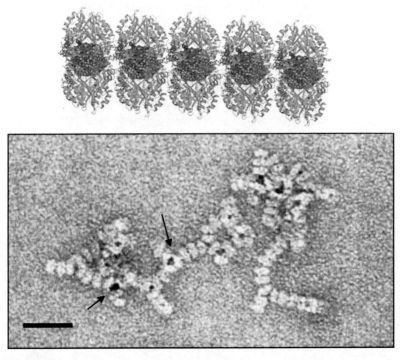

FIGURE 13.3   Synthesis of Pd NPs by SP1 protein with N-terminal hexahistidine tag. Arrows indicate Pd NPs within protein matrix. The scale bar equals 55 nm. Reprinted with permission from Reference 37.

of metalized SP1 chains (Fig. 13.3). Importantly, the Pd-SP1 nanowires were able to catalyze the reduction of 4-nitrophenol in the presence of $NaBH_4$, demonstrating the utility of SP1 as a scaffold for nanocatalysts [37]. A similar strategy for the generation of metalized nanocomposites was used with the TF55β complex. Instead of attaching a polyhistidine sequence to the amino-terminus, the TF55β apical loop was replaced by a ten-histidine stretch. As with the 6His-SP1 complex, the presence of the extra 180 histidine residues increased the binding of Pd^{2+} to the TF55β octadecamer specifically at the pore region. Bimetallic transition metal (TM) alloys of Pd-Ni and Pd-Co, with unique ferromagnetic properties, were then created using the imidazole-bound Pd^0 after reduction by dimethylamine borane. Using the self-assembling properties of the TF55β octadecamer, ordered 2D arrays of the TM bimetallic nanocomposites were created [38].

Ordered 2D arrays created from self-organizing materials with electric and magnetic properties may also be used in creating new classes of data storage devices in which the storage medium is dense, due to the nanometer length scale, but simple and inexpensive to manufacture. A number of studies have begun to look at the construction of such devices using chaperonins as templates. In one study, the amino-terminal hexahistidine tag in 6His-SP1 was replaced with a silica-binding peptide to create an Si-SP1 construct. Nanocomplexes were then formed by binding 5 nm, preformed silicon nanoparticles (with a presumed SiO_2 layer) to the silica-binding peptides at the SP1 pore. Once the silicon nanoparticles were bound, the SP1 ring structure acted as an insulator, shielding the nanoparticles from the underlying conductive surface. The result was a capacitive nanocomposite in which the individual hybrid nanoparticles could be stably and independently charged for use in binary logic functions and memory [39]. Another templating approach aimed at the organization of PbSe nanocrystals for flash memory applications involved the GroEL chaperonin. GroEL is isolated from *Escherichia coli*, and, unlike TF55β and SP1, is not thermoresistant. However, it has a quaternary structure that is similar to the thermoresistant chaperonins in that it is a homo-tetradecamer composed of two stacked rings of seven monomers each. Using commercially available wild-type GroEL, PbSe nanocrystals were trapped on ordered 2D arrays of the chaperonin. In this application, the proteinaceous template was burned away, resulting in direct annealing of the ordered PbSe nanocrystals to an SiO_2 substrate [40].

Many of the potential applications involving chaperonin nanocomposites require deposition of highly ordered composites onto a substrate. The ability of certain chaperonins to self-assemble into higher-ordered structures has proven to be a good starting point in the design of constructs and methods aimed at increasing the efficiency and ease of creating highly ordered nanocomposites. Early studies of TF55β observed that at high concentrations in the presence of Mg^{2+} and nucleotides, the octadecameric rings complexes could further spontaneously self-assemble into nanowire filaments and 2D crystals through interactions of the apical and equatorial surfaces, respectively [41]. A technique termed circular permutation was used to form mutants of TF55β in which the amino- and carboxy-termini were moved from the pore region to the apical or equatorial regions of the octadecamer. The positions of the amino- and carboxy-termini as well as incubation temperature were found to be important in whether the TF55β octadecamers self-assembled into 2D crystals or nanowire filaments. Placement of the termini at the apical region resulted in 2D crystal formation, whereas equatorial localization of the termini generated nanowire filaments or 2D arrays depending on the mutant [42]. Remarkably, the ability to self-organize into octadecamers and ordered arrays has shown significant tolerance to additional polypeptides placed at the apical region of TF55β (Fig. 13.4). For example, a mutant of TF55β self-assembled into an octadecamer with a 15 kDa cohesin sequence fused apically. Also, 2D crystal arrays have been observed to form in a TF55β mutant in which a 28 kDa yellow fluorescent protein sequence was attached to an apically localized carboxy-terminus [42].

Although a certain degree of ordering was observed with the spontaneous formation of metal-bound SP1 nanowires, high levels of SP1 ordering on 2D surfaces have largely been achieved through templating and packing techniques. One method of forming 2D SP1 arrays involves the spontaneous organization of a highly ordered, continuous SP1 monolayer on the surface of a planar phospholipid monolayer under aqueous conditions [35]. This method has also been successfully employed in the creation TF55β and GroEL 2D crystal arrays [40, 43]. However, the practicality

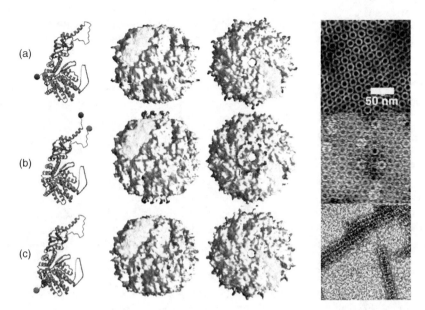

FIGURE 13.4   Circular permutation of TF55β and resulting structure. (a–c) Ribbon and space-filling structures (top and side views) of the (a) 153, (b) 267, and (c) 480 circular permutation mutants with the position of the amino-termini shown as dark dots and the carboxy-termini shown as light dots (left). TEM micrographs show the resulting structures formed by octadecamers of circular permutation mutants (right). Reprinted with permission from Reference 42. Copyright 2006 IOP Publishing Ltd.

of this method has been questioned due to potential disruption of surface contacts caused by the presence of phospholipids at the 2D protein array surface. To circumvent this limitation, an alternate technique was presented which combined forced formation of a 2D SP1 monolayer with a packing step to create a highly ordered 2D array of SP1 without the use of phospholipids. This method involved forcing the soluble SP1 dodecamer to the air/water interface by injecting it into a solution of high glucose and cadmium ion concentrations [44]. SP1 appears to be particularly suited for this approach as its inherent stability seems to protect it from denaturation as it is forced to the air/water interface. Additionally, the hexahistidine tag seems to allow for oligomerization in the presence of cadmium ions, since multimers of the dodecameric complex are observed at the air/water interface after injection into glucose/cadmium ion solution. Once an SP1 monolayer formed at the air/water interface, the dodecamers were then compressed using the Langmuir–Blodgett technique into highly ordered arrays. These arrays were then transferred directly to a mica surface for analysis [44].

One prerequisite for the ordering of nanocomposites on a surface is to confer binding affinity of the protein scaffold to that surface. This is readily achieved by substituting amino acids and/or peptide sequences into the coding region of each chaperonin monomer, and represents a powerful advantage that protein cages lend to the creation of nanocomposites. Surface-oriented amino acids in distinct areas of SP1 were identified and changed to a thiol-containing amino acid for increased binding to gold surfaces. Amino acid substitutions to cysteine were made at the inner pore and outer protein edge of the SP1 dodecamer where both substitutions increased the binding affinity of SP1 to a flat gold surface, with the cysteine substitution made at the outer protein edge, yielding a more efficient binding, most likely due to the increased accessibility of the SP1 protein edge to the gold surface [45]. SP1 fused to an amino-terminal silica-binding peptide (Si-SP1) possessed increased binding affinity to silica beads, but only in the presence of a chaotropic agent. The use of a chaotropic agent was proposed to increase the accessibility of the amino-terminal silica-binding peptide to the silica surface, and allowed the tuning of Si-SP1 affinity to the silica surface by adjusting the levels of chaotrope in the binding buffer [45]. The circular permutation technique would be applicable here as specifically positioned amino- and carboxy-termini of the chaperonin monomers would be easy to modify genetically with binding peptides directed against a target substrate material. Variants of chaperonins with differential surface-binding affinities will be important tools in the creation of complex arrays of scaffold-based nanocomposites that can be templated through self-assembly or nanolithography techniques.

13.5   POLYMERIC PROTEIN AND CARBOHYDRATE QUASI-CAGES

The protein cage systems discussed so far have involved highly structured proteins that form regularly ordered arrays around a sequestered inorganic nanoparticle. Beyond the discrete spherical protein cages with uniform sizes and fully enclosed symmetrical cavities such as ferritin and numerous viruses, there are many more quasi-cage proteins which possess a loosely defined spatial cavity or voids within the tertiary and quaternary structure that can be used to confine nanoparticle synthesis and/or encapsulate and trap inorganic materials. In these polymeric protein systems, the protein biopolymer forms multiple hydrophilic and hydrophobic domains, alternating crystalline and non-crystalline regions, and highly repeating secondary structures which provide a suitable environment for encapsulation of inorganic nanomaterials. Also, polymeric proteins can be modified with metal-binding peptides with the ability to sequester and disperse inorganic nanoparticles in a way that enhances both the properties of the protein matrix as well as the function of the nanoparticle. Here, we describe several non-cage polymeric proteins and polysaccharide matrices which have been successfully used to encapsulate and synthesize inorganic materials. Common polymeric proteins include silk fibroin and bovine serum albumin (BSA), and also encompass many more protein types which are not discussed.

Silk fibroin has enjoyed a history spanning thousands of years in the textile indus-try and still remains a sought-after commodity. Although silk-fibroin-based materials are well known for their remarkable mechanical properties, recently discovered opti-cal and biomedical properties promise significant applications of silk-based materials beyond textiles in the coming decades. Silk fibroin is a proteinaceous material which is produced in bulk from the cocoon of the silkworm, *Bombyx mori*. Silkworm silk fibers are composed of a 325 kDa heavy chain and a 25 kDa light chain in which the heavy chain is made up of numerous GAGAGS repeats. An intriguing feature of silk fibers is their quasi-ordered structure in which highly crystalline β-sheet domains are surrounded by an amorphous matrix of fibroin protein [46, 47]. For processing into films and other materials, it is common to denature silkworm cocoon fibers, creating a water-soluble form of fibroin. This form of fibroin consists of α-helices as well as random coils and has been termed silk I. Treatment of silk I preparations by dehydration generates an insoluble form of silk fibroin with high β-sheet structure, termed silk II. The transition from silk I to silk II can have a significant impact on the properties of silk-fibroin-based materials and how the materials form nanocomposites. Another type of silk studied for its materials properties is the dragline silk isolated from the orb-weaving spider, *Nephila clavipes*. Dragline silk fibers are secreted from the major ampulate gland of the spider and are largely made up of two different spidroin proteins MaSp1 and MaSp2 (for Major Ampulate Spidroin 1 and 2). MaSp 1 and 2 are made up of $(GA)_n/A_n$ repeats that make up a β-sheet structure, as well as GPGQQ repeats that form a β-spiral; and like silk worm silk, these contain crystalline β-sheet structures surrounded by non-structured regions [48]. Spider dragline silk is more mechanically robust than silkworm silk, but is only produced in small quantities in its native form. Therefore, bulk production of dragline silk has focused on recombinant and transgenic approaches. As a protein cage scaffold material, silkworm fibroin and spidroin do not provide a high degree of ultrastructure as do some of the geometrically ordered protein scaffolds (e.g., chaperonins, viruses, ferritin). However, these fibroins do provide many benefits as a scaffold for organic/inorganic nanocomposites in that they are malleable, functionalizable, biocompatible, and biodegradable. The last two features have led to a number of studies testing a role for the fibroins as starting materials for nanocomposites aimed at biomedical applications.

Gold nanoparticles have a number of features that support their use in biomedical applications in that they are stable, biocompatible, possess low toxicity, and can be functionalized with bioactive compounds—many properties that are shared with silk fibroin. One early study examined whether silk fibroin could be used to reduce gold ions to form nanoparticles [49]. Silkworm fibroin contains ∼5 mol% tyrosine, and tyrosine side chains are known gold reducing agents toward the formation of nanoparticles. Soluble, denatured silkworm fibroin was added dropwise to a solution of $HAuCl_4$ at an elevated pH. The result was a monodispersed, core-shell nanoparticle with a ∼15 nm gold core and a silk fibroin shell, resulting in a nanocomposite of ∼45 nm diameter [49]. A similar strategy was employed to create core-shell quantum dot nanoparticles with a CdSe/ZnS core surrounded by a silk fibroin shell. These fluorescent nanoparticles were monodispersed as well as stable and were used to label ovarian tumor cells with low cytotoxicity [50]. Additionally, Fe_3O_4 superparamagnetic nanoparticles were coated with a silk fibroin shell and were shown to be biocompatible with cultured cells [51]. Future studies will need to determine the advantages that silk fibroin shells confer on nanoparticles for imaging and tissue engineering applications.

It was speculated that the reduction of gold by silk fibroin in the core-shell nanoparticles resulted in a transition from an α-helical structure to a random coil whose presence was important for the formation of the fibroin shell [49]. A subsequent study examined the role of silk secondary structure in the formation of gold nanoparticles using silk fibroin films [52]. In this study, gold nanoparticles were formed on silk fibroin films either in the silk I or silk II conformation. Gold nanoparticles incorporated into silk I films were $\sim$17 nm and tended to aggregate into larger particles up to 100 nm diameter. However, silk II films generated smaller, well-dispersed gold nanoparticles of $\sim$7 nm, suggesting that the more ordered β-sheet structures in silk II facilitated the size constrain of gold nanoparticles. It was also observed that gold nanoparticle formation on silk I films stabilized the film conformation, preventing the transition of the fibroin into silk II during dehydration [52]. Together, the data indicate interplay between the organic and inorganic phases of silk nanocomposites where the quasi-ordered nature of silk fibroin participates in the scaffolding and ordering of inorganic nanoparticles which in turn play a role in the silk fibroin secondary structure.

The ability of silk fibroin to imbed functional nanoparticles has also been tested in the creation of broad-spectrum antimicrobial materials with self-sterilization properties that can be used as wound bandages, and surfaces in hospitals or water treatment filters. Work on antimicrobial nanoparticles has focused primarily on two elements: silver and titanium. Silver has been used throughout history as an antimicrobial agent and demonstrates efficacy against a broad array of bacteria with low human toxicity. One approach to create antimicrobial silver/silk nanocomposites involved dipping electrospun silk fibroin fiber mats in an $AgNO_3$ solution and forming silver nanoparticles through photoreduction, thereby generating a silver/silk nanocomposite with silver nanoparticles both in and around silk fibroin fibers [53]. In another approach, a silver-binding peptide was fused to a spidroin repeat and the resulting chimera was produced in bacteria, as demonstrated in Figure 13.5. Films of the chimera were then soaked in an $AgNO_3$ solution, which provided a nucleation point for the silver ions, and the resulting film demonstrated an ability to inhibit microbial growth [54]. Titanium possesses photocatalytic activity in its oxide form that can also be exploited for antimicrobial applications. Preformed TiO_2 nanoparticles (30–50 nm in size) have been added to aqueous preparations of denatured silk fibroin and cast into films [55]. It was observed that the addition of TiO_2 nanoparticles to the silk fibroin led to a secondary structure transition from random coil to β-sheet with an accompanying increase in film crystallinity and tensile strength as well as a decrease in film solubility. The TiO_2/silk nanocomposite films also demonstrated biocidal activity against three common species of bacteria. Titration studies determined that the optimal blend of TiO_2 in the silk fibroin films was 0.1% (w/w), with higher levels of dispersed TiO_2 nanoparticles resulting in decreased tensile strength and bacteriocidal activity [55]. Titanium oxide nanoparticles have also been formed *in situ* during casting of the silk fibroin nanocomposite [56]. This method utilized the sol-gel process and butyl titanate as a precursor. Tests of this material indicated well-dispersed TiO_2 nanoparticles of $\sim$80 nm and a concomitant decrease in film solubility. Mechanical strength of the silk/TiO_2 nanocomposites increased with increasing TiO_2 content and reached

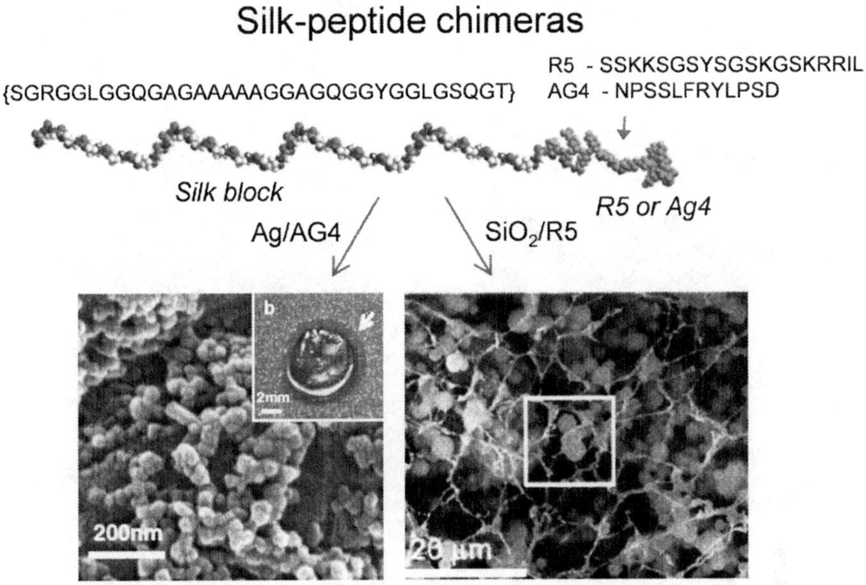

Silk-peptide chimeras

R5 - SSKKSGSYSGSKGSKRRIL

{SGRGGLGGQGAGAAAAAGGAGQGGYGGLGSQGT}   AG4 - NPSSLFRYLPSD

FIGURE 13.5 Silk-peptide chimeras for the synthesis of SiO_2 and Ag NPs. Space-filling model of silk-peptide fusion with either R5 or AG4 sequence added at C-terminus. (a) SEM image of Ag NPs produced by silk-AG4 chimera (inset shows optical image of Ag-loaded silk films). (b) SEM image of SiO_2 NPs encapsulated in silk-R5 chimera. SEM images are reprinted with permission from References 54 and 55.

a maximum at 0.4% TiO_2. X-ray diffraction analysis of silk fibroin films containing the *de novo* synthesized TiO_2 nanoparticles indicated an incomplete transition from silk I to silk II, as was observed with the addition of preformed TiO_2 nanoparticles. As concentrations of the newly formed TiO2 nanoparticles increased over 0.4%, a decrease in the amount of β-sheet structure was observed that was accompanied by a loss of mechanical strength and an increase in film solubility [56]. Thus, the addition of an inorganic component to the denatured silk fibroin preparations plays a dual role by increasing the functionality of the films as well as by increasing their mechanical strength. The latter role is an important aspect of the functional interaction between the silk fibroin and the organic phase, particularly in the areas of application where the mechanical parameters of the nanocomposite are critical.

The strength and biocompatibility of silk has led to the studies into silk fibroin as a scaffold for tissue engineering. For this application, the mechanical properties of the scaffold are important, so research has focused on enhancing the strength of silk-based films with inorganic nanoparticles. Clay platelets have attracted attention due to their high mechanical strengths and surface chemistries. Montmorillonite (MMT) is a layered aluminosilicate containing 1 nm thick platelets of 30–300 nm diameter and an anionic surface charge. A single-step fabrication protocol was created that involved a mixture of denatured, aqueous silk, MMT, and glutaraldehyde as a protein

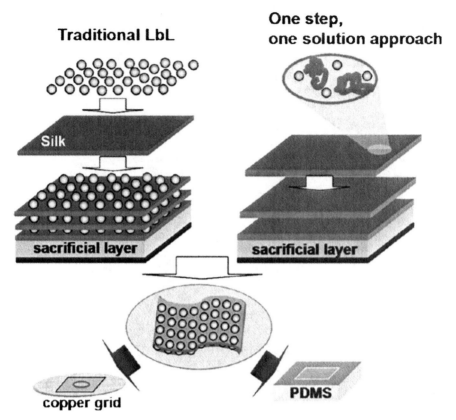

FIGURE 13.6   Layer-by-Layer (LBL) approaches for the fabrication of MMT- or POSS-reinforced silk nanocomposites. Traditional LBL assembly (left) and a one-step LBL approach (right). Reprinted with permission from Reference 57. Copyright 2010 American Chemical Society.

cross-linking agent that could be cast into thin films (Fig. 13.6) [57]. Mechanical tests of the silk fibroin films showed a 50% increase in Young's modulus and a threefold increase in toughness just from the glutaraldehyde cross-linking. This was consistent with structural studies that indicated glutaraldehyde cross-linking induced a transition from silk I to silk II. When MMT was added, the cross-linked silk/MMT films demonstrated well-dispersed MMT platelets and a high surface roughness due to poor orientation of the platelets. A further 3.7-fold increase in Young's modulus was observed in these silk fibroin/MMT films with no added benefit to toughness [57]. In addition to MMT, a synthetic polyhedral oligomeric silsesquioxane (POSS) nanoparticle was added to the cross-linked silk fibroin [57]. POSS nanoparticles are hybrids of silica and silicone, so they are rigid, but have organic side chains that can increase compatibility with the interfaces found in protein matrices. The 2 nm POSS nanoparticles were well dispersed in cross-linked silk fibroin films and demonstrated low surface roughness in contrast to what was observed with MMT. Young's modulus of the silk fibroin/POSS films increased ~2.7-fold over that of the cross-linked silk fibroin with a 1.4-fold increase in toughness also being observed [57].

As an example of how the chemical make-up of the protein matrix can have an effect on the organization of the inorganic phase in the nanocomposite, one study looked at the ability of a silk-elastin-based fusion protein to disperse MMT platelets [58]. Elastin is a structural protein found in vascular walls and skin and may have uses as a protein scaffold for tissue engineering. An artificial protein made up of silk repeats (GAGAGS), elastin repeats (GVGVP), and cationized elastin repeats (GKGVP) was created and termed SELP (for Silk and Elastin-Like Protein) and expressed recombinantly in bacteria. MMT platelets were found to disperse extremely well in the films made up of the SELP matrix with high alignment of the platelets as well as good local and global dispersal being observed. Mechanical studies of the SELP films indicated a Young's modulus of 2 GPa without addition of MMT; however, with the addition of MMT, Young's modulus of the nanocomposite films increased about 50%, rivaling that of denatured silk fibroin films [58]. Unlike what was observed in silk fibroin films, addition of the inorganic phase to the SELP matrix did not induce secondary structure in the resulting nanocomposite. This was speculated to be due to the suppression of secondary structure caused by the elastin repeats. To test the role of the positively charged lysines in the SELP protein, the lysine amino head groups were succinylated to reverse the head group charge to negative. When MMT was dispersed in the films made of the succinylated SELP (SELP$_{succ}$), local exfoliation and dispersion of MMT was observed. However, inhomogeneity was observed globally in the SELP$_{succ}$/MMT films with agglomerization of the MMT being observed in the regions of the film when viewed at low magnification [58]. These findings suggest that a protein matrix can be parsed into the regions that bind the inorganic nanoparticles and lead to local dispersion and the regions that are involved in long-range uniformity of the inorganic phase.

Chitosan is yet another biopolymer that forms a polymeric matrix for dispersion and entrapment of inorganic materials. Chitosan is the deacetylation product of chitin and is even more abundant than silk with sources derived from insects, fungi, and crustaceans (see molecular structure in Fig. 13.1). Much like silk, this simple biopolymer exhibits excellent biocompatibility, slow biodegradation, and high mechanical strength [59]. Consequently, chitosan has been used as a matrix to encapsulate optically responsive gold nanorods [59] and for the *in situ* synthesis of gold [60] and iron oxide nanoparticles [61]. Pini et al. fabricated 3.5% chitosan films doped with gold nanorods (~59 × 15 nm) by film casting in order to promote bioadhesion with arterial tissue upon laser light activation. Consequently, they demonstrated that irradiation of the chitosan-Au films using an 800 nm laser resulted in homogeneous adhesion of film and tissue by transmission electron microscopy (TEM). Mechanistically, as the nanorods are irradiated, they cause local heating within the chitosan matrix and induce the breakage of intermolecular and intramolecular H-bonds. These interchain dissociated H-bonds are then available to interact with the polar groups of the tissue surface, thereby promoting adhesion [59]. Chitosan has also been successfully used for the *in situ* synthesis of gold and iron oxide nanoparticles based on the coordination of metal ions with the amine and hydroxyl groups on chitosan. At 2 wt% chitosan, magnetite nanoparticles were formed by the simultaneous coprecipitation of iron oxide and gelation of chitosan in alkali media. Notably, the resulting ferrogels featured an increased viscoelastic modulus due to reinforcement of chitosan by magnetite nanoparticles [61]. Similarly, gold nanoparticles were synthesized in the presence of chitosan using an acidic solution of $HAuCl_4$ and at 40°C or 80°C. This resulted in the gelation of chitosan hydrogels and dispersion of NPs which exhibited increased activity in the catalytic reduction of *p*-nitrophenol [60].

Polymeric proteins can be molded and configured into micron-sized 3D shapes and structures with nanometer-scale precision and can also be used to template inorganic structures of the same shape. For example, Kaer et al. have demonstrated the fabrication of a 3D BSA protein hydrogel microstructure by using mask-directed multiphoton lithography to control protein cross-linking [62]. The resulting cross-linked BSA hydrogels were then used to direct silica condensation throughout the protein hydrogel template via flocculation of silica nanoparticles. After silica formation, exact silica replicas of the 3D protein scaffold were artificially created with nanoscale features of the protein bioscaffold preserved within the inorganic matrix. Consequently, silica biostructures were created to mimic the structures of diatom and radiolarian frustules. Ultimately, the combination of direct-write lithography techniques and the biomineralization activity of proteins allow the type, geometry, and composition of desired 3D structures to be programmed and defined by the user. This represents a step closer toward mimicking the complex hierarchical structures created in nature without much effort and from a few simple protein building blocks, and also makes it possible to generate new structures not created by nature.

13.6 SUMMARY AND PERSPECTIVES

The introduction of specific metal-binding peptides has greatly expanded the scope and functionality of protein cages for nanoparticle synthesis, while also aiding in the directed assembly of multicomponent nanomaterials (e.g., SWNTs and Co oxide NPs) and enabling multidimensional patterning on inorganic surfaces. Additionally, these materials have shown enhanced properties upon protein confinement in terms of optical, mechanical, electrical, and catalytic ability, which has led to promising new applications and significant gains in performance. For example, highly ordered 2D arrays of PbSe created by the GroEL chaperonin protein showed increased performance in flash memory applications because of the ability of the protein to self-organize into arrays and stabilize PbSe NPs. Also, substantial improvements in antimicrobial activity were achieved with numerous Ag-protein films. In total, these materials highlight the important role of the engineered protein template.

Biopolymers can also be used in the entrapment of inorganic nanomaterials. Either using the native biopolymer or by inserting metal- or metal-oxide-binding peptides, nanoparticles can be entrapped within the biological matrix. These hybrid polymers may find uses in biomedical and other advanced material applications.

REFERENCES

[1] Mann, S. *Biomineralization Principles and Concepts in Bioinorganic Materials Chemistry*; Oxford Univeristy Press: New York, 2001.

[2] Soto, C. M.; Ratna, B. R. *Curr. Opinion Biotechnol.* **2010**, *21*, 426–438.

[3] Jones, M. R.; Osberg, K. D.; Macfarlane, R. J.; Langille, M. R.; Mirkin, C. A. *Chem. Rev.* **2011**, *111*, 3736–3827.

[4] Slocik, J. M.; Govorov, A. O.; Naik, R. R. *Nano Lett.* **2011**, *11*, 701–705.

[5] Coppage, R.; Slocik, J. M.; Sethi, M.; Pacardo, D. B.; Naik, R. R.; Knecht, M. R. *Angew. Chem. Int. Ed.* **2010**, *49*, 3767–3770.

[6] Chen, C.-L.; Rosi, N. L. *Angew. Chem. Int. Ed.* **2010**, *49*, 1924–1942.

[7] Dickerson, M. B.; Sandhage, K. H.; Naik, R. R. *Chem. Rev.* **2008**, *108*, 4935–4978.

[8] Naik, R. R.; Stringer, S. J.; Agarwal, G.; Jones, S. E.; Stone, M. O. *Nat. Mater.* **2002**, 169–172.

[9] Brown, S.; Jesperson, T. S.; Nygard, J. *Small* **2008**, *4*, 416–420.

[10] Brown, S.; Mathiasen, T. *Adv. Mater.* **2011**, *23*, 132–135.

[11] Brown, S. *Nat. Biotechnol.* **1997**, *15*, 269–272.

[12] Slocik, J. M.; Naik, R. R. *Chem. Soc. Rev.* **2010**, *39*, 3454–3463.

[13] Kramer, R. M.; Li, C.; Carter, D. C.; Stone, M. O.; Naik, R. R. *J. Am. Chem. Soc.* **2004**, *126*, 13282–13286.

[14] Kobayashi, M.; Kumagai, S.; Zheng, B.; Uraoka, Y.; Douglas, T.; Yamashita, I. *Chem. Commun.* **2011**, *47*, 3475–3477.

[15] Sano, K. I.; Shiba, K. *MRS Bull.* **2008**, *33*, 524–529.

[16] Zheng, B.; Zettsu, N.; Fukuta, M.; Uenuma, M.; Hashimoto, T.; Gamo, K.; Uraoka, Y.; Yamashita, I.; Watanabe, Y. *Chem. Phys. Lett.* **2011**, *1–3*, 76–80.

[17] Zheng, B.; Uenuma, M.; Iwahori, K.; Okamoto, N.; Naito, M.; Ishikawa, Y.; Uraoka, Y.; Yamashita, I. *Nanotechnology* **2011**, *22*, 275312.

[18] Deng, Q. Y.; Yang, B.; Wang, J. F.; Whiteley, C. G.; Wang, X. N. *Biotechnol. Lett.* **2009**, *31*, 1505–1509.

[19] Fan, J.; Yin, J.-J.; Wu, X.; Hu, Y.; Ferrari, M.; Anderson, G. J.; Wei, J.; Zhao, Y.; Nie, G. *Biomaterials* **2011**, *32*, 1611–1618.

[20] Fan, R.; Chew, S. W.; Cheong, V. V.; Orner, B. P. *Small* **2010**, *6*, 1483–1487.

[21] Naito, M.; Iwahori, K.; Miura, A.; Yamane, M.; Yamashita, I. *Angew. Chem. Int. Ed.* **2010**, *49*, 7006–7009.

[22] Galvez, N.; Valero, E.; Ceolin, M.; Trasobares, S.; Lopez-Haro, M.; Calvino, J. J.; Dominguez-Vera, J.M. *Inorg. Chem.* **2010**, *49*, 1705–1711.

[23] Suzuki, M.; Abe, M.; Ueno, T.; Abe, S.; Goto, T.; Toda, Y.; Akita, T.; Yamada, Y.; Watanabe, Y. *Chem. Commun.* **2009**, 4871–4873.

[24] Jeong, G. H.; Yamazaki, A.; Suzuki, S.; Yoshimura, H.; Kobayashi, Y.; Homma, Y. *J. Am. Chem. Soc.* **2005**, *127*, 8238–8239.

[25] Kramer, R. M.; Sowards, L. A.; Pender, M. J.; Stone, M. O.; Naik, R. R. *Langmuir* **2005**, *21*, 8466–8470.

[26] Gobel, C.; Schuster, B.; Baurecht, D.; Sleytr, U.; Pum, D. *Coll. Surf. B: Biointerfaces* **2010**, *75*, 565–572.

[27] Mark, S. S.; Bergkvist, M.; Bhatnagar, P.; Welch, C.; Goodyear, A. L.; Yang, X.; Angert, E. R.; Batt, C. A. *Coll. Surf. B: Biointerfaces* **2007**, *57*, 161–173.

[28] Kagawa, H. K.; Osipiuk, J.; Maltsev, N.; Overbeek, R.; Qualite-Randall, E.; Joachimiak, A.; Trent, J. D. *J. Molec. Biol.* **1995**, *253*, 712–725.

[29] Grogan, D.; Palm, P.; Zillig, W. *Arch. Microbiol.* **1990**, *154*, 594–599.

[30] McMillan, R. A.; Paavola, C. D.; Howard, J.; Chan, S. L.; Zaluzec, N. J.; Trent, J. D. *Nat. Mater.* **2002**, *1*, 247–252.

[31] Wang, W.-X.; Pelah, D.; Alergand, T.; Shoseyov, O.; Altman, A. *Plant Physiol.* **2002**, *130*, 865–875.

[32] Dgany, O.; Gonzalez, A.; Sofer, O.; Wang, W.; Zolotnitsky, G.; Wolf, A.; Shohman, Y.; Altman, A.; Wolf, S. G.; Shoseyov, O.; Almog, O. *J. Biol. Chem.* **2004**, *279*, 51516–51523.

[33] Wang, W.-X.; Dgany, O.; Wolf, S. G.; Levy, I.; Algom, R.; Pouny, Y.; Wolf, A.; Marton, I.; Altman, A.; Shoseyov, O. *Biotechnol. Bioeng.* **2006**, *95*, 161–168.

[34] Xie, H.; Li, Y.-F.; Kagawa, H. K.; Trent, J. D.; Mudalige, K.; Cotlet, M.; Swanson, B. I. *Small* **2009**, *5*, 1036–1042.

[35] Medalsy, I.; Dgany, O.; Sowwan, M.; Cohen, H.; Yukashevska, A.; Wolf, S. G.; Wolf, A.; Koster, A.; Almog, O.; Marton, I.; Pouny, Y.; Altman, A.; Shoseyov, O.; Porath, D. *Nano Lett.* **2008**, *8*, 473–477.

[36] Khoutorsky, A.; Heyman, A.; Shoseyov, O.; Spira, M. E. *Nano Lett.* **2011**, *11*, 2901–2904.

[37] Behrens, S.; Heyman, A.; Maul, R.; Essig, S.; Steigerwald, S.; Quintilla, A.; Wenzel, W.; Burck, J.; Dgany, O.; Shoseyov, O. *Adv. Mater.* **2009**, *21*, 3515–3519.

[38] McMillan, R. A.; Howard, J.; Zaluzec, N. J.; Kagawa, H. K.; Mogul, R.; Li, Y.-F.; Paavola, C. D.; Trent, J. D. *J. Am. Chem. Soc.* **2005**, *127*, 2800–2801.

[39] Medalsy, I.; Klein, M.; Heyman, A.; Shoseyov, O.; Remacle, F.; Levine, R. D.; Porath, D. *Nat. Nanotechnol.* **2010**, *5*, 451–457.

[40] Tang, S.; Mao, C. B.; Liu, Y. R.; Kelly, D. Q.; Banerjee, S. K. *IEEE Trans. Electron Devices* **2007**, *54*, 433–438.

[41] Trent, J. D.; Kagawa, H. K.; Yaoi, T.; Zaluzec, N. J. *Proc. Natl. Acad. Sci.* **1997**, *94*, 5383–5388.

[42] Paavola, C. D.; Chan, S. L.; Li, Y.; Mazzarella, K. M.; McMillan, R. A.; Trent, J. D. *Nanotechnology* **2006**, *17*, 1171–1176.

[43] Ellis, M. J.; Knapp, S.; Koeck, P. J. B.; Fakoor-Biniaz, Z.; Ladenstein, R.; Hebert, H. *J. Struct. Biol.* **1998**, *123*, 30–36.

[44] Heyman, A.; Medalsy, I.; Dgany, O.; Porath, D.; Markovich, G.; Shoseyov, O. *Langmuir* **2009**, *25*, 5226–5229.

[45] Heyman, A.; Medalsy, I.; Or, O. B.; Dgany, O.; Gottlieb, M.; Porath, D.; Shoseyov, O. *Angew. Chem. Int. Ed.* **2009**, *48*, 9290–9294.

[46] Lotz, B.; Colonna, C. F. *Biochimie* **1979**, *61*, 205–214.

[47] Zhou, C. Z.; Confalonieri, F.; Jacquet, M.; Perasso, R.; Li, Z. G.; Janin, J. *Proteins* **2001**, *44*, 119–122.

[48] Lewis, R. V. *Chem. Rev.* **2006**, *106*, 3762–3774.

[49] Zhou, Y.; Chen, W.; Itoh, H.; Naka, K.; Ni, Q.; Yamane, H.; Chujo, Y. *Chem. Commun.* **2001**, 2518–2519.

[50] Nathwani, B. B.; Jaffari, M.; Juriani, A. R.; Mathur, A. B.; Meissner, K. E. *IEEE Trans. Nanobiosci.* **2009**, *8*, 72–77.

[51] Yin, G.; Huang, Z.; Deng, M.; Zeng, J.; Gu, J. *J. Coll. Inter. Sci.* **2011**, *363*, 393–402.

[52] Kharlampieva, E.; Zimnitsky, D.; Gupta, M.; Bergman, K. N.; Kaplan, D. L.; Naik, R. R.; Tsukruk, V. V. *Chem. Mater.* **2009**, *21*, 2696–2704.

[53] Kang, M.; Jung, R.; Kim, H.-S.; Youk, J. H.; Jin, H.-J. *J. Nanosci. Nanotechnol.* **2007**, *7*, 3888–3891.

[54] Currie, H. A.; Deschaume, O.; Naik, R. R.; Perry, C. C.; Kaplan, D. L. *Adv. Funct. Mater.* **2011**, *21*, 2889–2895.

[55] Xia, Y.; Gao, G.; Li, Y. *J. Biomed. Mater. Res Part B: Appl. Biomater.* **2009**, *90B*, 653–658.

[56] Feng, X.-X.; Zhang, L.-L.; Chen, J.-Y.; Guo, Y.-H.; Zhang, H.-P.; Jia, C.-I. *Intern. J. Biol. Macromol.* **2007**, *40*, 105–111.

[57] Kharlampieva, E.; Kozlovskaya, V.; Wallet, B.; Shevchenko, V. V.; Naik, R. R.; Vaia, R.; Kaplan, D. L.; Tsukruk, V. V. *ACS Nano* **2010**, *4*, 7053–7063.

[58] Drummy, L. F.; Koerner, H.; Phillips, D. M.; McAuliffe, J. C.; Kumar, M.; Farmer, B. L.; Vaia, R.; Naik, R. R. *Mater. Sci. Engineer. C-Biomim. Supramolec. Sys.* **2009**, *29*, 1266–1272.

[59] Matteini, P.; Ratto, F.; Rossi, F.; Centi, S.; Dei, L.; Pini, R. *Adv. Mater.* **2010**, *22*, 4313–4316.

[60] Hortiguela, M. J.; Aranaz, I.; Gutierrez, M. C.; Ferrer, L.; Monte, F. D. *Biomacromolecules* **2011**, *12*, 179–186.

[61] Hernandez, R.; Zamora-Mora, V.; Sibaja-Ballestero, M.; Vega-Baudrit, J.; Lopez, D. *J. Coll. Inter. Sci.* **2009**, *339*, 53–59.

[62] Khripin, C. Y.; Pristinski, D.; Dunphy, D. R.; Brinker, C. J.; Kaehr, B. *ACS Nano* **2011**, *5*, 401–1409.

PART VI

COORDINATION CHEMISTRY INSPIRED BY PROTEIN CAGES

14

METAL–ORGANIC CAGED ASSEMBLIES

Sota Sato and Makoto Fujita

14.1 INTRODUCTION

Self-assembly in nature produces huge, exquisite hollow structures like protein cages [1] and virus capsids [2] with a variety of biological functions such as recognition, storage, and organic/inorganic synthesis. Chemists have been inspired by the sophisticated biological systems with well-defined structures and have achieved to construct huge hollow structures self-assembled from a well-designed set of small subunits and to realize the analogous functions. Beyond the conventional synthetic compounds, these artificial cages will expand the chemistry where nanometer-sized scale exclusively shows characteristic structures, physical and chemical properties, and reactions.

Ferritin is a hollow, spherical protein cage with iron-storage functions to regulate the concentration of iron ions in living organisms [3, 4]. On the one hand, iron is an essential element of living system contributing to many metabolic processes; on the other hand, the control on the concentration of iron ion is necessary to avoid excess amount of it, which follows neurodegenerative diseases and generation of free radicals to make damages on proteins and nucleic acids [5, 6]. The cage structure is self-assembled from 24 polypeptide subunits with an outer diameter of 12 nm and an inner diameter of 8 nm, and the molecular weight of the cage is 450 kDa. The structure has been unambiguously revealed by X-ray crystallographic studies, which revealed distinctive structures to play a role in the iron-storage function of ferritin: two types of channels, threefold hydrophilic channel and fourfold hydrophobic channel,

Coordination Chemistry in Protein Cages: Principles, Design, and Applications, First Edition.
Edited by Takafumi Ueno and Yoshihito Watanabe.
© 2013 John Wiley & Sons, Inc. Published 2013 by John Wiley & Sons, Inc.

are constructed where subunits intersect each other [1, 7]. The threefold channel is a pathway to transport Fe(II) ions, and the amino acid residues on the internal surface of ferritin trap the ions. Finally, the Fe(II) ions are oxidized to Fe(III) ions, and up to 4500 iron atoms are stored as microcrystalline hydrous ferric oxide clusters ($5Fe_2O_3 \bullet 9H_2O$).

More extended examples of the strictly controlled cage like a protein cage, ferritin, self-assembled from just 24 polypeptide subunits are spherical virus capsids self-assembled from just 60 T protein subunits, where T is a mathematically defined triangulation number, and $T = 1, 3, 4, 7, 13$, and 16 are naturally occurring values. Virus is an infectious agent with 20–100 nm size in diameter and replicates themselves by infection to living organisms using DNA or RNA stored inside the hollow. One of the most important milestones in the earliest days to reveal the structures of virus is the structural hypothesis proposed by Crick and Watson in 1956 [8]. They assumed that virus is constructed from the self-assembly of a single, short protein to save genetic information and that the largest regular polyhedron, regular icosahedron, composed of 20 faces of regular triangles, should be the basic framework of virus. They also predicted the total numbers of subunits to be 60, because each triangular face of C_3 symmetry will be constructed from three P_1 symmetric protein subunits, and 20 faces consist of the whole shell structure of a capsid. In 1962, Casper and Klug generalized the hypothesis and proposed that the number of protein subunits is defined as 60 T to construct closed, discrete shell with icosahedral symmetry, assuming that each triangular face contains 3 T subunits ($T = h^2 + hk + k^2$, h and k: integers) [9, 10].

Experimentally, the three-dimensional structures of capsids were revealed by X-ray crystallographic analysis and cryo-TEM studies with an image reconstruction technique, and their geometrically restricted structures matched the rule of "number of subunits = 60 T": for example, satellite tobacco necrosis virus with 60 subunits ($T = 1$) [11], Cowpea chlorotic mottle virus (CCMV) with 180 subunits ($T = 3$) [12], and bluetongue virus with 780 subunits ($T = 13$) [13, 14]. Furthermore, multilayered capsids have been reported, for example, an Orthoreovirus, a member of the Reoviridae virus family, has a double capsid structure with an outer capsid ($T = 13$) and an inner capsid ($T = 2$) [15]. Although these virus capsids may seem to be rigid, some of the capsids show dynamic structural change depending on the external environment [16]. CCMV capsid is known to be in a native, closed-state at acidic conditions where acidic residues are protonated and to be in a swollen, open-state at neutral conditions, where Ca^{2+} ions are trapped by the acidic residues, and these structural changes are reversible.

Quite recently, an extremely huge self-assembled cage in nature was structurally determined. The huge protein, vault, consists of 78 protein subunits with the length of 67 nm and the diameter of 40 nm and has a huge hollow protected by strong protein wall with the thickness of 2–3 nm [17]. Although the function of vault has not been revealed clearly, nature indicates possibilities to construct huge self-assembled cages even beyond virus capsids.

The capsular structures of protein cages and virus are attractive skeletons for the applications to develop new functions beyond their inherent biological roles. A protein cage, ferritin, is one of the most promising candidates because the skeleton of ferritin is stable against a wide range of pH and temperature, and the mutation methods have been well established. The intrinsic function of ferritin to store iron(III) oxide was used to prepare size-controlled metal oxide or sulfide [18–20]. Furthermore, through the reduction reaction of corresponding metal oxides, Pd(0) or Au(0)-Pd(0) catalysts were prepared in ferritin, where it is notable that the substrates size selectively went through the protein shell and reacted on the catalysts [21, 22]. The residues of amino acids on the surface of the protein shell should have controls on traffic of metal ions and organic molecules, and the well-defined shell structures accessible by X-ray crystallographic analysis enabled to directly reveal the stepwise mechanism to accommodate palladium ions trapped at the specific recognition sites on the amino acid residues [23].

The virus capsids also have been applied to a versatile platform for synthesis and fabrication of nanomaterials. For example, the cavity of CCMV capsid was used as a reaction vessel to synthesize polyoxometalates, where cationic internal surface interacts with anionic starting materials and products, affording a single crystal of polyoxometalates with a diameter of 150 Å [24]. For rational improvement of the capsid surface, endohedral or exohedral functionalizations on the virus capsids have been extensively investigated by genetic modifications to introduce reactive groups like thiol or amine groups derived from residues of amino acids, followed by chemical modification of the reactive groups [25]. With these methods of chemical modification, a wide variety of organic or inorganic substituents have been introduced on the outer or inner surface of virus capsids [26, 27].

14.2 CONSTRUCTION OF POLYHEDRAL SKELETONS BY COORDINATION BONDS

For chemists, the insights about the mechanism of self-assembly of natural cages like ferritin or capsids and the insights about the relationship between the structures and the functions generate rich inspiration to develop artificial cage systems. There has been large effort to synthesize large-caged assemblies in artificial systems, but it is not still easy to construct well-defined, huge cages at will.

Quite recently, fusion proteins were prepared by connecting two oligomerization domains of proteins, and, in this semi-artificial system, cages with the diameter of 15 nm were successfully obtained [28]. Moreover, complete artificial cages with the diameter of 30–50 nm self-assembled from a 24-mer β-annulus peptide [29]. These achievements are very impressive, but the cages from peptide or protein motifs are difficult to design in general.

Various types of cages were synthesized by assembling building blocks through weak multiple interactions. One of the most efficient interactions employed as a driving force for the self-assembly of caged compounds is metal–organic coordination bond, and a wide variety of cages are reported. Furthermore, metal–organic cages are sometimes successfully functionalized to show their unique functions. In the following sections, the chemistry of metal–organic cages is reviewed.

14.2.1   Geometrical Effect on Products

The strengths of coordination bonds are suitable to construct stable self-assemblies, and the moderate rates of reversible formation and cleavage realize efficient shift of equilibrium from oligomeric intermediates to the most stable product. Also, the defined coordination numbers and directions are key to form well-defined product structures. In comparison, micelle and vesicle, which are well known self-assembled discrete structures in equilibrium, lack numerical and directional definitions in their interactions, and the size distributions of products composed of different number of subunit amphiphiles are inevitable.

In theory, the structures of metal–organic cages can be predicted based on the structural characteristics in the starting subunits. However, even we fully understand the building subunits in the chemical structures, the electronic properties, and the coordination geometry of metal ions and ligands, the product structures are frequently difficult to predict because molecular components are much more flexible than expected and metal centers can permit considerable deviation in their coordination angles.

A simplest example to indicate the difficulty in designing a product having a single structure is the self-assembly of mononuclear metal complex with two coordination sites in 90° and linear bridging ligand, which is one of the milestones to demonstrate metal–organic self-assemblies to form rings in the earliest era (Fig. 14.1). In addition to the easily expected square product, a triangular product also formed despite the hindered coordination angles from 90° [30, 31]. In a little more complex system, the self-assembly of an $M_{12}L_{24}$ complex from 36 building units and that of an M_6L_{12} complex from 18 building units are reported using the same bridging units and palladium (II) ion or platinum (II) ion, although both the metal ions are known to behave similarly in square planar coordination manner (Fig. 14.2) [32].

To understand the nature in self-assembly, simple and designable chemical systems are the desirable models to avoid complicated and unexplainable geometrical deviations in products. A robust self-assembly of a series of M_nL_{2n} cages has been reported from metal ions (M) with square planar coordination sphere and bent bidentate ligands (L). When the bidentate ligand is linear bearing coordinative sites at each opposite terminal, an infinite 2D grid with the composition of M_nL_{2n} ($n = \infty$) is formed where square complex is connected to each other [33–36]. To make cages defined as entropically favored regular or semiregular polyhedra with the metal ions

a square M_4L_4 complex   a triangle M_3L_3 complex

FIGURE 14.1   Self-assembly of a triangular complex with an easily expected square complex against angular hindrances.

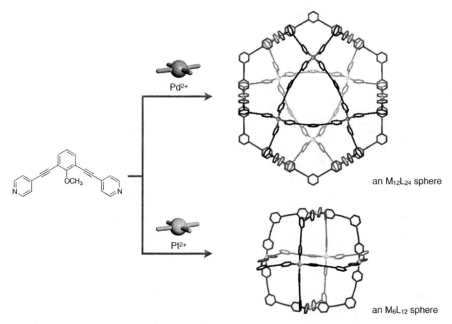

FIGURE 14.2   Self-assembly of spheres with different composition from the same ligand and Pd^{2+}/Pt^{2+} ions with square planar coordination sphere.

onto the vertices and the ligand onto the edges, the n of M_nL_{2n} cages is limited by geometrical restrictions to 6, 12, 24, 30, or 60, and the larger bend angle is required for a polyhedron with larger n by the geometrical requirements (Fig. 14.3). The synthesis of a series of M_nL_{2n} spherical cages for $n = 6$, 12, and 24 and M = Pd(II) ion have been achieved, where the mathematically calculated bend angle is 90° for M_6L_{12}, 120° for $M_{12}L_{24}$, and 135° for $M_{24}L_{48}$, and the angle of the used ligand was 90°

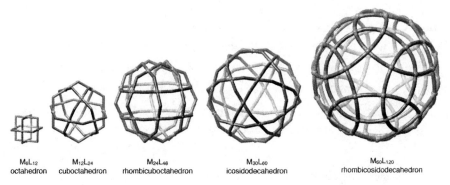

FIGURE 14.3   A family of M_nL_{2n} spheres self-assembled from bidentate ligands and metal ions with square planar coordination sphere.

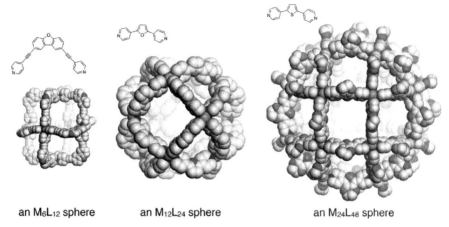

an M₆L₁₂ sphere   an M₁₂L₂₄ sphere   an M₂₄L₄₈ sphere

FIGURE 14.4 X-ray crystal structures of self-assembled M_nL_{2n} spheres ($n = 6, 12$, and 24).

for M_6L_{12}, 120°–127° for $M_{12}L_{24}$, and 135°–149° for $M_{24}L_{48}$ (Fig. 14.4) [37–40]. Further fine-tuning of the bend angle was performed by a unique method of mixed ligand system where two ligands with different bend angles, 127° and 149°, were mixed in a ratio and underwent the self-assembly with palladium (II) ions. The mean bend angle showed sharp threshold: the system where the angle smaller than 131.4° afforded an $M_{12}L_{24}$ sphere, and that where the angle larger than 133.6° afforded an $M_{24}L_{48}$ sphere [39]. On this empirically determined expectation, a series of $M_{24}L_{48}$ spheres were synthesized by a single kind of ligand which was rationally designed to have the angle larger than 135° [40].

The fact that only the distinctive polyhedron was obtained without structural deviation in n number in each self-assembly of M_nL_{2n} sphere, even in the largest number of self-assembly from 72 subunits, shows the strict rule of geometrical restriction on this artificial system works as well as that on protein cage systems. Although it might go against one's intuition, a single structure without structural deviations can be constructed from a large number of subunits, both in natural and artificial systems, when multiple interactions such as hydrogen bonds or ionic bonds, and geometrical and directional restrictions are employed.

14.2.2 Structural Extension Based on Rigid, Designable Framework

As overviewed in the above section, coordination bonds are suitable driving force to construct well-defined cages in artificial systems. If the coordination motifs are reserved, a structural extension of a caged complex can be achieved by the rational design of ligands.

One of the earliest examples of metal–organic cage is M_4L_6 type tetrahedron self-assembled from four Mg(II) ions and six ligands bearing two chelating sites [41]. The same coordination motifs were reserved in the tridentate ligands bearing three

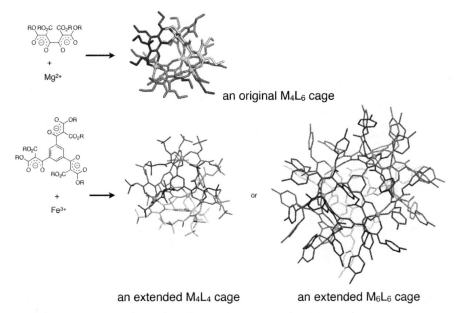

an original M$_4$L$_6$ cage

an extended M$_4$L$_4$ cage   an extended M$_6$L$_6$ cage

FIGURE 14.5   X-ray crystal structures of cages synthesized from bidentate or tridentate ligands and metal ions.

chelating sites, which afforded M$_4$L$_4$ or M$_6$L$_6$ cages upon complexation with Fe(III) ions (Fig. 14.5) [42].

A wide variety of tetrahedral coordination cages have been developed using ligands bearing two or three catecholate moieties as coordination sites and tri- and tetravalent metal ions such as Ga(III), Fe(III), Ti(IV), and Sn(IV), and so on. The compositions of these cages are M$_4$L$_4$ or M$_4$L$_6$ and the unique anionic frameworks suitable for the encapsulation of cationic species in water are designable and reproducible [43]. For an M$_4$L$_4$ cage, an original tetrahedron was synthesized from the smallest ligand bearing three catecholates and Ti(IV) ions, where the cavity size is 40 Å^3 determined by X-ray crystallography (Fig. 14.6) [44]. The central part of the ligand was extended by spacers so that the coordination catecholate are placed apart from each other and the threefold symmetry of the ligands is maintained, and the cavity size was extended up to 450 Å^3, whose structure was confirmed by ESI-MS [45]. The framework of M$_4$L$_6$ complexes has been also extended (Fig. 14.7). The mother M$_4$L$_6$-typed cage was formed from naphthalene-based ligand with two catecholate sites, which afforded homochiral ΔΔΔΔ and ΛΛΛΛ enantiomers upon complexation with Fe(III) or Ga(III) ions, which was later resolved into optically pure forms [46, 47]. The original cavity size was ~300–500 Å^3, and the introduction of a spacer in the ligand and the template synthesis with a suitably sized ammonium ion extended the cavity size at least up to 700 Å^3 [48].

One of the most useful coordination motifs is the use of [enPd(II)]$^{2+}$ (en = ethylenediamine) unit and ligands bearing several pyridyl groups as coordination

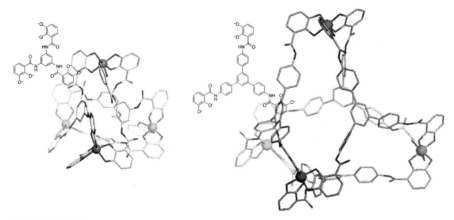

FIGURE 14.6 Extension of M_4L_4 cages from the original structure (left, X-ray structure) to the huger one (right) by the insertion of phenylene spacers to the core of the ligand.

sites. The simplest example is the square complex [30], whose two-dimensional structure was readily expanded into three-dimensional structure [49]. These methods were turned to be useful to synthesize a wide variety of hollow complexes, such as tube, bowl, and catenane structures by the rational design of ligands [50], and an M_6L_4 cage derived from a triangular ligand, 2,4,6-tris(4-pyridyl)-1,3,5-triazine, shows efficient encapsulation properties of neutral organic molecules within the hydrophobic cavity in water (Fig. 14.8) [51]. The encapsulation procedure is easy, and just suspending excess amounts of water-insoluble adamantane in the aqueous solution of an M_6L_4 cage is enough to prepare the solution of the inclusion complex of (adamantane)$_4 \subset M_6L_4$ [52]. There have been many reports on the application of the cage to accommodations of homo and hetero molecules and to controls on physical properties and reactivity of the encapsulated guests.

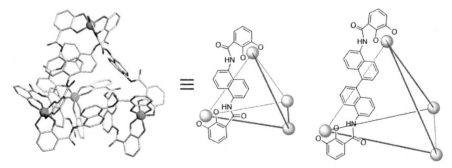

FIGURE 14.7 Extension of M_4L_6 cages. The original structure (left) was determined by X-ray crystallography, and the schematic representations for the original and extended complexes are shown with the chemical structures of a ligand out of six ligands for each complex.

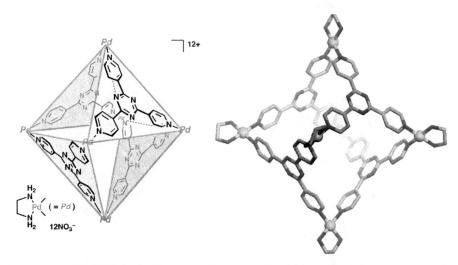

FIGURE 14.8   X-ray crystal structure of an M_6L_4 octahedral cage.

The structural extensions were also applied to the molecular systems employing the 90° end-capping unit, $[enPd(II)]^{2+}$, triangular ligand for the roof and floor parts, and linear pillar ligand, affording unique prism-like pillared cages. The use of different length of pillar ligand, the product structure has different number of π-stacked aromatic molecules with different height and the same diameter (Fig. 14.9) [53]. The roof and floor of the shortest cages accommodate one layer of planar molecule(s), such as triphenylene derivative [54], coronene [55], or a nucleotide duplex [56], and the simple extension of the length of pillar ligand constructed seven layers of stacked aromatic molecules at maximum [57]. In the structural extension, directional interactions between aromatic molecules as donor and acceptor play a dominant role to control the product structures. Furthermore, the stacked number was extended up to nine by an elaborated molecular design to realize interpenetrating stacked structures

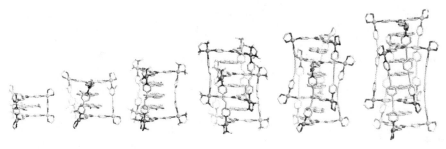

FIGURE 14.9   X-ray crystal structures of organic-pillared complexes with three-to-eight–stacked aromatic planes and MM-optimized structure of the highest complex with nine aromatic planes.

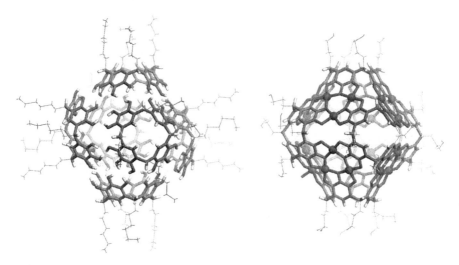

FIGURE 14.10   X-ray crystal structure of a L_6 sphere (left) and an $M_{24}L_6$ sphere (right).

[58]. The finite stacked structures are also interesting in their characteristic physical properties, and electron transport in vertical direction [59] and magnetic alignment in NMR magnetic field [60] have been reported recently.

Although the number of examples is limited, unique structural extensions are reported by exchanging the types of interaction used as driving force of the assemblies. A structurally similar sphere with diameter of 4 nm was synthesized through the use of coordination bonds or hydrogen bonds from the same molecule, pyrogallol[4]arene, affording L_6 or $M_{24}L_6$, respectively (Fig. 14.10) [61, 62]. Twelve hydroxyl groups on a ligand work as hydrogen donor–acceptor pair to form hexameric homo assembly of L_6 or as coordinating sites for Cu(II) ions in the $M_{24}L_6$ complex. The hexameric homo assembly was originally prepared by using calix[4]resorcinarene [63].

Another structural extension by the exchange between coordination bonds and hydrophobic effect, van der Waals forces, and CH−π interactions are reported. The same hexagram-shaped amphiphile molecule self-assembles into $[M_6L_8]^{12+}$ complex with metal ions or L_6 (Fig. 14.11) [64–66]. The interactions constructing the L_6 assembly are known to be weaker than coordination bond; in general, the multiple subunits are designed to fit well with each other to produce directional interactions to keep the discrete structure in both solution and the solid state.

In spite of the eager efforts to extend the framework based on established coordination motifs with geometrical restrictions as reviewed above, in the light of molecular size and weight, the straight strategy to obtain huge cages is to increase the number of subunits. The geometrical restriction in these assemblies tends to afford highly symmetric, spherical complexes [67–69]. The first example of spherical cage is the assembly of eight ligands bearing three chelating moieties and 12 Cu(II) ions, affording $M_{12}L_8$ composition (Fig. 14.12) [70]. The complex structure built

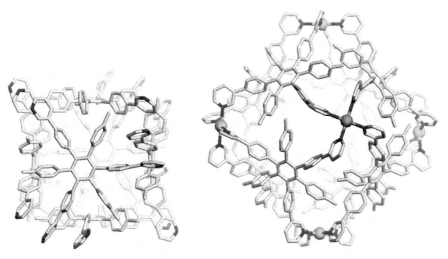

FIGURE 14.11   X-ray crystal structure of a L_6 sphere (left) and an M_6L_8 sphere (right).

from 20 subunits in total was revealed by crystal analysis to be a hollow, spherical framework with the diameter of 1.3 nm. Inside the hollow, 5–6 DMF molecules used as the solvent were accommodated, and two of them are well resolved in the crystal analysis.

The coordination bond between carboxylate and Cu(II) ion afforded the framework of $M_{24}L_{24}$ with the symmetry of cuboctahedron when we consider the polyhedron of the complex, mapping the ligands to edges and metal ions to vertices (Fig. 14.13).

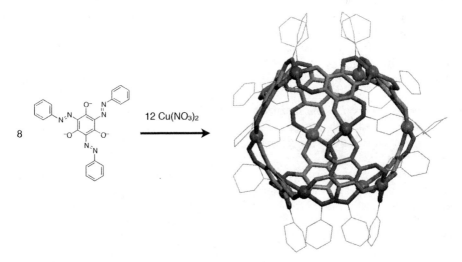

FIGURE 14.12   X-ray crystal structure of an $M_{12}L_8$ sphere.

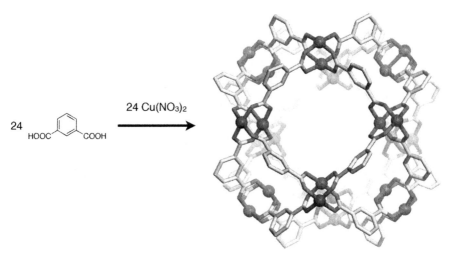

FIGURE 14.13   X-ray crystal structure of an $M_{24}L_{24}$ cuboctahedron sphere.

The sphere was synthesized from 1,3-benzenedicarboxylate ligands and Cu(II) ions and reported by two groups independently [71, 72].

Triangular ligands bearing three pyridyl groups and dinuclear Pt(II) complexes self-assemble into spherical complexes [73, 74]. Fifty subunits are connected by the metal–organic interactions, and nanoscopic dodecahedron was constructed.

As reviewed in the Section 14.2.1, the combination of bent ligands bearing two pyridyl groups and metal ions with square planar coordination sphere like Pd(II) and Pt(II) is reported to afford a series of geometrically restricted M_nL_{2n} spheres, where the structures were extended with $n = 6$, 12, and 24 [37–40]. Based on this rigid spherical frameworks, the first example of stellated polyhedron was demonstrated [75]. By attaching an additional pyridyl group on the ligand to introduce 24 coordinative sites on the periphery of an $M_{12}L_{24}$ sphere, six metal ions were trapped to form a complex with $M_{18}L_{24}$ composition (Fig. 14.14). Because both coordination sites for the construction of an $M_{12}L_{24}$ framework and for the trap of additional metal ions are the same pyridyl groups, competitive formation of coordination bonds can be expected to form indeterminate oligomers. But the flexible linker employed to introduce the additional site presumably made entropic differences in coordination nature, and the selective cleavage and formation of coordination bond at the additional metal sites are achieved.

The introduction of a spacer to a subunit ligand is a typical method to enlarge the product sizes, and an $M_{12}L_{24}$ sphere with a diameter of 3.5 nm was extended to that of 6.3 nm by the introduction of ethynylphenylene. The corresponding small and large ligands are linked together with a flexible triethylene glycol linker, and novel sphere-in-sphere molecule was synthesized on complexation with metal ions, where small ligands constructed the small sphere and large ligands constructed the large one independently without formation of oligomeric products (Fig. 14.15) [76].

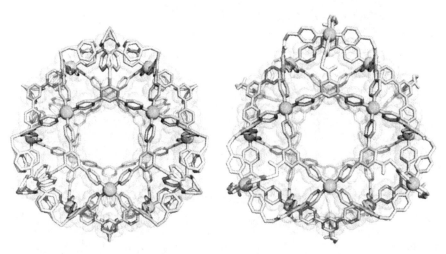

FIGURE 14.14   X-ray structure of an $M_{12}L_{24}$ sphere bearing free coordination sites (left) and a stellated $M_{18}L_{24}$ sphere by additional trap of six Pd(II) ions (right).

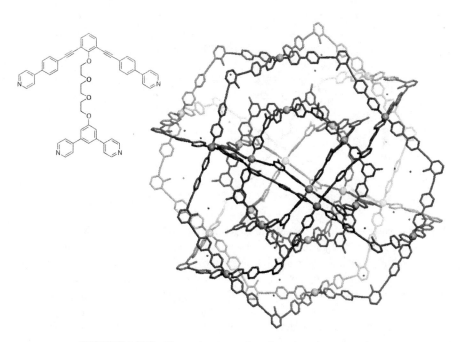

FIGURE 14.15   X-ray structure of a sphere-in-sphere complex.

14.2.3 Mechanistic Insight into Self-Assembly

Multiple weak interactions and their reversibility are characteristics in natural or artificial self-assemblies of many subunits to form huge, well-defined skeletons, whose high stability is not easy to expect from the simple sum of the weak interactions. Because the total system is very complex with many assembling steps from subunits to the final products through possible intermediates, rather simplified model studies are required to understand and prove the mechanism of the self-assemblies.

For virus self-assembly, the first theoretical approach was reported using molecular dynamics (MD) simulations for 125,000 molecules of solvent and subunit in a cubic region with periodic boundaries [77]. The importance of reversible processes to diminish unfavorable oligomers was revealed for the highly efficient self-assembly of the product with a single structure. An artificial self-assembly of dodecahedral capsids from molecules with pentagonal core of corannulene was simulated with physical models bearing magnets as interaction sites [78]. Then the model was translated into a theoretical simulation to show the pathway of the assembly.

Theoretical simulations become valuable only when the resultant predictions are evaluated in the light of experimental results. A simple experiment was designed to access the rate of ligand exchange reactions for self-assembly of $M_{12}L_{24}$ spherical complexes, where 48 Pd(II)-pyridine bonds are included in the structure of the final products. Monodentate Pd(II)-pyridine bonds are known to be kinetically labile and easily dissociate on equilibrium conditions. The kinetics of ligand exchange rates were measured by well-designed NMR and MS experiments, and it was revealed that $M_{12}L_{24}$ complexes are remarkably stable and the half-lives are much longer than those for comparable monodentate Pd(II)-pyridine complexes by a factor of approximately 10^5 [79]. These findings supported a stepwise mechanism of self-assembly where time scale of equilibrium differs and first experimentally proved the cooperative effect of multiple weak interactions to firmly keep the kinetically trapped framework of the thermodynamic products.

14.3 DEVELOPMENT OF FUNCTIONS VIA CHEMICAL MODIFICATION

The frameworks of artificial hollow cages themselves sometimes show functions to accommodate guest molecules and ions through rather hydrophobic environment surrounded by aromatic walls of the frameworks or through electrostatic effects. The number of the guests is typically 1–3, and they are tightly packed within the cavity, where the conformation of guests is sometimes fixed by the shape of the framework. The tight encapsulation of guest molecules controls the physical properties of the guests and catalyzes unique reactions of the guests, which sometimes can be achieved only within the cages.

Natural protein cages and their cavities are huger than those of typical artificial cages; therefore, tight packing way of encapsulation through direct interactions between the accommodated substrates and the framework are inefficient and insufficient. Instead, functional residues attached on the skeletal framework play dominant roles to guide substrates into the cavities and to store them firmly. In the following sections, developments of functions through designed functional groups on the cage frameworks are reviewed.

14.3.1   Chemistry in the Hollow of Cages

Chemical modification of internal wall of hollow cages is limited by obvious reasons that the cages must be large enough to accept the functionalization and to keep additional volume for substrates. The diameter of an original $M_{12}L_{24}$ spherical cage was extended from 3.5 nm to 4.6 nm by the introduction of acetylene spacers into the ligand to accept selective internal functionalizations [80]. The introduction of a residue to a ligand before complexation reaction enables quantitative and site-specific endohedral coating of the cavity (Fig. 14.16). Inside the spheres, 24 functional groups are densely accumulated to produce localized environment with the physical property derived from the residues, and still unoccupied volume filled with solvent remains to allow encapsulation of guests and to retain flexible motions of the attached residues.

The introduction of flexible fluorous chains made a corresponding fluorous nanophase, and phase separation between the nanophase and bulk organic solvent was achieved in the same manner of bulk phase separation between fluorous liquid and organic liquid (Fig. 14.17) [81]. The nanophase extracted up to eight perfluoroalkanes into the cavity, where the guests are not tightly packed in the cavity and actually dissolved at the center of the nanodroplet. The methodology to construct localized characteristic environments is applied to produce a hydrophobic nanophase to extract hydrophobic guests [82], an aromatic nanophase to extract aromatic molecules like C_{60} [83], and a chiral pocket surrounded by peptide residues like protein pockets

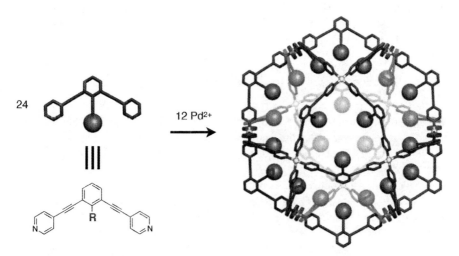

FIGURE 14.16   Schematic representation of an endohedral functionalization of an $M_{12}L_{24}$ sphere.

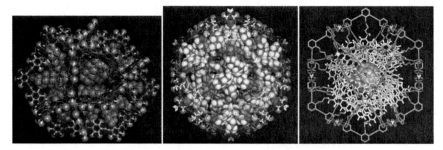

FIGURE 14.17 Fluorous(left), hydrophobic (middle), and aromatic (right) nanophases constructed in the hollow of an $M_{12}L_{24}$ spherical framework.

[84]. These artificial residues are highly designable based on organic chemistry, and the switching of the interior hydrophobicity was also achieved by isomerization of accumulated azobenzene chromophores [85].

The reactions within the cavities have distinctive characters: multiple reactants produce a well-controlled and size-limited product, whereas the reaction is essentially the same as that in bulk solvent. One strategy to introduce the substances is covalently anchoring reactive sites as a part of functionalized residues, and radical polymerization between polymerizable functional groups proceeded only in the cavity to afford oligomers [86]. The product was cleaved from the ligand and analyzed to show controlled extents of polymerization. Another more general strategy to accommodate the substances is the extraction of reactive species in the cavity by phase separation phenomenon. Organic monomers or inorganic precursors of sol–gel condensation were effectively extracted within the tailor-made environment inside the spheres, and controlled polymerized products were obtained [87–89]. Especially, thus prepared silica gel particles are notable in their well-controlled diameter of less than 3 nm and their narrow structural distribution (PDI < 1.01), which cannot be achieved by conventional synthetic and separation methods (Fig. 14.18).

14.3.2 Chemistry on the Periphery of Cages

Anchoring functional groups on the external periphery of cages is an attractive strategy to rationally develop designable functions without size limitation by the size of the framework. The original weak interaction of the attached residue can be cooperatively enhanced by the accumulation on the framework. For the $M_{12}L_{24}$ spheres, fragments of peptide, sugar, or DNA strand are delocated on the periphery to show distinctive functions.

A library of sugar balls was prepared by attaching a variety of sugar chains, affording huge molecular particles with molecular weights of up to over 20,000 and diameter of up to 7 nm [90]. Lectin, a class of proteins bearing multiple sugar-specific recognition sites, was mixed in the solution of the saccharide-coated spheres, and the formation of insoluble aggregate between the lectin and the sphere was observed specifically depending on the structure of the attached sugar. The analogous formation of bridging aggregate was formed between DNA-attached spheres and the

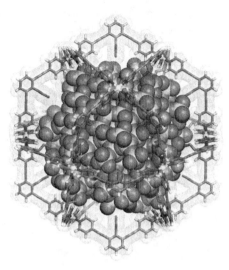

FIGURE 14.18 A monodisperse silica gel nanoparticle template synthesized within the hydrophilic hollow of a sphere.

complementary oligonucleotides, which was derived from larger density of accumulated DNA fragments on the sphere than that on conventional gold nanoparticles or micelles [91].

A peptide aptamer with recognition ability of titania is coated on the periphery of an $M_{12}L_{24}$ sphere, and the sphere showed irreversible fixation on titania surface (Fig. 14.19) [92]. Compared to the ferritin delocated with the aptamers showing weaker fixation property, the minimum distance between the aptamers controlled

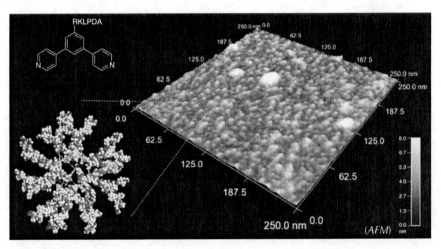

FIGURE 14.19 Irreversible adsorption of a sphere via functionalized peptide aptamers recognizing the titania plate.

by the framework of the sphere and the employed linker between the framework and the aptamer can be considered to be ideal for deriving the maximum fixation performance.

14.4 METAL–ORGANIC CAGES FOR PROTEIN ENCAPSULATION

Now that the sizes of artificial cages are comparable to those of biomolecules, protein encapsulation in the cages is expected to develop functions like molecular chaperone. Well-defined molecular sphere, $M_{12}L_{24}$ complex, was used for the encapsulation of a ubiquitin, where a bidentate ligand tethered with a ubiquitin was added to the solution of Pd(II) ions and additional ligands bearing a sugar chain (Fig. 14.20). Thus, the obtained sphere encapsulated one ubiquitin surrounded by hydrophilic sugar moieties, and the structure was well analyzed by NMR spectroscopy, ultracentrifugation, and X-ray crystallography [93].

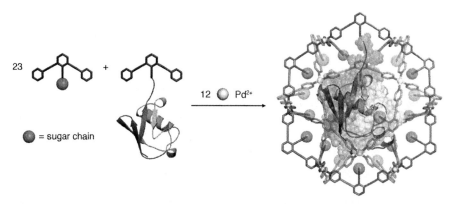

FIGURE 14.20 Synthesis of an $M_{12}L_{24}$ sphere accommodating a ubiquitin inside the hollow.

14.5 SUMMARY AND PERSPECTIVES

Metal-organic cages assembled from metal ions and organic ligands bearing multiple coordination sites are described. Coordination bond with defined direction and number of coordination has turned to be versatile to construct a well-defined product structure in a stable state and sometimes in predictable and designable ways. Inspired by natural protein cages, chemists have prepared many kinds of metal–organic cages and also developed rational strategies to extend the established cages in size and complexity of the frameworks. Development of functions by way of placing the functional residues inside or outside the three-dimensional cavities, which are frequently realized in natural protein cages, is an ongoing subject. Now that the scale of artificial metal–organic cages is reaching to that of huge biomolecules like proteins, sugar chains, and DNA strands, the next research area might be the artificial assemblies where efficient and well-defined interactions incorporating biomolecules and artificial caged frameworks in an even molecular size.

REFERENCES

[1] Lawson, D. M.; Artymiuk, P. J.; Yewdall, S. J.; Smith, J. M. A.; Livingstone, J. C.; Treffry, A.; Luzzago, A.; Levi, S.; Arosio, P.; Cesareni, G.; Thomas, C. D.; Shaw, W. V.; Harrison, P. M. *Nature* **1991**, *349*, 541–544.

[2] Speir, J. A.; Munshi, S.; Wang, G.; Baker, T. S.; Johnson, J.E. *Structure* **1995**, *3*, 63–78.

[3] Andrews, S. C. *Advan. Microb. Physiol.* **1998**, *40*, 281–351.

[4] Theil, E. C. Ferritin. In *Handbook of Metalloproteins*; Messerschmidt, A., Huber, R., Poulos, T., Wieghardt, K., Eds.; John Wiley & Sons: Chichester, 2001, pp 771–781.

[5] Christen, Y. *Am. J. Clin. Nutr.* **2000**, *71*, 621S–629S.

[6] Crichton, R. R.; Ward, R. J. *Analyst* **1995**, *120*, 693–697.

[7] Chasteen, N. D.; Harrison, P. M. *J. Struct. Biol.* **1999**, *126*, 182–194.

[8] Crick, F. H. C.; Watson, J. D. *Nature* **1956**, *177*, 473–475.

[9] Caspar, D. L. D.; Klug, A. *Cold Spring Harb. Symp. Quant. Biol.* **1962**, *27*, 1–24.

[10] Twarock, R. *Philos. Transact. A. Math. Phys. Eng. Sci.* **2006**, *364*, 3357–3373.

[11] Liljas, L.; Unge, T.; Jones, T. A.; Fridborg, K.; Lövgren, S.; Skoglund, U.; Strandberg, B. *J. Mol. Biol.* **1982**, *159*, 93–108.

[12] Speir, J. A.; Munshi, S.; Wang, G.; Baker, T. S.; Johnson, J. E. *Structure* **1995**, *3*, 63–78.

[13] Grimes, J. M.; Burroughs, J. N.; Gouet, P.; Diprose, J. M.; Malby, R.; Ziéntara, S.; Mertens, P. P. C.; Stuart, D. I. *Nature* **1998**, *395*, 470–478.

[14] Prasad, B. V.; Yamaguchi, S.; Roy, P. *J. Virol.* **1992**, *66*, 2135–2142.

[15] Zhou, Z. H. In *Segmented Double-stranded RNA Viruses: Structure and Molecular Biology*; Patton, J. T., Ed.; Caister Academic Press: Norfolk, UK, 2008; pp 27–43.

[16] Speir, J. A.; Munshi, S.; Wang, G.; Baker, T. S.; Johnson, J. E. *Structure* **1995**, *3*, 63–78.

[17] Tanaka, H.; Kato, K.; Yamashita, E.; Sumizawa, T.; Zhou, Y.; Yao, M.; Iwasaki, K.; Yoshimura, M.; Tsukihara, T. *Science* **2009**, *323*, 384–388.

[18] Douglas, T.; Dickson, D. P. E.; Betteridge, S.; Charnock, J.; Garner, C. D.; Mann, S. *Science* **1995**, *269*, 54–57.

[19] Meldrum, F. C.; Wade, V. J.; Nimmo, D. L.; Heywood, B. R.; Mann, S. *Nature* **1991**, *349*, 684–687.

[20] Meldrum, F. C.; Heywood, B. R.; Mann, S. *Science* **1992**, *257*, 522–523.

[21] Ueno, T.; Suzuki, M.; Goto, T.; Matsumoto, T.; Nagayama, K.; Watanabe, Y. *Angew. Chem. Int. Ed.* **2004**, *43*, 2527–2530.

[22] Suzuki, M.; Abe, M.; Ueno, T.; Abe, S.; Goto, T.; Toda, Y.; Akita, T.; Yamada, Y.; Watanabe, Y. *Chem. Commun.* **2009**, 4871–4873.

[23] Ueno, T.; Abe, M.; Hirata, K.; Abe, S.; Suzuki, M.; Shimizu, N.; Yamamoto, M.; Takata, M.; Watanabe, Y. *J. Am. Chem. Soc.* **2009**, *131*, 5094–5100.

[24] Douglas, T.; Young, M. *Nature* **1998**, *393*, 152–155.

[25] Mateu, M. G. *Protein Eng. Des. Sel.* **2011**, *24*, 53–63.

[26] Gillitzer, E.; Willits, D.; Young, M.; Douglas, T. *Chem. Commun.* **2002**, *21*, 2390–2391.

[27] Wang, Q.; Lin, T.; Tang, L.; Johnson, J. E.; Finn, M. G. *Angew. Chem. Int. Ed.* **2002**, *41*, 459–462.

[28] Padilla, J. E.; Colovos, C.; Yeates, T. O. *Proc. Natl. Acad. Sci. USA* **2001**, *98*, 2217–2221.

[29] Matsuura, K.; Watanabe, K.; Matsuzaki, T.; Sakurai, K.; Kimizuka, N. *Angew. Chem. Int. Ed.* **2010**, *49*, 9662–9665.

[30] Fujita, M.; Sasaki, O.; Mistuhashi, T.; Fujita, T.; Yazaki, J.; Yamaguchi, K.; Ogura, K. *Chem. Commun.* **1996**, 1535–1536.

[31] Lee, S. B.; Hwang, S.; Chung, D. S.; Yun, H.; Hong, J.-I. *Tetrahedron Lett.* **1998**, *39*, 873- 876.

[32] Fujita, D.; Takahashi, A.; Sato, S.; Fujita, M. *J. Am. Chem. Soc.* **2011**, *133*, 13317–13319.

[33] Gable, R. W.; Hoskins, B. F.; Robson, R. *J. Chem. Soc. Chem. Commun.* **1990**, 1677–1678.

[34] Fujita, M.; Kwon, Y. J.; Washizu, S.; Ogura, K. *J. Am. Chem. Soc.* **1994**, *116*, 1151–1152.

[35] Biradha, K.; Hongo, Y.; Fujita, M. *Angew. Chem.* **2000**, *112*, 4001–4003 *Angew. Chem. Int. Ed.* 2000, *39*, 3843–3845.

[36] Noro, S.-I.; Kitagawa, S.; Kondo, M.; Seki, K. *Angew. Chem.* **2000**, *112*, 2161–2164 *Angew. Chem. Int. Ed.* 2000, *39*, 2081–2084.

[37] Suzuki, K.; Tominaga, M.; Kawano, M.; Fujita, M. *Chem. Commun.* **2009**, 1638–1640.

[38] Tominaga, M.; Suzuki, K.; Kawano, M.; Kusukawa, T.; Ozeki, T.; Sakamoto, S.; Yamaguchi, K.; Fujita, M. *Angew. Chem. Int. Ed.* **2004**, *43*, 5621–5625.

[39] Sun, Q.-F.; Iwasa, J.; Ogawa, D.; Ishido, Y.; Sato, S.; Ozeki, T.; Sei, Y.; Yamaguchi, K.; Fujita, M. *Science* **2010**, *328*, 1144–1147.

[40] Bunzen, J.; Iwasa, J.; Bonakdarzadeh, P.; Numata, E.; Rissanen, K.; Sato, S.; Fujita, M. *Angew. Chem. Int. Ed.* **2012**, *51*, 3161–3163.

[41] Saalfrank, R. W.; Stark, A.; Peters, K.; von Schnering, H. G. *Angew. Chem. Int. Ed. Engl.* **1988**, *27*, 851–853.

[42] Saalfrank, R. W.; Glaser, H.; Demleitner, B.; Hampel, F.; Chowdhry, M. M.; Schünemann, V.; Trautwein, A. X.; Vaughan, G. B. M.; Yeh, R.; Davis, A. V.; Raymond, K. N. *Chem. Eur. J.* **2002**, *8*, 493–497.

[43] Caulder, D. L.; Brückner, C.; Powers, R. E.; König, S.; Parac, T. N.; Leary, J. A.; Raymond, K. N. *J. Am. Chem. Soc.* **2001**, *123*, 8923–8938.

[44] Brückner, C.; Powers, R. E.; Raymond, K. N. *Angew. Chem. Int. Ed.* **1998**, *37*, 1837–1839.

[45] Yeh, R. M.; Xu, J.; Seeber, G.; Raymond, K. N. *Inorg. Chem.* **2005**, *44*, 6228–6239.

[46] Caulder, D. L.; Powers, R. E.; Parac, T. N.; Raymond, K. N. *Angew. Chem. Int. Ed.* **1998**, *37*, 1840–1843.

[47] Davis, A. V.; Fiedler, D.; Ziegler, M.; Terpin, A.; Raymond, K. N. *J. Am. Chem. Soc.* **2007**, *129*, 15354–15363.

[48] Biros, S. M.; Yeh, R. M.; Raymond, K. N. *Angew. Chem. Int. Ed.* **2008**, *47*, 6062–6064.

[49] Fujita, M.; Umemoto, K.; Yoshizawa, M.; Fujita, N.; Kusukawa, T.; Biradha, K. *Chem. Commun.* **2001**, 509–518.

[50] Fujita, M.; Tominaga, M.; Hori, A.; Therrien, B. *Acc. Chem. Res.* **2005**, *38*, 369–378.

[51] Fujita, M.; Oguro, D.; Miyazawa, M.; Oka, H.; Yamaguchi, K.; Ogura, K. *Nature* **1995**, *378*, 469–471.

[52] Kusukawa, T.; Fujita, M. *J. Am. Chem. Soc.* **2002**, *124*, 13576–13582.

[53] Klosterman, J. K.; Yamauchi, Y.; Fujita, M. *Chem. Soc. Rev.* **2009**, *38*, 1714–1725.

[54] Kumazawa, K.; Biradha, K.; Kusukawa, T.; Okano, T.; Fujita, M. *Angew. Chem. Int. Ed.* **2003**, *42*, 3909–3913.

[55] Yoshizawa, M.; Nakagawa, J.; Kumazawa, K.; Nagao, M.; Kawano, M.; Ozeki, T.; Fujita, M. *Angew. Chem. Int. Ed.* **2005**, *44*, 1810–1813.

[56] Sawada, T.; Yoshizawa, M.; Sato, S.; Fujita, M. *Nature Chem.* **2009**, *1*, 53–56.

[57] Yamauchi, Y.; Yoshizawa, M.; Akita, M.; Fujita, M. *J. Am. Chem. Soc.* **2010**, *132*, 960–966.

[58] Yamauchi, Y.; Yoshizawa, M.; Fujita, M. *J. Am. Chem. Soc.* **2008**, *130*, 5832–5833.

[59] Kiguchi, M.; Takahashi, T.; Takahashi, Y.; Yamauchi, Y.; Murase, T.; Fujita, M.; Tada, T.; Watanabe, S. *Angew. Chem. Int. Ed.* **2011**, *50*, 5708–5711.

[60] Sato, S.; Morohara, O.; Fujita, D.; Yamaguchi, Y.; Kato, K.; Fujita, M. *J. Am. Chem. Soc.* **2010**, *132*, 3670–3671.

[61] Cave, G. W. V.; Antesberger, J.; Barbour, L. J.; McKinlay, R. M.; Atwood, J. L. *Angew. Chem. Int. Ed.* **2004**, *43*, 5263–5266.

[62] McKinlay, R. M.; Cave, G. W. V.; Atwood, J. L. *Proc. Natl. Acad. Sci. USA* **2005**, *102*, 5944–5948.

[63] MacGillivray, L. R.; Atwood, J. L. *Nature* **1997**, *389*, 469–472.

[64] Hiraoka, S.; Harano, K.; Shiro, M.; Shionoya, M. *J. Am. Chem. Soc.* **2008**, *130*, 14368–14369.

[65] Hiraoka, S.; Harano, K.; Shiro, M.; Ozawa, Y.; Yasuda, N.; Toriumi, K.; Shionoya, M. *Angew. Chem., Int. Ed.* **2006**, *45*, 6488–6491.

[66] Harano, K.; Hiraoka, S.; Shionoya, M. *J. Am. Chem. Soc.* **2007**, *129*, 5300–5301.

[67] Tranchemontagne, D. J.; Ni, Z.; O'Keeffe, M.; Yaghi, O. M. *Angew. Chem. Int. Ed.* **2008**, *47*, 5136–5147.

[68] Leininger, S.; Olenyuk, B.; Stang, P. J. *Chem. Rev.* **2000**, *100*, 853–908.

[69] Seidel, S. R.; Stang, P. J. *Acc. Chem. Res.* **2002**, *35*, 972–983.

[70] Abrahams, B. F.; Egan, S. J.; Robson, R. *J. Am. Chem. Soc.* **1999**, *121*, 3535–3536.

[71] Eddaoudi, M.; Kim, J.; Wachter, J. B.; Chae, H. K.; 'Keeffe, M. O.; Yaghi, O. M. *J. Am. Chem. Soc.* **2001**, *123*, 4368–4369.

[72] Moulton, B.; Lu, J.; Mondal, A.; Zaworotko, M. J. *Chem. Commun.* **2001**, 863–864.

[73] Olenyuk, B.; Whiteford, J. A.; Fechtenkötter, A.; Stang, P. J. *Nature* **1999**, *398*, 796–799.

[74] Olenyuk, B.; Levin, M. D.; Whiteford, J. A.; Shield, J. E.; Stang, P. J. *J. Am. Chem. Soc.* **1999**, *121*, 10434–10435.

[75] Sun, Q.-F.; Sato, S.; Fujita, M.; *Nature Chem.* **2012**, *4*, 330–333.

[76] Sun, Q.-F.; Murase, T.; Sato, S.; Fujita, M.; *Angew. Chem. Int. Ed.* **2011**, *50*, 10318–10321.

[77] Rapaport, D.C.; Johnson, J.E.; Skolnick, J. *Comput. Phys. Commun.* **1999**, *121*, 231–235.

[78] Olson, A. J.; Hu, Y. H. E.; Keinan, E. *Proc. Natl. Acad. Sci. USA* **2007**, *104*, 20731–20736.

[79] Sato, S.; Ishido, Y.; Fujita, M. *J. Am. Chem. Soc.* **2009**, *131*, 6064–6065.

[80] Tominaga, M.; Suzuki, K.; Murase, T.; Fujita, M. *J. Am. Chem. Soc.* **2005**, *127*, 11950–11951.

[81] Sato, S.; Iida, J.; Suzuki, K.; Kawano, M.; Ozeki, T.; Fujita, M. *Science* **2006**, *313*, 1273–1276.

[82] Suzuki, K.; Iida, J.; Sato, S.; Kawano, M.; Fujita, M. *Angew. Chem. Int. Ed.* **2008**, *47*, 5780–5782.

[83] Suzuki, K.; Takao, K.; Sato, S.; Fujita, M. *J. Am. Chem. Soc.* **2010**, *132*, 2544–2545.

[84] Suzuki, K.; Kawano, M.; Sato, S.; Fujita, M. *J. Am. Chem. Soc.* **2007**, *129*, 10652–10653.

[85] Murase, T.; Sato, S.; Fujita, M. *Angew. Chem. Int. Ed.* **2007**, *46*, 5133–5136.

[86] Murase, T.; Sato, S.; Fujita, M. *Angew. Chem. Int. Ed.* **2007**, *46*, 1083–1085.

[87] Kikuchi, T.; Murase, T.; Sato, S.; Fujita, M. *Supramol. Chem.* **2008**, *20*, 81–94.

[88] Suzuki, K.; Sato, S.; Fujita, M. *Nature Chem.* **2010**, *2*, 25–29.

[89] Suzuki, K.; Takao, K.; Sato, S.; Fujita, M. *Angew. Chem. Int. Ed.* **2011**, *50*, 4858–4861.

[90] Kamiya, N.; Tominaga, M.; Sato, S.; Fujita, M. *J. Am. Chem. Soc.* **2007**, *129*, 3816–3817.

[91] Kikuchi, T.; Sato, S.; Fujita, M. *J. Am. Chem. Soc.* **2010**, *132*, 15930–15932.

[92] Ikemi, M.; Kikuchi, T.; Matsumura, S.; Shiba, K.; Sato, S.; Fujita, M. *Chem. Sci.* **2010**, *1*, 68–71.

[93] Fujita, D.; Suzuki, K.; Sato, S.; Yagi-Utsumi, M.; Yamaguchi, Y.; Mizuno, N.; Kumasaka, T.; Takata, M.; Noda, M.; Uchiyama, S.; Kato, K.; Fujita, M. *Nature Commun.* **2012**, *3*, 1093.

INDEX

Note: Page number followed by f and t indicates text in figure and table respectively.

Coordination Chemistry in Protein Cages: Principles, Design, and Applications, First Edition.
Edited by Takafumi Ueno and Yoshihito Watanabe.
© 2013 John Wiley & Sons, Inc. Published 2013 by John Wiley & Sons, Inc.